THE PHYSICS OF THREE-DIMENSIONAL RADIATION THERAPY

Conformal Radiotherapy, Radiosurgery and Treatment Planning

T0358887

Series in Medical Physics

THE PHYSICS OF THREE-DIMENSIONAL RADIATION THERAPY

Conformal Radiotherapy, Radiosurgery and Treatment Planning

Steve Webb

Professor of Radiological Physics,
Head, Joint Department of Physics,
Institute of Cancer Research and
Royal Marsden NHS Trust,
Sutton, Surrey, UK

I*o*P

Institute of Physics Publishing
Bristol and Philadelphia

British Library Cataloguing-in-Publication Data

A catalogue record for this book is available from the British Library.

ISBN 0-7503-0247-X
ISBN 0-7503-0254-2 Pbk

Library of Congress Cataloging-in-Publication Data are available

Reprinted with corrections 2001

The author has attempted to trace the copyright holder of all the figures and tables reproduced in this publication and apologizes to copyright holders if permission to publish in this form has not been obtained.

Cover picture depicts a multileaf collimator being used to shape an x-ray field during radiotherapy. The collimator has many pairs of small lead jaws that generate a field matched to the view an observer would have of the tumour from the source position.

Series Editors:
C G Orton, Karmanos Cancer Institute and Wayne State University, Detroit, USA
J A E Spaan, University of Amsterdam, The Netherlands
J G Webster, University of Wisconsin-Madison, USA

Published by Institute of Physics Publishing, wholly owned by The Institute of Physics, London

Institute of Physics Publishing, Dirac House, Temple Back, Bristol BS1 6BE, UK

US Office: Institute of Physics Publishing, The Public Ledger Building, Suite 1035, 150 South Independence Mall West, Philadelphia, PA 19106, USA

Typeset in LaTeX using the IOP Bookmaker Macros
Printed in Great Britain at the University Press, Cambridge

CONTENTS

PREFACE

Conformal radiotherapy is tailoring the high-dose volume to the target volume in a patient, simultaneously delivering a low dose to non-target tissues. By and large the term is associated with external-beam radiotherapy although there is no reason why this distinction should necessarily be drawn. Brachytherapy is in a sense the ultimate in conformal therapy but has its own rich literature and is not covered in this volume. What makes conformal radiotherapy so novel and exciting is that, whilst the ambition to achieve it is as old as x-rays themselves, it is only in very recent years that the prospect is becoming reality. This is because of the synergistic confluence of progress in radiation acceleration technology, in computational power for three-dimensional dose planning, in three-dimensional imaging for tumour localization and therapy monitoring and in techniques for ensuring and assessing the accuracy of positioning at the time of treatment. Developments in automatic techniques for shaping the beam, tools for characterizing three-dimensional dose distributions and methods of predicting biological response from dose plans are all new and contribute to this synergism.

It might be argued that it is too soon to prepare a work reviewing this field. All these aspects of conformal therapy are in a fluid state, continually being developed. In this volume I have tried to extract the key elements which will probably stand the test of time; but it is inevitable that conformal therapy will continue to develop and this book is prepared in the hope of being useful during this development. Some of what is presented is moderately controversial and where this is the case I have tried to document the elements of doubt and debate.

Why select high-energy photons as the tool for conformal therapy? By the very nature of photon–tissue interactions, they have an in-built disadvantage in scattering and depositing dose proximal and distal to the target volume. In an ideal world they would not be the radiation of choice for this challenging problem. Protons are inherently 'better behaved' since, to first order, they travel in moderately straight lines and they can be tailored to deposit none of their energy distal to the target. Protons have been irradiating patients for close on 40 years yet the worldwide patient totals are still only some 10^4 compared with millions of photon treatments. Until now the technology has been simply too costly, yet now this modality is receiving renewed interest and is reviewed here.

Conversely the unwanted photons distal to the target volume and exiting from the patient can be used to advantage in the developing field of portal imaging.

Traditionally, film has been the medium of choice, but in the last ten years new detectors have been developed with many advantages over film as detector, a possibility obviously not available to proton irradiation whose quality control has to be performed differently. Megavoltage imaging, both planar and tomographic, is a growing activity, largely, though not exclusively, non-commercial-based at present.

In the last ten years magnetic resonance imaging has been added to the arsenal of imaging modalities which have been applied to planning and monitoring radiation therapy. These include x-ray computed tomography, digital subtraction angiography, single-photon emission computed tomography, positron emission computed tomography and ultrasound. Organizing tomographic imaging into truly three-dimensional treatment planning has become another major growth area both within university and hospital centres and in industry.

Techniques for treatment plan optimization have been available for many decades. In particular, optimization by rotation techniques were once very much in vogue. By drawing analogies with image reconstruction methods, dose planning by inverse computed tomography has attracted interest and become a very real possibility with modern, fast digital computers. Using the multileaf collimator to create beams with irregular geometric shapes, tailored to the projection area of the target volume, it is now possible to optimize the arrangement of a finite number of such fields, creating irregular and indeed concave high-dose volumes. The multileaf collimator is over ten years old, but only now coming into use in more than just a few centres. Photon dose planning based on point-dose kernels (the distribution of dose around a photon interaction at a point) is also just becoming possible.

Stereotactic radiotherapy and radiosurgery have been practised for four decades but with the realization that linear accelerators can potentially do the work of the special-purpose gamma knife machines, there is a resurgence of interest. The underlying physics requires a very special confidence in the use of medical images to guide the precise position of the small-area radiation beams. Along with brachytherapy this is also in a sense the ultimate in conformation. It finds use not only in treating tumours but also benign disease such as arteriovenous malformations.

Many physicists and radiotherapists content themselves with dose distributions, relying on clinical experience to translate these into the predicted biological response of tumours and normal tissue. Work is underway to put the translation from the surrogate, dose, to biological outcome onto a firmer footing. The field is controversial, not least because the models are complicated and much of the basic radiobiological data are poorly understood or disputed. Statements about biological outcome made here should be treated with caution.

The process of treating a patient is a chain involving all these activities. Any weak link in the chain reduces the efficacy of making the other steps more precise (figure P.1). Hence although most of us can only expect to contribute to one small part of the process, a perspective on the wider context is needed. Some go so far

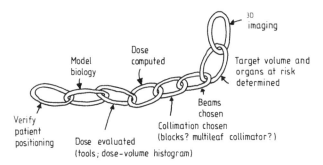

Figure P.1. *Showing the 'chain' of processes leading to conformal radiotherapy and illustrating that the physical basis of radiotherapy depends on satisfactory performance at all stages.*

as to say that an effective radiotherapy centre with its physics support must be able to demonstrate that attention has been paid to all parts of the chain. This is not easy in a world of limited resources, both human and material, but it should be a goal.

Steve Webb
April 1992

ACKNOWLEDGEMENTS

The Institute of Cancer Research and the Royal Marsden Hospital collaborate on a large research programme working towards improving the physical basis of radiotherapy leading to achieving conformal radiotherapy. Clinical research is conducted under Alan Horwich in the Department of Radiotherapy and, alongside, the Joint Department of Physics under Bob Ott is developing new techniques to improve the physical basis of conformal radiotherapy. In turn, this research fuelled interest in a postgraduate teaching course on the physics of conformal radiotherapy, which led to the production of this book. I am enormously grateful to all those whose enthusiasm for this subject has supported this research and teaching.

I am particularly grateful to my colleagues who are experts in specific aspects of radiotherapy physics who have shared their knowledge with me. These include Bill Swindell (Head of the Radiotherapy Physics Team), Alan Nahum, Philip Mayles, Philip Evans, Mike Rosenbloom, Roy Bentley, Glyn Shentall, Margaret Bidmead , Jim Warrington, Sarah Heisig and Penny Latimer. Several PhD students including Glenn Flux, Michael Lee and Rong Xiao, who attended the course, have also provided helpful comments.

I am most grateful to Sue Sugden and Pru Rumens in the Institute of Cancer Research Library for helping me to obtain copies of references and to Ray Stuckey and his photographic staff for help with half-tone illustrations.

The research work in conformal radiotherapy is concentrated in so few centres worldwide that in writing these reviews, some 'household names' in the radiotherapy physics world keep appearing. It would be invidious to single out some subset for listing here but the reader will have no difficulty identifying them from the fruits of their research and my greatest debt in preparing this work is to them. I hope this compendium will make their work even better known and that they will approve, even though inevitably this can be only a précis of their achievements.

I should like to thank all the publishers who have allowed figures to be reproduced. The authors are acknowledged in the figure captions. All publishers were contacted with request for permission to reproduce copyright material.

The reference lists are up to date as of May 1992.

The work of the Institute of Cancer Research and Royal Marsden Hospital is supported by the Cancer Research Campaign.

I thank Sean Pidgeon (Commissioning Editor) at Institute of Physics Publishing for his enthusiasm for this project.

I am also grateful to Mark Telford and Jenny Pickles of Institute of Physics Publishing for their excellent work in the publication of this book.

This book is dedicated to Linda and the fight against chronic disease.

CHAPTER 1

THREE-DIMENSIONAL RADIATION-THERAPY TREATMENT PLANNING

1.1. CONFORMAL RADIOTHERAPY TREATMENT PLANNING

1.1.1. What is conformal radiotherapy?

The aim of radiation therapy is to tailor a tumourcidal dose envelope to a target volume and to deliver as low a radiation dose as possible to all other normal tissues. This is what is meant by conformation or conformal radiotherapy—the high-dose volume 'conforms' to the target (figure 1.1). By target volume is implied the full extent of the tumour including any marginal spread of disease (sometimes called the biological target volume (BTV)) and a 'safety' margin (extending the BTV to the so-called mobile target volume (MTV)) (Urie *et al* 1991). A simultaneous goal is to minimize the dose to organs at risk. Against the background of these aims, what is perhaps surprising is that so little practical radiotherapy completely achieves this goal. The reasons are not hard to establish. Difficulties in determining the three-dimensional target volume, in designing appropriately shaped and oriented radiation ports, in calculating the dose distribution from complex superpositions of radiation fields in inhomogeneous tissue and in establishing that the patient is correctly positioned at the time of treatment delivery, all conspire to defeat the aims. It is now recognized that only by improving the physical basis of radiotherapy at *all* of these stages can the problem of accurately establishing the physical basis of radiotherapy be said to be properly tackled. As may be expected there are several groups of workers attacking separate aspects of the problem but few taking on the full burden of solving the complete chain of tasks (Schlegel 1987, Suit and Verhey 1988). In the meantime practical radiotherapy is a patchy activity varying at worst from techniques little changed for decades, through to the most ambitious protocols, with a wide spectrum inbetween. Optimizing radiotherapy with the goal of conformation is highly desirable to remedy this situation. This is not meant to imply that when the physical basis of radiotherapy has been optimized all the problems magically

1

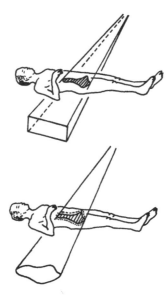

Figure 1.1. *Illustrates the difference between 'conventional' and 'conformal' radiotherapy. In the upper figure a rectangular beam is shown irradiating a target volume (shaded) which changes its cross section with longitudinal position. Radiotherapy might comprise a number of such beams, coplanar, with at best some blocking. Inevitably this irradiates some normal tissue in an unwanted manner. In the lower figure, one of many, possibly non-coplanar, beams is shown with the shape of the beam tailored to the projection of the target in that direction. The result of using a number of such beams should be a dose distribution (an isodose surface is represented dotted) which conforms to the surface of the target volume.*

vanish. There will still be difficulties associated with incomplete radiobiological knowledge—i.e. how does tissue damage correlate with radiation dose—and of course this whole discussion can only apply to treating well-differentiated local primary disease; metastases are another matter (Horwich 1990).

It has been estimated (Suit and Westgate 1986) that improved local control is so important that thousands of lives per year could be saved by improving the physical basis of radiotherapy and this has become the driving force behind the efforts to achieve conformal therapy.

Smith (1990) voices concerns about how the new technology for conformal radiotherapy should be evaluated. In his view many authors have not reported quantitative evaluations of the real clinical importance of their physics developments. In particular whilst claims that conformal therapy reduced the dose to normal tissues, it was felt necessary to prove that the tissue otherwise irradiated in non-conformal therapy was clinically damaged. Conformal techniques should be compared with the best available non-conformal techniques. Techniques improving dose calculation should be quantified in terms of tumour control probability and normal tissue complication probability with the dose normalization changed to increase the latter to tolerance and consequently improve the tumour control. Attention should be given to the possible problem of 'good' dose distributions being spoilt by unwanted patient movement.

Conformal radiotherapy is technologically complex and would appear to be justified by evidence that improved dose distributions would lead to improved local control. Lichter (1991) argues that these technological improvements should now be given the opportunity to show whether real clinical advantages can be produced compared with conventional technology. He raises several, almost moral, dilemmas. Firstly, researchers should take time to define the clinical experiments, the end points and the potential outcomes. Are we interested in more cures or better local control with reduced morbidity, or the same cure with fewer acute side effects? Secondly, Lichter questions the timing of such trials. Should these perhaps wait until the technology of multileaf collimators (chapter 5), optimization (chapter 2), megavoltage imaging (chapter 6) and 3D treatment planning are all well established? Thirdly, would patients tolerate being randomized into low-technology versus high-technology trials?

Since three-dimensional treatment planning (3DTP) is central to conformal therapy we begin with discussing planning tools.

1.1.2. The planning tools required for conformal radiotherapy

Fraass (1990) and Fraass *et al* (1990) have broken down the planning problem into its component physical considerations. The improvement offered by the most recent treatment planning computational tools arises from the use of fully three-dimensional data to define anatomy, design beams, calculate doses and evaluate plans. These tools are now becoming available in some centres as computational hardware becomes fast enough and as the problems of utilizing three-dimensional medical images are solved. Some three-dimensional planning systems have been developed 'in house' by university hospitals and commercial three-dimensional planning systems are also becoming available.

1.1.2.1. Conventional 2D treatment planning.

Let us recall what we might call the 'conventional' approach to treatment planning (and in doing so make some broad generalizations). Rosenman (1991) provides a useful thumbnail historical account of the development of radiotherapy planning and the subject is given tutorial coverage by Bleehan *et al* (1983). Before the advent of commercially available computerized tomography (CT) around 1972, the target volume was estimated from planar radiographs. With the patient lying in the treatment position, a few radiation fields or ports were determined at the simulator with the gantry at maybe two, three or four orientations spaced on the circumference of a single circle. The ports could be of different sizes at each orientation but were generally rectangular in shape since this is all the field collimators could achieve (figure 1.2). The use of a single circle was regulated by the treatment machine, which could only deliver coplanar fields. Non-coplanar fields were possible by

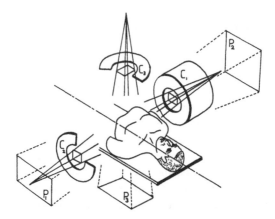

Figure 1.2. *An illustration of some of the features of 'conventional' (i.e. non-conformal) radiotherapy. The patient is being treated with a pair of parallel-opposed lateral x-ray fields and an anterior–posterior field. The three fields are coplanar, i.e. their central axes all lie in a plane with the fields being set up by rotating a gantry about a stationary patient lying on the patient support system. Each field is collimated to a rectangle by two pairs of orthogonal lead jaws. C_i $(i = 1, 2, 3)$ represents the collimation. The high-dose volume, where the three fields overlap in the patient, is approximately a parallelpiped (not exactly because of beam divergence). Unless the tumour happened to be this shape, which is unlikely, much normal tissue is unnecessarily irradiated. Even if it were this shape, the dose from just a few beams would have inhomogeneities. The fields might be partially blocked to change this shape. P_i $(i = 1, 2, 3)$ are the three exit portals. Conventionally a film will record the exit radiation and be compared with a simulator film. The portals are not shown equidistant from the isocentre to avoid cluttering the diagram. In practice the film would be placed closer to the patient than shown. The treatment couch is flat.*

twisting the couch but treatment planning was difficult if dose calculation was not confined to a single plane. The assumption was made that the patient was cylindrically uniform over the axial extent of the radiation fields. An outline of the external contour was made in the midplane of the fields using a lead wire or a plaster-of-paris cast. A treatment plan was calculated—in early days by hand but later by computer—using this outline, and appropriate orientations and beam weights for the ports were found. The dose plan was generally represented by a series of isodose lines in this single transaxial plane. A decision on the suitability of the plan was made by eye based on rules of thumb and experience.

Of course the above description is arguably too simple but serves as the basis of explaining the improvements which are now being introduced with three-dimensional treatment planning. Decades of experience honed a variety of techniques tailored to specific tumour sites and ingenious attempts were made to improve matters. Before CT, simple blurred 'classical' tomograms were used to visualize transaxial anatomy; couch and head twists were designed to avoid the restriction to coplanarity; estimates of air spaces were made and simple tissue-inhomogeneity corrections were attempted. Many hours at the planning computer were devoted to trying different beam positions to avoid sensitive structures and to improve the dose distributions. Blocks were designed at the simulator. However, all up, the available technology provided the limit which no amount of human ingenuity could overcome.

Let us critically list the limitations of the classical approach:

1. Planar x-rays often did not show clearly the tumour extent.

2. The tumour extent was determined anatomically—functional imaging with good spatial resolution was often unavailable and, where available, was often not correlated with anatomy.

3. The correct 3D shape of the tumour was not determined.

4. The 3D location of sensitive structures was not determined.

5. The assumption that the patient was uniform axially over the length of the treatment field was wrong.

6. Planning was made in a single axial slice, usually a mid-field slice—there was no distinction between the behaviour in this slice and that in adjacent slices.

7. The use of a few beams limited the possibilities for tailoring the dose to the target region. The high-dose volume was far larger than the target volume and contained normal and potentially sensitive structures—so-called 'organs at risk'. Tumours with irregular shapes were 'boxed in' to cubic or rectangular–parallelpiped high-dose volumes.

8. The beams were axially coplanar.

9. The body contour was difficult to determine—it may even have been altered by the method of measurement (e.g. pressed by lead wire).

10. The internal tissue inhomogeneities were hard to determine accurately.

11. The dose plans were computed by an algorithm which was too simple.

12. The dose planning took no account of the 3D nature of the body surface and changes in density in 3D.

13. The doses were displayed as isodose lines in a plane. There was no quantitative method of determining how much of the tumour received dose within specified ranges. Even if this were worked out for the single plane in which the doses were computed there was no way of finding out what happened in the adjacent tissue.

14. The beam ports were rectangular, not tailored to the projection of the treatment volume.

...and so one could go on. Perhaps the most serious deficiency was that one simply had no way of knowing what dose had really been delivered.

Notice in this catalogue of difficulties that there were several 'missing technologies' which all conspired to produce these limitations. Medical imaging was planar and largely anatomical. Treatment machines had very limited geometrical movements (rotational movements and rectangular fields). Computers were too slow for accurate dose planning (see, for example, Cunningham 1988) and graphic display of dose and tissues was limited.

1.1.2.2. The impact of x-ray CT. In 1972 one of these limitations was dramatically removed with the advent of x-ray computed tomography (CT). In a very short time, accurate 3D tomographic information on both diseased and normal tissues became available and was quickly harnessed to treatment-planning computers (Parker and Hobday 1981, Goitein 1982, Lichter *et al* 1983). However, no such widely available dramatic improvement followed in the design of treatment machines, which were still only able to deliver rectangular (maybe blocked) fields from a limited number of angular directions. There were still no 3D computer planning systems. So by no means all the limitations listed above disappeared. It was still the norm to make use of a mid-field axial slice and imagine the patient to be uniform axially for the purposes of treatment planning. However, it was possible to superpose the central slice plan on the true anatomical adjacent slices and make some qualitative observations. There was generally no attempt to compute the dose–volume histogram (i.e. the histogram showing the fractional volume of tissue raised to specific dose levels—see section 1.1.9), partly because, of course, the non-central dose distributions were incorrect and partly because such calculations were still prohibitively time consuming. In short, 3D dose calculations were generally not performed despite the availability of 3D anatomical information. 3D display and plan evaluation had to be sacrificed as a result. The limitations of the treatment machines prohibited imaginative choices of port shapes and the use of a large number of beam orientations. It is for this reason that, although in terms of visualization, CT had a dramatic impact on radiation treatment planning, the methods used were still essentially *two* dimensional. Most of these requirements for 2D imaging were met by so-called 'turn-key' systems, marketed commercially, widely tested, but somewhat limited in scope.

Lichter (1986) has pointed out that the clinical results obtained using CT-based treatment planning have never been evaluated in comparison with conventional planning methods for the very good reason that CT enabled clinicians to define anatomy so much better that it would have been unethical not to use the data in one arm of a trial (data on the frequency of *changing* plans with CT knowledge is available but this is not the same as assessing outcome).

1.1.2.3. Three-dimensional multimodality medical imaging and 3D treatment planning. Several attempts were made soon after CT data became available to use it properly to compute 3D dose distributions. However it is only now that computers are sufficiently fast to allow this in sensible times. With the passage of time since the early seventies, further 3D imaging modalities have become available, such as emission tomography (SPECT and PET) and magnetic resonance imaging (MRI). For an overview of the physics and engineering of 3D medical imaging the reader could choose from (among other books) Moores *et al* (1981), Hamilton (1982), Wells (1982), Gifford (1984), Guzzardi (1987), Aird (1988) and Webb (1988). Aspects of medical imaging relevant to planning and evaluating radiotherapy are discussed in chapter 8 along with some brief details of how these technologies work. There are special requirements for CT scanning producing images for radiotherapy treatment planning (Lichter *et al* 1983, McCullough 1983, Goitein 1983, Sontag and Purdy 1991) including: a flat couch, patient in the treatment position with quiet respiration, contiguous thin slices (3–5 mm in the head; 5–10 mm in the body), large patient aperture, scout film indexed to scans etc. MRI allows imaging in arbitrary planes, can differentiate certain tissues better than CT and is useful for monitoring response to treatment, but may suffer image geometrical distortion and does not give electron densities directly (Sontag and Purdy 1991). Moerland *et al* (1991) have suggested distortion could be as much as 5 mm. Ultrasound does not find great use except in determining chest wall thickness and external contours. Emission computed tomography is also beginning to play a part in treatment planning and monitoring. Techniques to merge and combine 3D image data sets are being developed (see section 1.2); non-axial slices may be viewed. Using the beam's-eye-view of 3D structures created from 3D image datsets, multiple non-coplanar geometrically shaped fields can be defined at arbitrary orientations (figure 1.3) (McShan *et al* 1990). 3D dose calculations can be performed including 3D inhomogeneity corrections and the results may be shown superposed on 3D anatomy either as sequential slices or as shaded-surface distributions. These facilities enable development of tools for plan analysis and comparison whereby it can be precisely established what fraction of the treatment volume receives dose in prescribed ranges. The scene is set for removing all of the physical limitations listed in section 1.1.2.1 and to permit true 3D planning (Sontag and Purdy 1991, Laughlin *et al* 1991). Two major obstacles remain. Firstly, the computational hardware is expensive and the software environments to achieve these goals are only developed at a small number of centres. Some are available commercially, but at a high price. For the most advanced dose-calculation techniques, computer times remain high. Secondly, the treatment-delivery machine hardware has simply not kept up and is still geared to the 'conventional' era or at best the '2D planning with CT' era. Some imaginative workers are developing dynamic therapy (see chapter 7) or multileaf collimators (MLC) and there is a need to automate the computation of the settings of the leaves (see chapter 5). In the early 1990's these are not yet the normal modes of treatment.

At least in a few centres, planning treatments with full use of three-dimensional information has been achieved and attention can turn to questions of plan evaluation. Once again the tools may be available to quantify dose–volume histograms, but there is still considerable debate over how to turn this physical data into a prediction of biological response (Tait and Nahum 1990) (see section 1.1.11). This requires knowledge of tumour control probability and normal tissue complication probability, and whilst some data are available (e.g. Lyman 1985), and methods of using it (Kutcher and Burman 1989), the data are incomplete and open to questioning (McShan 1990b). Against this background it is not surprising that most workers still interpret the treatment plan in terms of dose assessment rather than biological effect. Coupled with this is the development of methods to ensure that the patient is correctly positioned at the time of treatment. Digital portal imaging (see chapter 6) and megavoltage computed tomography can gather

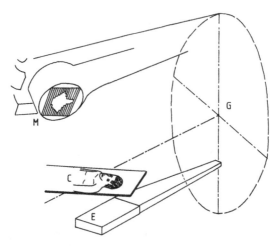

Figure 1.3. *An illustration of some of the features of 'conformal' radiotherapy. The patient is being treated by a number of beams which have been shaped to the projected area of the target by a multileaf collimator (M). This enables the fields to depart from rectangular and take an irregular shape. The central axis of the one beam shown does not lie in the transaxial plane of the patient because the longitudinal axis of the couch (C) is not orthogonal to the plane of rotation of the gantry (G) (couch twist). One might imagine a number of beams whose central axes do not lie in a single plane, a situation described as 'non-coplanar'. The image of the portal radiation is captured by an electronic portal imager (E) diametrically opposite the x-ray source.*

the data to assess how well this is done with present positioning techniques and allow repositioning, should errors be thought excessive (Wong and Purdy 1990, Evans *et al* 1990, Wong 1991).

If the 2D distribution of portal intensity is measured by a digital scanner, the intensity can be backprojected into the megavoltage CT imaging data (Ishigaki *et al* 1990). This gives a first-order map of primary dose. If the dose kernel (see chapter 2) were to be folded into this, the result would be a map of delivered dose including scatter. Such 'transit dosimetry' is still in its infancy and not a technique in the repertoire of most radiotherapy departments (see also chapter 6, section 12).

It would not be particularly instructive in a teaching text to describe in detail specific treatment-planning systems since these are continually developing and not all the features required of a 3D treatment planning system are available in all systems. Also this level of detail is only really needed by the user of a particular system who will want to consult the papers of the groups concerned. The main 3DTP systems currently in use are at:

- the Department of Radiation Oncology, University of Michigan at Ann Arbor (McShan and Fraass 1987, Fraass and McShan 1987, Fraass 1986, Fraass *et al* 1987, McShan 1986, 1990a,b, Kessler *et al* 1991) (*SCANDIPLAN; Scanditronix);
- the Joint Centre for Radiation Therapy, Massachusetts General Hospital, Boston

(Goitein *et al* 1983b, Kijewski 1987);
- the German Cancer Centre at Heidelberg (Schlegel *et al* 1987, Boesecke *et al* 1987) (*part system/ STP, Fischer);
- the University of North Carolina at Chapel Hill (Sherouse *et al* 1990);
- the University of Tampere, Finland (Vitaenen) (*CADPlan; Varian);
- the Karolinska Hospital, Stockholm, Sweden (Brahme 1992);
- the Memorial Sloan Kettering Cancer Centre, New York (Laughlin *et al* 1991);
- the University of Pennsylvania School of Medicine and Fox Chase Cancer Centre, Philadelphia (Sontag *et al* 1987, Sontag and Purdy 1991);
- the Mallinckrodt Institute of Radiation, Washington University School of Medicine, St Louis (Purdy *et al* 1987, Sontag and Purdy 1991).

The systems asterisked have been marketed commercially (names shown in brackets). Other commercial systems include the AB Helax system.

Instead of describing these 3D planning systems in detail, the principal requirements for 3DTP and how these might be met are discussed in this chapter.

Sontag and Purdy (1991) emphasize the need for good documentation for new 3D treatment-planning systems. Jacky and White (1990) have pointed out the requirement for careful validation of the accuracy of 3D planning software. By comparing the predictions of computer calculations and the mathematics which is supposedly coded, the programs are validated independent of the weaknesses of any particular beam model. Verification by not comparing with measurements can show up programming errors which might otherwise masquerade as inaccuracies of the model. Smith (1990) summarizes the two major initiative in 3D radiation therapy planning covered by the Nordic CART (Computer aided radiotherapy) initiative, started in 1985 and by the American Centres supported by NCI contract, started in 1982 and reported in a special issue of the International Journal of Radiation Oncology, Biology and Physics in May 1991 (see also section 1.3).

Brahme (1992) has re-emphasized that fully 3D treatment planning requires *all* of the following to be performed three-dimensionally: diagnosis, biological response monitoring, dose calculation (including optimization and possibly the use of 3D dose kernels), treatment simulation, display, dose delivery, fixation and dose verification.

Achieving 3D treatment planning does not necessarily in itself imply that conformal therapy will be the result. A poor set of choices for the beams would certainly not achieve this objective. However, 3D treatment planning is such an important component of the chain of events leading hopefully to conformal therapy, that it is the first part of that chain that is discussed in some detail.

1.1.3. The computer system

Among the earliest 3D planning systems making use of extensive computer graphics were those of Reinstein *et al* (1978) and McShan *et al* (1979). As computer facilities have become faster and cheaper, more sophisticated features have been developed. A 3DTP system requires a computer with a large storage capacity (several gigabytes) for 3D images and dose distributions together with some hundred megabytes of memory. Graphics display systems supported by device-independent GKS graphics software need to be attached. These should preferably support multiple-window displays, since many people find it easier to build up a 3D impression from simultaneously viewing multiple 2D slices than from true (shaded-surface) 3D displays of anatomy and dose. DEC VAX systems were chosen by the Ann Arbor team for example to meet these requirements. The system should be menu driven with interaction by mouse and keyboard. 3D dose calculations take a long time and for these batch computing facilities are needed (Tepper and Shank 1991).

1.1.4. Localization and coordinate systems

Functional imaging modalities such as SPECT and PET, and anatomical imaging by x-ray CT and MRI can all provide data for input to the treatment-planning process. There is no real concensus yet on the impact and usefulness of MRI data in treatment planning, but this will surely come in the next few years. Margulis (1992) has stressed that medical images, including MRI, assist accurate staging of disease as well as the determination of the target volume. Unfortunately the data obtained from these imaging modalities are not usually in the same format. Pixel/voxel sizes, the number of pixels per frame (tomographic slice), conventions for labelling and sequencing frames, are all potentially different. Also the patient may not have been in the same position when each of the studies was performed. Consequently the first requirement of a treatment-planning system is that it should be able to accept a variety of data input types and reformat data into a common arrangement of voxels. Secondly, it should provide for converting these into a single common reference system. The first requirement is not difficult to meet provided the manufacturer of the imaging equipment will divulge the data format. The second is far more difficult. Rotations, translations and changes of scale are relatively easy to accomodate by matrix algebra, but these only apply if the relative movements between datasets can be considered rigid transformations. If one set is warped relative to another the problem is more complicated and many techniques have been tried to overcome this (e.g. Kessler 1989). In view of the fundamental importance of image registration in three-dimensional radiation treatment planning, this topic is treated at length in section 1.2. The end result of these transformations is that the image capability of the treatment-planning system should become independent of the origin of the imaging studies. The planning system should use a coordinate system tied to the treatment room and the images should be locked into this system. Fraass (1986), Fraass *et al* (1990) and McShan and Fraass (1990) have called this the 'dataset concept' and implemented it in the University of Michigan planning system 'UMPLAN'. The use of coordinate transformations also allows positioning differences between the simulator and any single imaging modality (such as the CT scanner) to be observed and corrected. If imaging is available on the treatment machine then positioning errors at the time of treatment could be similarly corrected (see chapter 6). The imaging coordinate systems may be defined relative to some marks on the patient's skin or attached markers at the time of scanning or with reference to bony landmarks or external patient restraint systems (see section 1.2).

1.1.5. Structures from images

The staple input is x-ray CT, generally a large number of transaxial slices together with a scout view (see also chapter 8). The scout view is a digital planar x-radiograph taken on the CT scanner without gantry rotation and showing the longitudinal positions of the transverse sections (which are sometimes called slices or cuts) (figure 1.4). It is not the same as a digital planar radiograph taken with the beam diverging in all directions; in the scout view the beam diverges only in the transaxial direction. The treatment planner should be able to request non-axial slices including oblique cuts. Structures can be outlined by autotracking, i.e. the computer will automatically threshold the CT numbers to create a contour. Where this is inappropriate the contours of structures may be drawn by hand, usually using a mouse or trackball. If a large-screen digitizer is available, this process is more accurate. In particular, parallax errors can be reduced (McShan *et al* 1987). 3D treatment planning requires large numbers of contiguous CT slices and there is clearly a tradeoff between wanting thin slices and wanting high patient throughput on the CT scanner (Sontag and Purdy 1991). From the ensemble of contours, 3D solid surface views can be created. Some (e.g. Gildersleve 1991, Wong and Purdy 1990, Tepper and Shank 1991) regard this aspect of treatment planning as the most time consuming, generating a bottleneck in the series of processes from imaging to treatment setup. There is thus growing interest in 'image segmentation' and also in 'artificial intelligence', whereby a computer might be able to recognize specific anatomy from its shape, density or location in relation to other structures (Sontag and Purdy 1991). This has presently not been achieved and would be a difficult computer problem to solve. The basic need is to define (i) the 'target volume' being the tumour volume, margins of spread, safety margins, in fact all the tissue which it is required to raise to high dose, and (ii) 'organs at risk', being the sensitive structure which must be kept at low dose. If these are next to each other high dose gradients are the aim. There is plenty of evidence that the most important strategy is to achieve superior dose distributions in terms of increasing tumour control probability and also reducing treatment related morbidity. Obviously the worst possible error is to improperly define the target volume; this can never be corrected by later adjustments. For this reason conservative estimates of the target volume are often made.

Wambersie *et al* (1988) stress the over-riding importance in treatment planning of correctly determining the target volume. This is a clinical decision and no amount of quality control of other parts of the chain of activities involved in delivering radiotherapy can correct for wrong specifications at this stage.

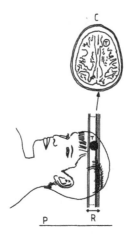

Figure 1.4. *Radiotherapy planning (RTP) makes use of the three-dimensional imaging modalities, x-ray computed tomography (CT), magnetic resonance imaging (MRI), single-photon emission computed tomography (SPECT) and positron emission tomography (PET). X-ray CT was historically first used in RTP and is still the most common. Tomograms (slices) are made over a range R spanning the expected location of the tumour T. P is the plane of the scout view. An outline of the target T is made on each slice C. Other normal tissue and organs at risk may also be contoured, together with the patient external outline. These data form the geometrical input to treatment planning. Oblique slices may be made by reformatting transaxial data. MRI is capable of directly generating non-transaxial images. 3D imaging may also be used to monitor the outcome of therapy.*

1.1.6. Beams

A truly 3DTP facility will allow beams to be specified from any orientation, be able to block such fields and then show the resulting dosimetry superposed on any cut through the 3D images. This requires that if multiple windows are in operation and the beam positions are changed in one window (say, a transaxial cut) the beams are immediately updated in the other windows displaying different cuts.

Sailer *et al* (1990) have provided a very neat solution to the problem of labelling beams which can lie anywhere in 3D. Their paper is entitled '3D beams need 3D names'. Radiation beams have traditionally taken their names, such as anterior, posterior, lateral etc, from radiology. These names are unambiguous. However extensions such as 'oblique' or 'left anterior oblique' fail to convey 'how much left?' and 'how much anterior?' The solution lies in using the three primary axes in the patient. These are left(L)–right(R), anterior(A)–posterior(P), and inferior(I)–superior(S). Generally oriented beams can then be labelled: $A_1\theta_1 A_2\theta_2 A_3$ where A_1 is the nearest primary axis to the axis of the beam, θ_1 the angle to the second-closest primary axis A_2 and θ_2 the angle to the third-closest primary axis A_3. For example the beam shown in figure 1.5 is 'L40A20I'. By definition the larger angle must appear first in the specification and this is a useful check. Beams

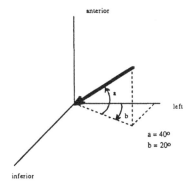

Figure 1.5.   *A suggestion for how the directions of beams might be labelled. A 'complex beam' (i.e. one not along a principal axis) is located in one octant of three-dimensional space. This beam, which is closest to the left principal axis, is named 'L40A20I' (from Sailer et al (1990)) (reprinted with permission from Pergamon Press Ltd, Oxford, UK).*

lying in a primary plane—one of transaxial, sagittal or coronal—have a simpler three-element specification.   For example a beam labelled 'A20L' lies in the transaxial plane, whilst 'A30S' lies in the sagittal plane. An example of a 'simple beam' lying along an axis would be 'L'. This method should give unambiguous communication of field directions, relative to patient axes. The transformation between patient axes and equipment (gantry/couch/collimator) axes would be needed to calculate the beam orientations in these coordinates. The terminology was extended by Goitein (1990).

1.1.7.   Blocked fields and the beam's-eye-view

The beam's-eye-view (BEV) of the target geometry is particularly useful for making decisions concerning the shape of blocks or irregular multileaf-collimator-defined fields to shield sensitive structures (figure 1.6).   The concept was introduced by Goitein *et al* (1983a).   A popular form of BEV is to display a perspective view, from the beam's source position, of stacked contours (figure 1.7). This is especially effective when depth cueing (closer regions made brighter and vice versa) is used. Shaded surface BEV display is also possible (Tepper and Shank 1991). Sontag and Purdy (1991) emphasize the improved usefulness of the BEV if it can show the isothickness lines of any wedge in the field. The shape of blocks should be able to be drawn directly on the BEV display using a joystick or trackball. This can sidestep the practice of defining blocks at the simulator which might only be changed when verified at the BEV stage. Block design becomes part of the treatment planning rather than the simulation process. The use of the beam's-eye-view can also effectively determine the optimal direction of beam entry (Wong and Purdy 1990). Where the flexibility exists it is obviously good to choose views of the target volume which exclude as much nearby sensitive tissue as possible (Reinstein *et al* 1983). Indeed, Goitein (1982) argues that determining

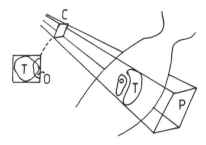

Figure 1.6.   *Schematic showing an x-ray beam collimated (non-conformally) to a rectangular cross section intersecting a patient. T represents the target volume; O represents an adjacent organ at risk; C is a plane orthogonal to the beam axis at which the 'beam's-eye-view' of the patient may be constructed. This (as the name suggests) is the view of the patient from the source—it is shown schematically connected to C with a dotted line. Note in this example the view of the target volume overlaps the view of the organ at risk. It might be possible to remove the overlap by orienting the beam differently. A conformal field would shape the radiation to the outline of the target as seen at C by a multileaf collimator (see chapter 5). P represents the plane in which a portal image can be formed (see chapter 6).*

the best geometrical field shape (to exclude as far as possible organs at risk) is a matter of higher priority than paying undue attention to the algorithm for calculating the dose to the target volume. This philosophy explains why many commercial treatment planning computers put great emphasis on geometrical precision and still use relatively uncomplicated photon algorithms. The use of non-coplanar beam portals is clearly fundamental (Laughlin *et al* 1991). It turns out that the optimum set of geometrically shaped beams defined by multi-leaf collimators (see chapter 5) can be determined entirely computationally so as to achieve this aim (Webb 1991; see also chapter 2). The process of planning by use of beam's-eye-view is sometimes called 'virtual simulation' (Sherouse *et al* 1990, Sherouse and Chaney 1991).

The 3D radiation therapy treatment planning system at the University of North Carolina is known as the 'virtual simulator' (Rosenman 1991). The essential feature of this 3D treatment-planning system is that all planning is done on the computer, based on tomographic (CT) data. A screen displays wire-frame 3D images of the patient and small versions of selected CT slices are also on screen. As the planner manipulates the beams looking at the 3D image, the positions of the beams in the 2D images is continually updated. When the final choice of beam positions has been made and the dose calculation done, the beam's-eye-view and portal radiograph for each field (sometimes called the 'digitally reconstructed radiograph (DRR)') are computed (figure 1.8). Only then does the patient go to the physical simulator for real x-ray images to be made for quality assurance purposes. Some other treatment planning computers (e.g. IGE TARGET and THERATRONICS CT–SIM) also have this facility. The DRR becomes the standard against which portal images taken on the treatment machine

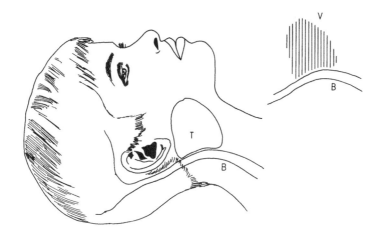

Figure 1.7. *The beam's-eye-view of the target T may be derived from* CT *data. Each slice allows one contour to be defined. In the beam's-eye-view V of the target the contour is seen edge-on and can be represented by a thin looped line. Each slice gives rise to one thin loop so in the example shown 17 slices span the target. The brainstem B can also be seen in the beam's-eye-view. It is clear in this example that a rectangular field enclosing the target would also enclose part of the brainstem. The beam's-eye-view thus allows the treatment planner to define the position of blocks. Alternatively the field could be shaped by a multileaf collimator. Because of beam divergence each contour in reality appears in perspective as a thin closed curve. Only the central axis of the beam sees the contour as a line. For simplicity single lines have been shown here instead.*

can be compared (Tepper and Shank 1991). However, it is resolution limited because of the fairly large slice thickness of the CT data from which it is generated (Sontag and Purdy 1991).

McShan *et al* (1990) discuss the great importance attached to the beam's-eye-view facility in the UMPLAN 3D treatment-planning system. By establishing the matrix transformation between the treatment reference system and the beam coordinate system, structures can be mapped from the beam's-eye-view projection space to the tomographic image space and vice versa. Hence blocks can be designed at the planning stage as in the 'virtual simulator'. A digitally reconstructed radiograph can be matched to the portal image. Indeed the DRR can sometimes provide the *only* portal image in circumstances where the practical portal imaging system would collide with the couch.

The beam's-eye-view can also provide a measurement of the volume of normal tissue irradiated by a portal, even when it is shaped to the projected area of the target volume; this will occur if some or all of the organ-at-risk is also in the line of sight of the irradiation. Chen *et al* (1992) have created plots of such a volume

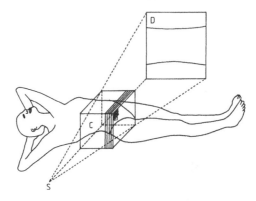

Figure 1.8. *A truly 3D treatment planning system can generate a digital portal radiograph (DPR) D by ray tracing from the x-ray source S through the CT data C. This ray tracing takes account of the ray divergence and differential photon attenuation. A series of CT slices are shown in the pelvic region of a patient. Because the in-plane pixel size will be smaller than the slice width, some interpolation is needed to obtain square pixels in the DPR, sometimes also called a digital reconstructed radiograph (DRR). The simplest DRR attenuates the primary radiation along each ray path. An experimentally determined portal radiograph at the simulator has contributions from scattered radiation also. Either the simulator radiograph or the DRR can be compared with a portal image at the time of treatment on the linear accelerator. If this is also digital, the comparison may be quantitative.*

(or fractional volume of the total organ-at-risk) as a function of beam orientation for a very large number of potential orientations including non-coplanar portals. They make the very plausible argument that these plots can help select suitable views. They do not claim this is an optimization technique since no doses have been calculated at this stage of the planning process. However this method of 'volumetrics' can be a good way to select an initial set of orientations from the large number of possibilities and then beamweights/ wedges etc can be optimized subsequently (see chapter 2). Another proposal from Chen *et al* (1992) is to display 'bitmaps' showing what fraction of the volume of an organ-at-risk receives primary dose from m fields when there are $n > m$ fields in all.

The BEV, useful though it is, has one major limitation. It shows the perspective of just one beam at a time. Sontag and Purdy (1991) have discussed the possibilities for 'physician's-eye-view' (PEV) whereby wire frames for the anatomy and *all* the beams are displayed as if the observer's eye is in the treatment room. PEV would help evaluate field abutments and gaps.

1.1.8. Dose calculation

The ideal dose calculation method should:

1. include all 3D geometry information;
2. make use of an accurate 3D map of electron density;

3. use a 3D description of the beam;
4. include 3D scatter and interface effects within the patient;
5. calculate the dose at all points in the volume;
6. be fast.

There is still no agreement on the 'best' photon algorithm (Cunningham 1982, Laughlin *et al* 1991). Computationally requirements 5 and 6 clash. Generally a crude matrix of dose points in selected planes are calculated fast allowing scope for changes and optimization. Once it is believed the beams are optimized, full fine-matrix calculations can be made. Perhaps surprisingly, for photon dose computation, the majority of commercial planning systems still use methods that were developed many years ago (Cunningham 1982, Sontag and Purdy 1991). The subject of dose optimization is considered more fully in chapter 2. It is now just becoming accepted that dose calculation based on the integration over dose-kernels is the most accurate way to properly compute three-dimensional distibutions (see chapter 2). However, elaborate though such methods be, it is presently not possible to perform such calculations as fast as needed for routine treatment planning. Hence for the time being they are generally reserved for benchmark computations, leaving day-to-day calculations still based on suboptimal dose-calculation models (Wong and Purdy 1990).

It is important to separate the problems of determining the dose *level* from the dose *distribution*. Obtaining the most uniform distribution of dose to the target volume whilst simultaneously sparing organs at risk is generally the physicist's problem (see chapter 2). The physicist must optimize the free variables (such as the number of fields, the position and shape of the fields etc) to try to achieve this. Deciding the target dose *level*, is a clinical decision which must be taken with adequate knowledge of the biological response of both normal and pathological tissue (see section 1.1.11). Dobbs (1992) has discussed the need for clarity in specifying the location within the target volume where the dose level should be set. A new ICRU report (ICRU 1993), to be published in 1993, will emphasize the need to clarify whether this is at the point of maximum, minimum, modal, mean or median dose value.

1.1.9. Dose display and evaluation of plans

1.1.9.1. Methods to display dose distributions; dose–volume histograms. Traditionally the dose matrix has been displayed as a series of isodose lines. The reason for this was threefold. Firstly, it emulated the output from 'hand planning'; secondly, in the early days of computer planning, grey-scale display of dose as a series of pixel values was not possible; thirdly, smoothly varying isodose lines could be interpolated from a smallish number of data points calculated on a coarse grid. Today the digital 3D dose matrix can be interrogated in many more ways.

Once the 3D dose matrix has been computed, the results may be displayed superposed on grey-scale anatomical images from CT or MRI. Generally it is

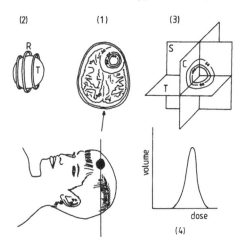

Figure 1.9. *Ways of displaying the results of a 3D dose calculation: (1) isodoses are drawn in a plane (here transaxial) and superposed on the grey-scale image (CT, MRI, SPECT or PET); (2) isodose ribbons wrap around a 3D shaded-surface image of the target; (3) axonometric display of transaxial (T), sagittal (S) and coronal (C) images with superposed isodose lines in all three planes—the images need not be from the same modality; (4) differential dose–volume histogram.*

adequate to evaluate the correspondence between isodose contours and target volumes by visually inspecting selected tomographic planes (figure 1.9). A useful facility is the display of dose with colour wash on the grey-scale anatomy. If the colour wash can be independently windowed, it is straightforward to visually perceive what fraction of a target volume has received dose within specified limits. If sharp and identifiable transitions in colour are used at discrete doses, these transitions define isodoses in the colour-wash display. Similar superposition of dose on anatomy may be performed with shaded surfaces in three dimensions (e.g. Rosenman 1991, Boesecke *et al* 1988, Tepper and Shank 1991). A facility is needed to view dose on anatomy in sagittal, coronal and oblique as well as the more usual transaxial slices. Some systems (e.g. UMPLAN) can produce an 'axonometric' display whereby selected slices in different directions (and if required even from different imaging modalities) can be displayed as a single perspective view. Isodose lines may be overlayed. Facility to retrieve 'dose at a point' and to display confidence limits are also desirable. Experience seems to have shown that 'movie loops', showing sequential slices of dose and anatomy, are not very successful (Tepper and Shank 1991).

However, it is in recognition that both the geometry and dose are known in complete 3D digital form that numerical evaluation of plans can be made with 3DTP. The cumulative (or integral) 'dose–volume histogram' (DVH) can specify the fraction of the target volume raised to specific dose values. This

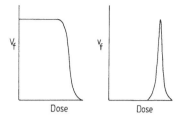

Figure 1.10. *Left—a schematic integral dose–volume histogram. The volume V_f is the fraction of the total volume irradiated to the stated dose or more. Right—a schematic of the corresponding differential dose–volume histogram. The volume v_f is the fraction of the total volume irradiated to the stated dose. The left curve is the integral of the right curve. An 'ideal' integral dose–volume histogram for a target volume would be a square box showing all the volume receiving a uniform single-valued dose. The corresponding differential histogram would be a delta function.*

may be computed from a differential dose–volume histogram (figure 1.10)†. This is, however, one of the most time-consuming tasks computationally and there is interest in questioning whether the information from coarser grids might be adequate. The DVH is computed in its integral (cumulative) form from the differential histogram showing the number of dose voxels raised to each dose level. Wong and Purdy (1990) argue that the DVH is an effective tool in evaluating rival 3D plans. Indeed it is, but even single multi-planar 3D plans computed by optimization techniques such as 'inverse computed tomography' (see chapter 2) can be nicely quantified via the DVH for normal and pathological tissues. Kessler *et al* (1991b) show how those voxels in a dose distribution corresponding to a specific part of the dose–volume histogram can be interactively displayed, thus recovering the spatial information otherwise lost.

Drzymala *et al* (1991) have summarized some good practices when constructing and using the DVH:

1. It is convenient to include both absolute and relative dose and volume by labelling all four sides of the plot.

2. When comparing rival plans, the DVHs should be on the same plot for visual comparison.

3. Multiple differential histograms on the same plot are much more difficult to interpret than multiple integral histograms, which are preferred.

4. Histograms can be smoothed by drawing a line through the right-most corners of each bin.

---

† The term differential dose–volume histogram is used here to mean (as the mathematics says) the histogram of fractional volume receiving stated dose. This is to contrast with the commonly used term 'integral dose–volume histogram' meaning the fractional volume receiving *the stated dose or more.*. Clearly then the former is the derivative of the latter and the latter is the integral of the former. It is perhaps unfortunate that the word 'differential' has entered the jargon here because it is strictly redundant. It could be wrongly interpreted that these two histograms were *two* differentiations (or integrations) apart.

5. Using a non-linear axis for volume may help observe fine structure in the curves and aid an interpretation of the prediction of normal tissue complication probability and tumour control probability from the curves (see section 1.1.11). Small differences in the shape of the DVH curve can have large effects on these and predictions should be treated with caution.

In summary, the DVH should never be used as the sole method for assessing a treatment plan but as an adjunct to visually inspecting dose distributions. Neither should any single parameter derived from it by a biological model be implicitly trusted.

If a digital dose prescription is available then a dose-difference matrix can be computed by subtracting the optimized plan from the prescription. Such a dose difference can then be displayed in the same ways described above as the dose matrix itself (e.g. by colour wash on selected slices or shaded-surface anatomy) (Tepper and Shank 1991).

1.1.9.2.   Scoring dose plans by 'regret'.   Shalev *et al* (1987) and Viggars and Shalev (1990b) have addressed the problem of basing a decision on the degree of acceptability of a treatment plan in terms of simple parameters. They specify the maximum and minimum dose (dose limits) for the target volume and the 'tolerance volume'—the volume of such targets which may go above or below these values, respectively. Maximum dose to non-target volumes and a corresponding tolerance volume are also specified. Let T_i be the prescribed tolerance volume for the ith region, D_i the dose limit for this region and $V(D_i)$ the actual volume which falls outside the prescribed limit, known as a 'volume of regret'. The score function for the ith constraint is

$$S_i = 10\left[1 - V(D_i)/T_i\right]. \tag{1.1}$$

A score of 10 is ideal; no volume falling outside the constraint; a score below zero is unacceptable, more volume than tolerable falling outside the constraint. Scores between 0 and 10 are acceptable. Viggars and Shalev (1990a) and Shalev *et al* (1991a) compare rival plans by observing the score functions as changes are made. For example conventional plans were compared with conformal plans (Shalev *et al* 1991b). The interactive system is called OSCAR, standing for 'objective scoring with coloured areas of regret', and implemented on an AECL Theraplan system.

As well as generating numerical scores, OSCAR combines the advantages of viewing 2D isodose distributions with the advantages of the dose–volume histogram. From the DVH the spatial regions in which dose falls outside the acceptable limits (volumes of regret) are colour-shaded to show the spatial locations where such regret occurs. OSCAR utilizes six colours; mild target overdose = light orange, severe target overdose = dark orange, mild target underdose = light blue, severe target underdose = dark blue, non-target tissue overdose = red, specific organ overdose = purple. This is *not* the same as colour washing grey-scale images. Colour is applied only where there is regret and it

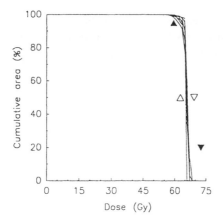

Figure 1.11. *Cumulative dose–area histogram for a mid-plane target region in seven patients with adenocarcinoma of the prostate. The inverted triangles represent tolerance points for target overdose and the upright triangles are tolerance points for target underdose. Open symbols are mild; closed symbols are severe tolerances. Note the histogram weaves between these constraints. This demonstrates a satisfactory treatment plan has been found (from Hahn et al (1990)) (reprinted with permission from Pergamon Press Ltd, Oxford, UK).*

is specifically coded to the type of regret (Viggars and Shalev 1990b). These are 'images of regret'.

It is important to remember that the areas shown by colour-wash are not washed by dose values. Parts of the washed regions may well be within tolerance. The wash is applied everywhere to a volume where the overall score function for that volume lies outside tolerance.

The score function can lead to a different way of organizing treatment planning. The clinician is asked to provide a prescription, a series of targets and organs at risk together with the fractional volumes which may or may not be raised to certain tolerances. Provided the treatment planner derives a scheme with positive score function, the plan can be considered acceptable and in principle need not be shown to the clinician—the prescription having been met. Another way of interpreting this is that if the DVH follows a certain track in space, weaving between limits set by the clinician, then the plan has to be acceptable. It may not however be entirely clear how to clinically set the dose limits or how to interpret the regret if dose limits cannot be met.

This is illustrated in figure 1.11. The clinician requested no more than 20% of the target area (this was a 2D plan) received dose above 73 Gy and no more than 50% above 69 Gy. Conversely no more than 5% receives 59 Gy or less (i.e. no *less* than 95% should receive 59 Gy or more) and no more than 50% should receive 63 Gy or less (these are the T_i and D_i in equation (1.1)). These four points are shown in figure 1.11. The curves represent successful plans (positive score function) because they weave between the constraint points (Hahn *et al* 1990).

Shalev *et al* (1991b) measured changes in the DVH and in the score functions as a result of movement at the time of treatment and thus quantified the importance of movement. They found for example in a prostate study that ±0.5 cm movement led to changes in score function of about 5 units.

1.1.10.   Verifying the treatment plan

After treatment planning is completed, the patient may return to the simulator where the optimized fields can be set up. The DRR can be checked against the corresponding physical radiographs. Similar checks can be made at the time of treatment if imaging is available on the treatment machine. It is apparent that the overall precision in radiotherapy is compounded of potential errors at all stages of the process and errors at delivery have possibly received until now too little attention. One must remember the maxim that 'a chain is only as strong as its weakest link'. This subject is considered in depth in chapter 6.

1.1.11.   Biological response of normal and pathological tissue

It is at this stage that attention must be turned to folding in the biological response data to arrive at some statement of the anticipated effectiveness of treatment. A display of the volumetric distribution of the tumour control probability (TCP) and the normal tissue complication probability (NTCP) would be most useful (Wong and Purdy 1990). Until comparatively recently, treatment planning was confined to establishing acceptable dose distributions, normalized to some '100%' value which was then assigned to some value in Gy by clinical decision. Sometimes this has been called the modal dose, but this may be confusing as it could be interpreted to mean the most frequently occurring dose value. The basis for deciding this value lies in what is known of the way the probability of tumour control (TCP) varies with dose and the corresponding way the normal tissue complication probability (NTCP) varies with dose.

The tumour control probability is the probability of eradicating all tumour cells plotted as a function of dose. The normal tissue complication probability is the probability, as a function of dose, of inducing some particular complication (a collective word for describing a variety of conditions such as nausea, vomiting etc) in a normal (ie non-tumour-bearing) organ.

1.1.11.1.   Homogeneous irradiation.   For radiotherapy to be effective at all it is obvious that the curve for TCP must lie to the left of the curve for NTCP when target dose is plotted along the abscissa (see figure 1.12). The closer the curves, the more difficult it becomes to select a dose level which is likely to control the tumour without causing normal tissue damage. This is particularly problematic if the target volume and organs at risk are sufficiently close geometrically that they receive similar dose levels (for example if the organ at risk lies within a concavity of the outline of the target volume).

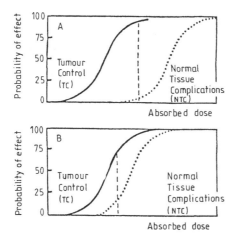

Figure 1.12. *A schematic showing the tumour control probability and normal tissue complication probability as a function of dose. Curves A show a favourable situation where a high tumour-control probability can result with a small complication probability. Curves B show a less favourable situation where a high tumour-control probability would result in a larger complication probability (from Wambersie et al (1988)).*

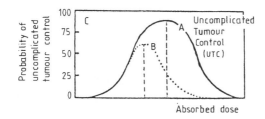

Figure 1.13. *The 'mantelpiece clock' overlap curve of TCP and (1-NTCP). Curves A and B correspond with using the curves from figure 1.12. This shows the probability of uncomplicated tumour control (P_{UTC}) (from Wambersie et al (1988)).*

The probability of uncomplicated tumour control is simply given by $P_{UTC} = TCP\,(1 - NTCP)$. If the individual probability curves are sigmoidal in shape, as they generally are, then the shape of P_{UTC} is like the outline of a mantlepiece clock (see figure 1.13) (Wambersie 1988). The optimum dose level is sometimes defined as the dose level which maximizes P_{UTC}. We may see that the closer the curves for TCP and NTCP, the lower is the probability of uncomplicated tumour control if target volume and organs at risk receive the same dose. It is again a clinical decision how to balance the risks. Often the (sic) modal dose is determined not so much to get a high probability of tumour control but to keep the probability of normal tissue complications very low indeed (some 1–2%) on the basis that quality of life counts high (see also section 1.1.9). This is referred to as 'treating to tolerance doses' (Emami *et al* 1991).

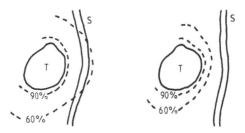

Figure 1.14. *A schematic showing the benefit of conformal versus conventional therapy. In the left-hand diagram showing conventional therapy a 90% isodose contour has been shaped to the target volume T, but the non-conformal technique leaves an organ at risk, such as the spinal canal S, within the 60% isodose contour. The inclusion of the spinal chord within the 60% contour is an example of 'regret'. In the right-hand diagram a conformal technique has reshaped the isodose contours so that the organ at risk is outside the 60% line. At constant TCP this would lead to a lower NTCP. Alternatively the dose could be increased to give the same NTCP as in the left-hand example with a correspondingly higher TCP.*

One way round the problem is to design therapy so that the dose to organs at risk is much lower than to the target volume—the goal of modern conformal therapy (figure 1.14). This gets us away from the simple situation where a single vertical line in the plot of probabilities gives the TCP and the NTCP and thus lessens the need for the two curves to be wide apart.

1.1.11.2. Inhomogeneous irradiation. The above description of how dose maps to biological effect is too simple. In all treatment plans there is an inevitable variability of dose to the target volume and a separate variation in dose to the organs at risk. It makes no sense to speak of *the* dose level to the target volume and *the* dose level to the organs at risk. Rather we should map the *distribution* of dose into a distribution of biological effect, dose being simply a 'surrogate' for biological effect (Goitein 1991). But here begins the difficulty. Whilst some data are available (e.g. Lyman 1985, Emami *et al* 1991) and methods of using it (Kutcher and Burman 1989) the data are incomplete and open to questioning (McShan 1990b). It is not too difficult with state-of-the-art computer planning systems to generate dose–volume histograms for target volume and normal tissue; converting these into an integral probability of biological effect is more controversial (Nahum and Tait 1992). It is sometimes not too difficult to do the mathematics, but more difficult to accept the predictions of the results. Let us first look at methods of evaluating the biological effect of inhomogeneous irradiation which have an intuitive appeal.

The problem is how to convert from a DVH to a measure of risk. In particular a method to choose between plans which create intersecting dose–volume histograms is essential to understanding the trade-off between plans which create small volumes raised to high dose compared with those which create large volumes raised to lower doses (see e.g. Kutcher *et al* 1991). In the literature can be found curves for the complication probability as a function of dose for different tissues and complications. These are generally for *whole organ* irradiation. The differential dose volume histogram shows the fractional volume of tissue raised to each dose level. Essentially the differential DVH has to be folded into the risk curve to compute the cumulative risk. In order to do this properly, account must be taken of the dose–volume relationship. If (in the differential DVH) the fractional volume v_i receives dose D_i, then the equivalent whole-organ (WO) dose (for that fractional volume) is

$$D_i^{WO} = D_i \, (v_i)^n \tag{1.2}$$

where n is a tissue-specific parameter.

(Tissues with a small value for n are said to have a 'small volume effect' and vice versa. As n tends to zero, the complication probability becomes independent of volume. Burman *et al* (1991) have fitted the biological data of Emami *et al* (1991) and provided values of the parameter n for 28 normal tissues with defined endpoints. These values range from $n = 0.01$ (ear) to $n = 0.87$ (lung).)

Using this whole-organ dose, it is now allowable to read off the $NTCP\left(1, D_i^{WO}\right)$ for that partial volume where the 1 implies the risk curve for the whole organ and D_i^{WO} is the dose adjusted to the equivalent whole-organ dose from equation (1.2). The total risk is then (Shalev *et al* 1987)

$$NTCP_{\text{tot}} = \left[1 - \Pi_i \left(1 - NTCP\left(1, D_i^{WO}\right)\right)\right]. \tag{1.3}$$

(See, however, equation (1.20): equations (1.2) and (1.3) hold only for small values of the NTCP. This is not a major limitation because clinical radiotherapy aims to work at the low end of the NTCP curve.) Support for the power-law model comes from Kutcher *et al* (1990), who fitted the radiation-response data of Hopewell *et al* (1987) for white marrow necrosis in the rat but the model's general validity is doubted by Yaes (1990).

Another way to achieve the same end is to assume the risk behaves as some power law of volume and, from the risk curve for full volume irradiation $NTCP(1, D)$, derive the set of risk curves for partial volumes $NTCP(v, D)$ (see figure 1.15 and equation (1.14), where the power-in-volume law is given). The differential dose–volume histogram can then be folded fractional volume by fractional volume into the appropriate curve, without the need to convert the dose D_i to an equivalent whole-organ dose, to derive an overall risk, i.e.

$$NTCP_{\text{tot}} = [1 - \Pi_i (1 - NTCP(v_i, D_i))]. \tag{1.4}$$

Yet another way is to use the whole-organ NTCP curve, but not convert the dose D_i to whole-organ dose D_i^{WO}, in which case a power of v appears in the equation, namely

$$NTCP_{\text{tot}} = \left[1 - \Pi_i \left[(1 - NTCP(1, D_i))\right]^{v_i}\right] \tag{1.5}$$

(see also equation (1.11)).

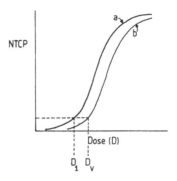

Figure 1.15. *Schematic showing the relationship between curve (a) for NTCP(1, D), giving the normal tissue complication probability for the whole organ irradiated at uniform dose D, and curve (b) for NTCP(v, D) giving the normal tissue complication probability for fractional volume v irradiated at the same uniform dose D. As v decreases, curve (b) moves to the right giving a lower NTCP at any given dose D. Curves (a) and (b) are related through equation (1.14). Also shown is a dotted horizontal line of constant NTCP. The doses D_1 and D_v are linked through equation (1.20), the 'iso-effect dose–volume relationship'. This holds for all (not just small) NTCP provided the power law model expressed by equation (1.14) holds.*

One way round these apparent computational difficulties is to *reduce* the cumulative DVH to an effective dose to the whole volume through one of the schemes proposed (e.g. Lyman and Wolbarst 1987, 1989). Then one may genuinely use the risk curve for whole-volume irradiation and read off the risk at that effective dose. The other popular scheme (Kutcher and Burman 1989) is to reduce the differential DVH to an effective volume at the maximum dose and use the appropriate risk curve for that volume. We shall go into more detail in section 1.1.11.4.

So we see in general there is a great deal of difficulty converting from inhomogeneous irradiation DVHs into cumulative risk. For this reason the use of score functions, as in section 1.1.9.2, can be a useful rule-of-thumb indicator.

It is also important to realize that clinicians may well be so keen to set a very low level of risk of normal tissue damage that all plans accepted will in practice give risks which are below clinical detection. If this is the case, the data to relate normal tissue risk against dose delivered are never accumulated! In the study of Shalev *et al* (1987) the predicted risks of rectal stricture were only a few % and in practice these were not seen indicating that a conservative set of risk data had been used.

In the study of Shalev *et al* (1991b) it turned out that the score function was linear with the risk. Since the score function is linear with volume beyond tolerance, this means the risk was linear with volume beyond tolerance (at least for the prostate treatments and at low risk values). It should not be interpreted that this is always the case. For example, irradiating just a small part of the spinal chord beyond tolerance will clearly render the whole motor system ineffective.

This is an example of a 'series organ' in the language of Brahme *et al* (1992) and Källman *et al* (1992a,b).

1.1.11.3. The dose-versus-biological-effect controversy. Wheldon (1988) has discussed the philosophy of applying over-rigorous mathematics to biological data in some depth.

Why are the biological data for converting dose to a probability of damage so controversial? Three reasons can be identified:

1. Populations vary. Not all patients respond the same, but are assumed to do so. It is convenient to discuss statistical ensembles rather than individuals (Wheldon 1988). Emami *et al* (1991) have recently gathered together what is probably the most comprehensive survey so far of normal tissue tolerance to radiation. They tabulated the doses giving $NTCP = 5\%$ and $NTCP = 50\%$ for whole-organ, 2/3-organ and 1/3-organ irradiation for 28 critical tissues. This establishes just two points on the NTCP curve at each volume from which the dose variation of NTCP can be fitted. The volume dependence is also established. Each set of data is accompanied by a detailed discussion of its source. Burman *et al* (1991) fitted the data of Emami *et al* (1991) and provide very useful tables and graphs showing the dose and volume dependences of the NTCP.

2. In documenting the effect of radiation on tissues, the dose is often assumed to be uniform in the target volume or is simply measured at one particular point (e.g. Hanks' (1985) 'pattern of care study'). In practice, the dose varies in the target volume tissue and so it is hard to establish accurate dose–response data from in-vivo irradiations.

3. Animal and cellular model extrapolations to humans are risky.

Goitein (1991) has warned of the maxim, well known in the computer field, known as 'garbage in—garbage out'; in other words the predictions of biological effect are only as good as the data on which they are based and the models in which these data are used. Cunningham (1983) wrote of the hope that one of the major contributions of CT to treatment planning might be its ability to generate accurate dose–response data for specific tissues. In principle, mapping 3D dose to anatomical CT data could indeed generate curves of NTCP versus dose, provided an acceptable way were found to reduce the inhomogeneous irradiation to an equivalent uniform dose. At present these techniques are still controversial. Also in-vivo NTCP is kept low in clinical practice so the data at high NTCP are not gathered.

However Goitein (1991) succinctly sums up why the future of radiotherapy planning must take account of biological models:

1. They can give a 'back of the envelope' estimate of what improvements might result from a change in dose level or distribution. They can show what effort in optimizing dose planning is worthwhile.

2. They can help choose between rival plans with different DVH; indeed they may show that both DVH curves for rival plans might be below the level at which

biological effect would be noticed and in this respect they would show that rival different dose plans were biologically indistinguishable. If one rival were more easily achieved practically, this would be useful information to know.

What inaccuracy actually matters in radiotherapy? This is a difficult question and there is no one answer. The way towards an answer lies in the data for the relative steepness of the dose response curves for the tissues irradiated, both target and normal. For normal tissue the relative steepness can be expressed as that percent change in dose changing the TCP from 50% to 75% and for normal tissue the steepness is the percent change in dose changing the NTCP from 25% to 50%. The former vary from 5 to 50% depending on tissue and the latter from 2 to 17% (Wambersie *et al* 1988). In principle such figures provide a means to answer the question whether geometric positioning errors at the time of treatment are or are not important. Another way is to determine restrospectively what effects can be detected clinically. Goitein (1985) suggested computing 'extreme error plans' corresponding to the worst-case geometric errors with the patient in worst-case positions.

Goitein (1982) discussed at length the question of accuracy in dosimetry in radiotherapy in terms of the accuracy required of CT data used in treatment planning. Quoting Geise and McCullough (1977), 2% dose uncertainty is associated with a spatial resolution of 5 mm in the assessment of organ position. Since all diagnostic CT scanners have much better spatial resolution than this, CT resolution is not a limitation on dose accuracy. Even the so-called 'non-diagnostic CT scanners' (Webb 1990) achieve a spatial resolution of 5 mm. Patient *movement* of the order of 5 mm would however lead to this kind of uncertainty.

It can be easily shown that *random* errors in attenuation coefficient measurements in CT are not an important practical consideration for the calculation of dose. On the other hand a *systematic* error could be very important. 3.5% error in the CT number would give a dose uncertainty of 2% (Goitein 1982). Another cause for concern is the proper determination of the linear or bilinear relationship between CT scanner number and electron density, which should be measured experimentally for each CT scanner (see e.g. Parker *et al* 1979, Ten Haken *et al* 1991). Other uncertainties in 3D radiation therapy planning have been considered by Urie *et al* (1991).

1.1.11.4. Formal equations of the biological model. The formal equations giving the normal tissue complication probability (NTCP) when a partial (fractional) volume v is irradiated to a uniform dose D have been given by Lyman (1985). This is sometimes called the integrated normal (or probit) model (Schultheiss *et al* 1983):

$$NTCP = 1/\sqrt{2\pi} \int_{-\infty}^{t} \exp\left(-x^2/2\right) dx \qquad (1.6)$$

with x a free variable and

$$t = [D - D_{50}(v)] / [m D_{50}(v)] \qquad (1.7)$$

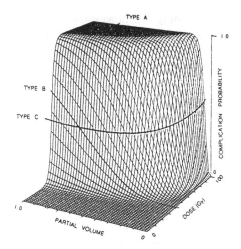

Figure 1.16. *Complication probability for liver, displayed as a three-dimensional surface as a function of dose and partial volume. For this organ the volume dependence is represented by $n = 0.32$. $D_{50} = 40$ Gy for whole organ irradiation with liver failure as the end-point. The slope parameter is $m = 0.15$. Type-A curves show the complication probability as a function of dose at constant partial volume. Type-B curves show the complication probability as a function of partial volume at constant dose. Type-C curves show the partial volume as a function of dose at constant complication probability (from Burman et al (1991)) (reprinted with permission from Pergamon Press Ltd, Oxford, UK).*

and

$$D_{50}(v) = D_{50}(v = 1) v^{-n}. \tag{1.8}$$

Here D_{50} is the dose which results in a 50% complication probability for some specified complication or end-point; $D_{50}(v = 1)$ is the value appropriate to irradiating *all* the volume and $D_{50}(v)$ is the value for partial volume irradiation. These two are related through a power law (with index n) of the partial volume v, this being the fractional volume of the whole organ. m is a slope parameter which characterizes the shape of the NTCP curve, which is sigmoidal, being the integration of a Gaussian with standard deviation $m D_{50}(v)$. For any particular tissue and specified end-point for normal tissue damage, the values of the parameters $D_{50}(v = 1)$, n and m must be determined from clinical observation. Burman *et al* (1991) have tabulated these three parameters for 28 tissues whose D_5 (the dose which results in a 5% complication probability) and D_{50} were tabulated by Emami *et al* (1991). For a discussion of the difficulties of the fitting procedure, see Burman *et al* (1991). Some tissues demonstrate an additional threshold behaviour, there being a partial volume below which no damage appears to occur. The model expressed by equations (1.6)–(1.8) connect complication probability, dose and partial volume. The interrelation between these can be represented by a surface, as shown in figure 1.16.

In general the treatment planning will, however, not yield a uniform distribution of dose either through the target volume or the normal structures (organs at risk), the fraction of volume raised to each dose level being characterized instead by the differential or integral dose–volume histogram (DVH). For the Lyman equations to be easily called into play, a *reduction scheme* to reduce the DVH to an effective fractional volume v_{eff}, uniformly irradiated to the maximum dose D_{max}, must be found. One such scheme has been given by Kutcher and Burman (1989):

$$v_{eff} = v_{max} + v_1 (D_1/D_{max})^{1/n} + v_2 (D_2/D_{max})^{1/n} + \ldots\ldots = \sum_i v_i (D_i/D_{max})^{1/n}$$

(1.9)

where volume v_i receives dose D_i in the *differential* DVH and D_{max} is the maximum dose to part of the normal tissue under consideration; n has the same meaning as before. It has been suggested that dose steps of some 1 Gy are adequate to compute v_{eff}. This effective volume method proceeds from the conjecture that each volume element of the differential dose–volume histogram independently obeys the same dose–volume relationship (equations (1.2) and (1.8)) as the whole organ (Kutcher *et al* 1991). Using D_{max} as the dose to which all elements are transformed ensures that v_{eff} is always less than unity. It is the fraction of the whole organ volume. If the whole organ is not covered in the CT scanning process (and hence in construction of the DVH), a so-called 'standard' ICRP volume should be used. The method is somewhat empirical but Kutcher and Burman (1989) showed it had many desirable features, for example:

1.   it reduces to the same complication probability for uniform irradiation as was used for the input data, and

2.   when small hot or cold spots are introduced to an otherwise uniformly irradiated volume, the NTCP marginally rises or falls respectively, the amplitude of the change depending on n.

These equations enable the NTCP for inhomogeneous irradiation of any normal tissue whose biological parameters are known to be calculated from the DVH. One may write the normal tissue complication probability (equation (1.6)) as $NTCP (v, D, m, n, D_{50} (v = 1))$ to indicate the full dependences. Using the reduction method one looks up $NTCP (v_{eff}, D_{max}, m, n, D_{50} (v = 1))$. If there is more than one such structure the overall $NTCP_{tot}$ may be found from

$$NTCP_{tot} = 1 - (1 - NTCP_1) (1 - NTCP_2) \ldots$$

(1.10)

where there are as many producted terms as there are normal structures considered. For example in the study of Nahum and Tait (1992) calculating the NTCP for pelvic irradiation, the rectum and the small bowel were considered to be the dose-limiting structures of importance and so two terms were used. Schultheiss *et al* (1987) argue that the aim of clinical radiotherapy should indeed be to maximize $\Pi [1 - w_i NTCP_i]$, where w_i is the morbidity of some complication whose probability is $NTCP_i$. This statement is consistent with attempting to minimize $NTCP_{tot}$ in equation (1.10) with the addition of the *importance* of the separate risks.

The tumour control probability (TCP) is also a sigmoidal curve with increasing dose and can be determined experimentally or by biological model. Nahum and Tait (1992) have provided such a model for bladder tumours (see equation (1.32)). As discussed above, the clinical practice is generally to prescribe a target dose which keeps the NTCP very small (Emami *et al* 1991) rather than optimize the probability of uncomplicated tumour control, simply because patients can often come to terms with a life-threatening tumour, whereas their quality of life is threatened by normal tissue damage as part of treatment. Nahum and Tait's study showed that the mean TCP could be improved if (for pelvic irradiation) all treatments were customized to give an overall NTCP of 5%. However, the magnitude of the improvement over globally setting the same target dose for all patients depended critically on the volume parameter *n* and indicated the potential dangers of relying on precise mathematical modelling with poorly understood biological data—as Goitein put it there is great potential for the 'garbage in—garbage out' syndrome.

The NTCP can be deduced from the DVH without reduction. The differential DVH is the set of fractional volumes v_i each with dose D_i. Suppose $NTCP(1, D)$ is the NTCP when the *whole organ* is irradiated to dose D, then the NTCP for an inhomogeneous irradiation specified by this differential DVH is given by

$$NTCP_{inhom} = 1 - \Pi_{i=1}^{M} [1 - NTCP(1, D_i)]^{v_i} \tag{1.11}$$

where there are M subvolumes in the differential DVH, each of which may be considered uniformly irradiated. This equation expressing the dose–volume relationship as a power-in-volume law was derived by Schultheiss et al (1983).

Niemierko and Goitein (1991) have shown that there is actually no need to compute the DVH in order to obtain $NTCP_{inhom}$. Suppose a normal tissue is specified by dose D_i to N *points*, then each point can be thought of as representing a small volume in which the dose is uniform and so it follows that

$$NTCP_{inhom} = 1 - \Pi_{i=1}^{N} [1 - NTCP(1, D_i)]^{1/N}. \tag{1.12}$$

Niemierko and Goitein (1991) argue that equations of this form for $NTCP_{inhom}$ characterize tissues with 'critical element architecture', that is comprising 'functional subunits (FSU)', any one of which receiving damage implies damage to the whole organ function (such as the spinal chord, nerves etc). If P_i is the probability of damage to the ith FSU then the NTCP for the whole organ is given by

$$NTCP = 1 - \Pi_{i=1}^{N} [1 - P_i]. \tag{1.13}$$

This equation and those of a similar form in this section give the probability of damage when tissues have a 'series' organization. Källman *et al* (1992a,b) give corresponding equations when the structure is partially series and partially parallel.

From equation (1.11) follows the functional form of the volume dependence of NTCP (see figure 1.15). Suppose dose D is given to fractional volume v only, the rest of the organ being virtually unirradiated, then the NTCP for the organ is given by

$$NTCP\,(v, D) = 1 - [1 - NTCP\,(1, D)]^v \qquad (1.14)$$

i.e. the $NTCP\,(v, D) < NTCP\,(1, D)$ for all fractional volumes and for all values of $NTCP\,(1, D)$†.

If the value of $NTCP\,(1, D)$ is very small, equation (1.14) may be expanded to

$$NTCP\,(v, D) = vNTCP\,(1, D) \qquad (1.15)$$

i.e. for small complication probabilities, the complication probability is a linear function of the (uniformly) irradiated volume for constant dose. This linearity obviously does not hold for higher values of the complication probability, when the full equation (1.14) applies.

An alternative form for the NTCP for full-volume irradiation to that given in equation (1.6) is

$$NTCP\,(1, D) = \frac{1}{1 + (D_{50}/D)^k} \qquad (1.16)$$

with

$$k = 1.6/m \qquad (1.17)$$

giving the relation between the parameter k and the shape parameter m. k is also equal to $1/n$ where n is defined by equation (1.8). The arguments have been attached to NTCP to clarify that this is an NTCP for uniform irradiation to dose D for the whole organ ($v = 1$).

This is sometimes called the logistic model and has the advantage of being in closed form. It differs from the integrated normal model by less than 1% (which is clinically indistinguishable) with suitable adjustment of parameters (Schultheiss *et al* 1983). These authors also made use of the logistic formula together with the two known doses giving $NTCP = 5\%$ and 50% to deduce the parameter $k = 1/n$.

---

† Equation (1.14) is from Schultheiss *et al* (1983) and was derived as follows. Imagine irradiating a full volume $v = 1$ at uniform dose D. The probability of escaping injury is $[1 - NTCP\,(1, D)]$, but it could also be expressed as $[1 - NTCP\,(1/A, D)]^A$ if A subvolumes were all irradiated to the same dose D. So

$$[1 - NTCP\,(1, D)] = [1 - NTCP\,(1/A, D)]^A. \qquad (1.\text{i})$$

If only B parts were irradiated ($v = B/A$), we have

$$[1 - NTCP\,(v = B/A, D)] = [1 - NTCP\,(1/A, D)]^{B=vA}. \qquad (1.\text{ii})$$

Substituting the RHS of equation (1.i) into the RHS of equation (1.ii) gives equation (1.14). Equation (1.11) is the general form for inhomogeneous irradiation.

This should be contrasted with the fitting method of Burman *et al* (1991) described earlier.

The logistic formula may be used directly in equations (1.11), (1.12), (1.14) and (1.15). For example, putting equation (1.16) into equation (1.12) we have

$$NTCP_{inhom} = 1 - \Pi_{i=1}^{N} \left[1 + (D_i/D_{50})^k \right]^{-1/N}. \tag{1.18}$$

Hence the $NTCP_{inhom}$ follows directly from a list of N points with their dose values D_i provided the D_{50} and k for the complication are known.

From equations (1.14) and (1.16) we may deduce the dose D_v required to be given to partial volume v to give the *same* NTCP as dose D_1 to the whole organ, the so-called 'iso-effect dose–volume relationship' (see figure 1.15). We set

$$NTCP(v, D) = NTCP(1, D_1) \tag{1.19}$$

and substitute from equations (1.14) and (1.16) to get

$$D_v = D_{50} \left(\left[1 + (D_1/D_{50})^k \right]^{1/v} - 1 \right)^{1/k}. \tag{1.20}$$

For very small doses (low complications)

$$D_v = D_1 v^{-1/k}. \tag{1.21}$$

Equation (1.21) is the well known power law between iso-effect dose and volume (see also equations (1.2) and (1.8)). For example, (with k=10) $D_{0.5} = 1.07D_1$, $D_{0.1} = 1.26D_1$ and $D_{0.05} = 1.35D_1$. However the FSU architecture model predicts significant departure from this 'law' when the NTCP is high. Niemierko and Goitein (1991) give experimental data showing better fit to equation (1.20). Schultheiss *et al* (1983) also present equation (1.20) and discuss the approximations leading to equation (1.21).

The histogram reduction scheme published by Kutcher and Burman (1989) (see equation (1.9)) is just one of several analytic reduction schemes proposed. It essentially reduces the differential DVH for an inhomogeneous distribution of the dose to an equivalent volume v_{eff} uniformly receiving the maximum dose. Equations (1.14) and (1.16) then give the NTCP with D set to D_{max} and v set to v_{eff}. A second reduction scheme has been proposed by Lyman and Wolbarst (1987, 1989) which reduces the cumulative DVH to a single *effective dose* D_{eff} through the whole volume such that this would give the same NTCP as the inhomogeneous irradiation. It does this by starting with the right-most element of the *cumulative* DVH, combining the effect of this bin into the bin immediately to the left of it (figure 1.17). Let us for the moment consider what happens when there are only two such elements in the cumulative DVH. (When there are more, the process of combining the right-most two bins simply sequences until there is only one bin left). A new dose D_{eff} is created intermediate between the last bin at D_1 and the

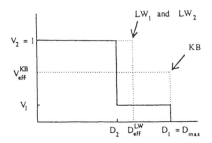

Figure 1.17. *A model dose–volume histogram with just two elements for discussion purposes. The* DVH *is 'reduced' by the method of Lyman and Wolbarst (LW) to give an effective dose* $D_{\text{eff}}^{\text{LW}}$ *to all the volume; and by the method of Kutcher and Burman (KB) to give a fractional volume* $V_{\text{eff}}^{\text{KB}}$ *raised to the maximum dose* D_{max}. *The biological effect of the two are supposedly the same (from Niemierko and Goitein (1991)).*

second from last bin at D_2 where the corresponding volumes are V_1 and V_2. We have

$$NTCP\,(V_2, D_{\text{eff}}) = \frac{V_2 - V_1}{V_2} NTCP\,(V_2, D_2) + \frac{V_1}{V_2} NTCP\,(V_2, D_1). \quad (1.22)$$

This equation states that the probability as a function of dose can be linearly interpolated in volume. The partial volume complication probabilities are computed from equations (1.6) to (1.8), which include the volume dependence of the D_{50}. In equation (1.22) upper-case V is used to signify the partial volume in the *cumulative* DVH to distinguish from lower-case v used earlier for the partial volume in the *differential* DVH. When the cumulative DVH has a large number of elements, equation (1.22) is repeatedly used to combine data from the right until a single D_{eff} results for the total volume.

Niemierko and Goitein (1991) have very elegantly shown the relationship between the result in equation (1.22) and the predictions of the critical element model. Let us apply equation (1.11) of this model to the two-element cumulative DVH. We get

$$NTCP_{\text{inhom}} = 1 - [1 - NTCP\,(1, D_2)]^{v_2}\,[1 - NTCP\,(1, D_1)]^{v_1} \quad (1.23)$$

where, as before, lower-case v represent fractional volumes in the *differential* DVH. But

$$v_2 = V_2 - V_1 \quad (1.24)$$

and

$$v_1 = V_1 \quad (1.25)$$

and so we have

$$NTCP_{\text{inhom}} = 1 - [1 - NTCP\,(1, D_2)]^{V_2 - V_1}\,[1 - NTCP\,(1, D_1)]^{V_1}. \quad (1.26)$$

But from equation (1.14) expressing the volume dependence of NTCP at any given dose (upper- or lower-case V may be used here)

$$1 - NTCP(1, D) = [1 - NTCP(V, D)]^{1/V}.\qquad(1.27)$$

Combining equations (1.26) and (1.27) we have

$$NTCP_{inhom} = 1 - [1 - NTCP(V_2, D_2)]^{\frac{V_2 - V_1}{V_2}}[1 - NTCP(V_2, D_1)]^{\frac{V_1}{V_2}}.\qquad(1.28)$$

Now expanding binomially *for small NTCP* we have

$$NTCP_{inhom} = \frac{V_2 - V_1}{V_2}NTCP(V_2, D_2) + \frac{V_1}{V_2}NTCP(V_2, D_1)\qquad(1.29)$$

which is exactly the same as equation (1.22) of Lyman and Wolbarst (1987, 1989). So we have the important result that *this Lyman and Wolbarst reduction scheme is identical to the critical element scheme of Niemierko and Goitein for small normal tissue complication probabilities.* It is *not* the same otherwise. This identity arises because both histogram reduction schemes incorporate the power-law dose–volume dependence. As we have seen this power-law dependence breaks down in the critical element model for large NTCP and so the methods would in these circumstances not be expected to give the same results. However since in clinical practice most dose planning is worked out so that the NTCP is very low, it is clear that both methods are, with this approximation, the same and could be used interchangeably. If NTCP is not small one must not make this assumption.

Niemierko and Goitein's critical element model can predict a D_{eff} and v_{eff} like the Lyman and Wolbarst and Kutcher and Burman terms which in the limit of small NTCP reduce to their results exactly but which are different for high NTCP. These results are (see Niemierko and Goitein 1991):

$$D_{eff} = D_{50}\left[\Pi_{i=1}^{N}[1 + (D_i/D_{50})]^{v_i} - 1\right]^{1/k}\qquad(1.30)$$

and

$$v_{eff} = \sum_{i}v_i\frac{\ln[1 - NTCP(1, D_i)]}{\ln[1 - NTCP(1, D_{max})]}.\qquad(1.31)$$

Let us work a simple numerical example to show the similarity of all the methods at low normal tissue complication probability. Suppose some single normal organ at risk receives 30 Gy to 0.4 of its volume and 40 Gy to the other 0.6. (It would be hard to see how such a simple distribution would arise in practice but this is just to illustrate the point.) From this differential DVH we see that the cumulative DVH has all of the volume at 30 Gy or more and 0.6 of the volume at 40 Gy. Let us suppose that for this organ $D_{50} = 50$ Gy and $n = 0.1$ (so $k = 10$). We shall use equation (1.16) to deduce the NTCP for the whole organ irradiated to some dose (the same qualitative conclusions would be obtained if we used equations

(1.6) and (1.7)). From equations (1.2) and (1.3) (the method converting each partial volume dose to an equivalent whole-body dose) we find $NTCP = 0.0628$. The methods using the doses from the differential DVH without converting them to whole-body equivalents, represented by equations (1.4) and (1.5) (or (1.11)), give the same result since one follows from the other through the use of equation (1.14) and $NTCP = 0.0616$. The Lyman and Wolbarst method (equation (1.22)) gives $NTCP = 0.0606$. The Kutcher and Burman method (equation (1.9) with use of equation (1.14)) gives $NTCP = 0.0615$.

We must still deal with the question of how to calculate the tumour control probability (TCP). This might be determined experimentally but in the absence of experimental data Nahum and Tait (1992) have provided one model. This is based on the survival of clonogenic cells (Källman *et al* 1992b). We have

$$TCP = (1/K) \sum_{i=1}^{K} \exp\left(-N_s\left(i\right)\right) \tag{1.32}$$

where
$$N_s\left(i\right) = N_0 \exp\left(-\alpha_i D\right) \tag{1.33}$$

and N_0 is the initial number of clonogenic cells (taken to be $\rho_c V_t$ where $\rho_c = 10^7$ cm^{-3}, V_t is the target volume in cc, D is the dose and α is a value taken from a Gaussian distribution with mean 0.35 and a standard deviation of 0.08. (K is the number of samples for α and should be a large number like 10^3 or 10^4.) Nahum and Tait (1992) found this gave a good fit to some biological data (figure 1.18). It generates a characteristically sigmoidal curve for TCP as a function of dose. The question still remains of what volume to use and what dose when the DVH for the target volume exhibits inhomogeneous irradiation. One option is to use the 90% volume and the dose to which 90% of the volume is raised. This ignores the 'rind' of low dose (which would pull down the TCP) and also ignores the larger doses given to small parts of the volume (which would increase the TCP). This is an open question.

A potential danger of taking the biological model too seriously is that since most irradiations are set to give very low NTCP, the conditions for determining the precise form of the NTCP curve up to values of $NTCP = 1$ are not achieved clinically. Also not enough time has yet elapsed for the predictions of even small NTCP to be checked out with real future damage. There may be considerable uncertainty in the values of the volume dependence parameter n by ± 0.08 (Burman *et al* 1991). However, Kutcher *et al* (1991) have shown that rival dose plans still *rank in the same order*, as n is varied within its range of uncertainty. Hence the problem of not knowing n accurately affects absolute calculations of NTCP rather than relative calculations. For this reason Sontag and Purdy (1991) and Webb (1992) felt confident in using TCP and NTCP to rank treatment plans. Also there may not be agreement about how to *weight* the importance of different complications. It is all very well to know the NTCP but how should one decide the

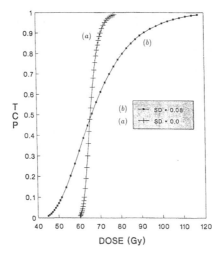

Figure 1.18. *Showing the tumour-control probability as a function of dose based on the clonogenic cell-kill model with* $\alpha = 0.35$, $\rho_c = 10^7$ *and* $V_T = 320$ cc *with* α *(a) not randomized and (b) randomized (from Nahum and Tait (1992)).*

importance of a complication independent of its probability of occurrence? Until these problems are dealt with some workers will prefer to stay with considerations of dose optimization rather than optimization of biological outcome.

On the other hand, there are several practical advantages accruing from the use of biological models:

1.   They can give simple numerical data (the NTCP values) which help answer the question 'which is the better of two cumulative DVH's which cross-over?';

2.   They reduce the staggering amount of data in treatment plans that have to be analysed to manageable proportions; and

3.   They provide a measure on a biological rather than a dosimetric scale.

1.1.11.5.   The time factor and inhomogeneous irradiation.   Time also enters the consideration of biological effect. This is a big subject, well treated by others (e.g. Wheldon 1988) and here we are brief. So-called 'iso-effect formulae' predict the way in which the dose per fraction, the fraction size and the total dose may be manipulated to give the same recognizable biological effect. There are many ways of providing a mathematical treatment. One is to compute a time–dose factor (TDF) given by

$$TDF = k_1 N d^\delta \, (T/N)^{-\tau} \, v^\phi \qquad (1.34)$$

where N is the number of fractions, d is the dose per fraction, (T/N) is the time in days between fractions, v is the partial volume of tissue irradiated (a fraction of the reference volume $v = 1$), k_1 is a scaling factor making $TDF = 100$ for the reference volume of tissue irradiated to tolerance and δ, τ and ϕ are tissue-specific exponents.

However, once again this oversimple equation ignores the inhomogeneity of dose throughout a three-dimensional target volume. Orton (1988) has proposed that target volumes might be considered broken into subvolumes v_i in which the dose per fraction d_i is uniform. Then because of the power law, the individually calculated $(TDF_i)^{1/\phi}$ are linearly additive and one can write an integral TDF factor (ITDF) as

$$ITDF = \left[\sum_i (TDF_i)^{1/\phi} \right]^{\phi} \tag{1.35}$$

where

$$TDF_i = k_1 N d_i^{\delta} (T/N)^{-\tau} v_i^{\phi}. \tag{1.36}$$

A second way to include time is the use of a linear quadratic factor (LQF) given by

$$LQF = k_1 N d \, (1 + d/(\alpha/\beta)) \, (T/N)^{-\tau} v^{\phi} \tag{1.37}$$

where the terms have the same meaning as for TDF and (α/β) is the well known ratio for the survival curve (Wheldon 1988). High values of the ratio indicate tissues relatively insensitive to changes in fraction size and vice-versa.

As for TDF, account for dose inhomogeneity through the target volume can be made by defining an integral LQF (ILQF) via

$$ILQF = \left[\sum_i (LQF_i)^{1/\phi} \right]^{\phi} \tag{1.38}$$

where

$$LQF_i = k_1 N d_i \, (1 + d_i/(\alpha/\beta)) \, (T/N)^{-\tau} v_i^{\phi}. \tag{1.39}$$

Orton (1988) tabulates the tissue-specific factors for skin, stroma, brain, spinal chord, lung, gut and kidney.

1.2.   REGISTRATION OF TWO IMAGE DATASETS FOR 3D TREATMENT PLANNING

In recent years the proliferation of medical imaging modalities has had a dramatic effect on the expectations of performing good radiotherapy. An immediate consequence has been the need to develop techniques for image registration (Chen *et al* 1990). The use of medical images to assist the radiotherapy treatment planning process is so important that the second part of this chapter develops the subject of image registration in some detail. In the early days of using images in radiation therapy planning it was easier and was traditional to merge data on sheets of paper. For example although CT data might be accessed by computer for treatment planning, the same software might not be able to accept SPECT or PET data. Goitein (1982) argues that in these early days the computer was

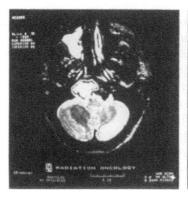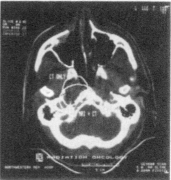

Figure 1.19. *Left—MRI image of a patient with a clival chordoma. The treatment volume has been outlined by the therapist. Right—the CT image of the patient with outlines from CT only and CT+MRI (from Kessler et al (1991)) (reprinted with permission from Pergamon Press Ltd, Oxford, UK).*

a limitation in using multi-modality images for treatment planning. In early times anatomical structures were translated from the hard copy of one imaging modality to another by pencil (Sontag and Purdy 1991). These practices are fortunately fast disappearing.

Imaging methods may be classified in a number of ways; e.g. those

- generating anatomical data: planar x-radiographs, CT, MRI, ultrasound,
- generating functional data: MRI, SPECT, PET,
- generating tomographic volume datasets: SPECT, PET, MRI, CT,
- with fine scale of resolution: MRI, CT, x-radiographs, some PET,
- with coarse scale of resolution: SPECT, some PET.

The need to register data arises for several reasons. Sometimes anatomical detail is more easily seen in MRI than in CT (e.g. differentiating tumour and oedema from grey and white matter in the brain. Bony structure is best visualized by CT whereas soft-tissue pathology and nervous tissue at the base of the skull is best seen by MRI (Hill et al 1991a)). Sometimes the target volume may be defined differently in two modalities. Correlation of MRI with CT is needed for isodose calculations which require knowledge of the tissue electron densities. This creates the problem of registering two different modalities but both with high spatial resolution (Chen *et al* 1985, Hill *et al* 1991a) (figure 1.19). Additionally one might like to make use of functional data, such as SPECT or PET, to determine the target volume and once again correlation with CT is needed to obtain the necessary tissue electron densities.

Serial CT or MRI scanning is common, for example pre- and post-operatively; prior to, and after, radiation therapy. The clinician needs to assess response to therapy. Hence there is a need to correlate high-resolution data from the same modality at two or more different scanning times.

Functional data from SPECT and PET is widely available. Radiation treatment planning is assisted by image combination; e.g. functional PET data may allow the treatment planner to avoid healthy brain tissue when planning, using CT data, to treat a brain tumour. The need to match functional data on to corresponding anatomical images (CT or MRI) creates the problem of correlating two modalities, one with a high and the other with a lower spatial resolution. The mismatch can be by a factor 10 (SPECT with CT say)(see e.g. Roeske *et al* 1991). Also the need arises to display high-resolution isodose distributions superposed on to (possibly low-resolution) functional images when monitoring response.

Used alongside anatomical data, diagnosis may be illuminated by functional imaging (Bidaut 1991). Conversely the anatomical data can provide the basis for photon attenuation corrections in forming the functional images. Yu and Chen (1991) have shown how this improves contrast. Chen *et al* (1991) used the 'hat-head' method (see section 1.2.7) to register MRI and CT data with PET data to provide anatomical boundary information which improved image reconstruction.

Correlation of patient position at the time of imaging for planning, derived from CT data, and, at the time of treatment, from planar radiographs (either port films or electronic portal images—see chapter 6), is also important, for example to verify that positioning at the time of treatment is the same as at the planning stage. Registration of digital radiographs artificially generated from the CT data with the planar radiograph at the time of treatment is required. Comparison of simulator films with portal images at the time of therapy also requires registration techniques. Megavoltage portal images (digital or film) can be correlated with simulator images or digital projection radiographs from CT data and studies using such correlations are underway (Swindell and Gildersleve 1991).

Matching different modalities is complicated because each has its own geometrical description with different pixel sizes, slice thicknesses, data orientations, magnifications and possibly non-aligned longitudinal axes (tilted transaxial planes). All these problems have to be addressed by image-correlation methods.

The choice of technique for correlating images has much to do with whether the intention to do so exists *a priori* before either scanning session takes place, in which case external markers may be used, or whether the images to be correlated already exist, maybe from different centres or departments. The 'strongest method' of ensuring precise image registration is the use of the stereotactic frame. Some types of frame are however somewhat invasive and they are generally reserved for head imaging only where precision is most important. Stereotactic image registration is discussed in a separate chapter along with stereotactic radiotherapy and radiosurgery. *A posteriori* image registration obviously cannot use applied fiducials—marks which appear on both sets of images. If both sets of data have high resolution, skilled observers may be able to identify corresponding anatomical landmarks which may then take the place of external markers (Hill *et al* 1991a). For example bony structure, relative to which the target volume may not move, can serve this purpose. If however one set of data has much lower spatial resolution, then this kind of *a posteriori* identification is generally not possible.

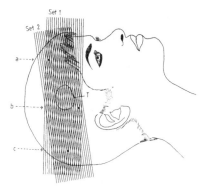

Figure 1.20. *Two* 3D *imaging modalities have been used to take a series of tomographic sections through a brain tumour 'T' but the slices are not parallel to each other. Imagine the most general situation where they are related by an unknown rotation and translation in three dimensions. The transformation can be determined by the use of external fiducial markers which appear in both sets of images. Markers a,b and c appear in slices 2, 18 and 13 in set 1, respectively, and in slices 6, 18 and 9 in set 2, respectively. By measuring their coordinates in each set, the transformation may be found.*

The only solution is to arrange attached external landmarks *a priori*. These two methods (use of external fiducials and correlative anatomical recognition) are collectively referred to as 'landmark-based image registration' (figure 1.20). In the absence of these, one particular *a posteriori* method may still work for the head—the so-called 'hat-head fitting method' or surface-based registration—see section 1.2.7.

Image correlation techniques can be divided into 2D and 3D. What follows is a general 3D description which can easily be collapsed to the corresponding 2D method.

1.2.1. *3-vector description*

Let us assume that two 3D digital image datasets p and p' are related to each other simply by a 3D rotation, 3D translation and a scale change. For the moment disregard any warping or elastic transformations. Each dataset comprises points assumed to be on 3D rectangular cartesian coordinates. A representative point in each set shall be r and r' which may be represented by 3-element column vectors, i.e.

$$r = \begin{pmatrix} x \\ y \\ z \end{pmatrix} \tag{1.40}$$

and

$$r' = \begin{pmatrix} x' \\ y' \\ z' \end{pmatrix} \tag{1.41}$$

Then

$$r' = \mathbf{S}\mathbf{R}r + b \tag{1.42}$$

where $\mathbf{R}$ is a (3×3) rotation matrix, $\mathbf{S}$ is a diagonal (3×3) scaling matrix and b represents the translation vector. Equation (1.42) implicitly operates for *all* points in the sets p and p'. The rotation matrix $\mathbf{R}$ can be broken down as the product of three separate rotations. Let $\mathbf{R}_x (\theta_x)$ be the rotation by θ_x about the x-axis, $\mathbf{R}_y (\theta_y)$ be the rotation by θ_y about the y-axis and $\mathbf{R}_z (\theta_z)$ be the rotation by θ_z about the z-axis. Then (and implicitly defining an order of sequential rotations)

$$\mathbf{R} = \mathbf{R}_z (\theta_z) \, \mathbf{R}_y (\theta_y) \, \mathbf{R}_x (\theta_x) \,. \tag{1.43}$$

We can write

$$\mathbf{R}_x (\theta_x) = \begin{pmatrix} 1 & 0 & 0 \\ 0 & \cos\theta_x & \sin\theta_x \\ 0 & -\sin\theta_x & \cos\theta_x \end{pmatrix}$$

$$\mathbf{R}_y (\theta_y) = \begin{pmatrix} \cos\theta_y & 0 & \sin\theta_y \\ 0 & 1 & 0 \\ -\sin\theta_y & 0 & \cos\theta_y \end{pmatrix}$$

$$\mathbf{R}_z (\theta_z) = \begin{pmatrix} \cos\theta_z & \sin\theta_z & 0 \\ -\sin\theta_z & \cos\theta_z & 0 \\ 0 & 0 & 1 \end{pmatrix}. \tag{1.44}$$

The scale matrix in general may be written

$$\mathbf{S} = \begin{pmatrix} s_x & 0 & 0 \\ 0 & s_y & 0 \\ 0 & 0 & s_z \end{pmatrix} \tag{1.45}$$

where s_x, s_y and s_z are the scaling factors in the x, y, z directions. If the scaling is isotropic, $\mathbf{S}$ collapses to $s\mathbf{I}$ where $\mathbf{I}$ is the identity matrix and s is a scalar constant.
Combining equations (1.42), (1.43), (1.44) and (1.45), we have

$$r' = \begin{pmatrix} x' \\ y' \\ z' \end{pmatrix}$$

$$= \begin{pmatrix} s_x C_z C_y & s_x S_z C_x - s_x C_z S_y S_x & s_x S_z S_x + s_x C_z S_y C_x \\ -s_y S_z C_y & s_y C_z C_x + s_y S_z S_y S_x & s_y C_z S_x - s_y S_z S_y C_x \\ -s_z S_y & -s_z S_x C_y & s_z C_y C_x \end{pmatrix} \begin{pmatrix} x \\ y \\ z \end{pmatrix} + \begin{pmatrix} b_x \\ b_y \\ b_z \end{pmatrix} \tag{1.46}$$

where $S_{x,y,z} = \sin\theta_{x,y,z}$ and $C_{x,y,z} = \cos\theta_{x,y,z}$.
Note the rotations and scaling are not commutative. They must be performed in the order written.

1.2.2. 4-vector description

It is rather clumsy to have the rotation and the scaling as a matrix operation and yet the translation as an additional (rather than multiplicative) vector. This can be overcome by defining r and r' to be 4-element column vectors:

$$r = \begin{pmatrix} x \\ y \\ z \\ 1 \end{pmatrix} \qquad (1.47)$$

and

$$r' = \begin{pmatrix} x' \\ y' \\ z' \\ 1 \end{pmatrix}. \qquad (1.48)$$

We now write

$$\mathbf{R}_x\,(\theta_x) = \begin{pmatrix} 1 & 0 & 0 & 0 \\ 0 & \cos\theta_x & \sin\theta_x & 0 \\ 0 & -\sin\theta_x & \cos\theta_x & 0 \\ 0 & 0 & 0 & 1 \end{pmatrix}$$

$$\mathbf{R}_y\,(\theta_y) = \begin{pmatrix} \cos\theta_y & 0 & \sin\theta_y & 0 \\ 0 & 1 & 0 & 0 \\ -\sin\theta_y & 0 & \cos\theta_y & 0 \\ 0 & 0 & 0 & 1 \end{pmatrix}$$

$$\mathbf{R}_z\,(\theta_z) = \begin{pmatrix} \cos\theta_z & \sin\theta_z & 0 & 0 \\ -\sin\theta_z & \cos\theta_z & 0 & 0 \\ 0 & 0 & 1 & 0 \\ 0 & 0 & 0 & 1 \end{pmatrix}. \qquad (1.49)$$

The scale matrix becomes

$$\mathbf{S} = \begin{pmatrix} s_x & 0 & 0 & 0 \\ 0 & s_y & 0 & 0 \\ 0 & 0 & s_z & 0 \\ 0 & 0 & 0 & 1 \end{pmatrix} \qquad (1.50)$$

and a translation matrix is written as

$$\mathbf{T} = \begin{pmatrix} 1 & 0 & 0 & b_x \\ 0 & 1 & 0 & b_y \\ 0 & 0 & 1 & b_z \\ 0 & 0 & 0 & 1 \end{pmatrix}. \qquad (1.51)$$

If the scaling parameters s_x, s_y, s_z can take negative values, this will also take care of the 'handedness' of the datasets, i.e. the possibility that the planes are numbered in reverse order in one set from another.

Now r and r' are linked by

$$r' = \mathbf{T} \mathbf{S} \mathbf{R}_z \left(\theta_z\right) \mathbf{R}_y \left(\theta_y\right) \mathbf{R}_x \left(\theta_x\right) r. \tag{1.52}$$

It is readily apparent that the (4×4) operator $\mathbf{R}_z \left(\theta_z\right) \mathbf{R}_y \left(\theta_y\right) \mathbf{R}_x \left(\theta_x\right)$ acting on r, as defined by equation (1.47) rather than (1.40), performs the 3D rotation as in (1.42) but leaves a 1 as the fourth element in the rotated vector (see equation (1.56)). The left-most operator $\mathbf{T} \mathbf{S}$ is

$$\mathbf{T} \mathbf{S} = \begin{pmatrix} s_x & 0 & 0 & b_x \\ 0 & s_y & 0 & b_y \\ 0 & 0 & s_z & b_z \\ 0 & 0 & 0 & 1 \end{pmatrix}. \tag{1.53}$$

When this operates on the rotated 4-vector it scales and translates the data leaving a 4-vector r' in which the fourth element is unity (see equation (1.57)). To appreciate this better, create the operation in equation (1.52) in stages: first

$$\mathbf{R}_x \left(\theta_x\right) r = \begin{pmatrix} x \\ y \cos \theta_x + z \sin \theta_x \\ -y \sin \theta_x + z \cos \theta_x \\ 1 \end{pmatrix} \tag{1.54}$$

then

$$\mathbf{R}_y \left(\theta_y\right) \mathbf{R}_x \left(\theta_x\right) r = \begin{pmatrix} x \cos \theta_y - y \sin \theta_y \sin \theta_x + z \sin \theta_y \cos \theta_x \\ y \cos \theta_x + z \sin \theta_x \\ -x \sin \theta_y - y \cos \theta_y \sin \theta_x + z \cos \theta_y \cos \theta_x \\ 1 \end{pmatrix} \tag{1.55}$$

and

$$\mathbf{R}_z \left(\theta_z\right) \mathbf{R}_y \left(\theta_y\right) \mathbf{R}_x \left(\theta_x\right) r$$
$$= \begin{pmatrix} x C_z C_y - y C_z S_x S_y + z C_z S_y C_x + y S_z C_x + z S_x S_z \\ -x S_z C_y + y S_z S_y S_x - z S_z S_y C_x + y C_z C_x + z C_z S_x \\ -x S_y - y C_y S_x + z C_y C_x \\ 1 \end{pmatrix} \tag{1.56}$$

where $S_{x,y,z} = \sin \theta_{x,y,z}$ and $C_{x,y,z} = \cos \theta_{x,y,z}$ and finally

$$r' = \mathbf{T} \mathbf{S} \mathbf{R}_z \left(\theta_z\right) \mathbf{R}_y \left(\theta_y\right) \mathbf{R}_x \left(\theta_x\right) r = \begin{pmatrix} a_1 \\ a_2 \\ a_3 \\ a_4 \end{pmatrix}$$

where

$$a_1 = s_x x \cos\theta_z \cos\theta_y + s_x y \left(\sin\theta_z \cos\theta_x - \cos\theta_z \sin\theta_y \sin\theta_x\right)$$
$$+ s_x z \left(\cos\theta_z \sin\theta_y \cos\theta_x + \sin\theta_x \sin\theta_z\right) + b_x$$
$$a_2 = -s_y x \sin\theta_z \cos\theta_y + s_y y \left(\sin\theta_z \sin\theta_y \sin\theta_x + \cos\theta_z \cos\theta_x\right)$$
$$+ s_y z \left(\cos\theta_z \sin\theta_x - \sin\theta_z \sin\theta_y \cos\theta_x\right) + b_y$$
$$a_3 = -s_z x \sin\theta_y - s_z y \cos\theta_y \sin\theta_x + s_z z \cos\theta_y \cos\theta_x + b_z$$
$$a_4 = 1. \tag{1.57}$$

We see that equation (1.57) is (of course) identical to (1.46). The advantage of using equation (1.52) or (1.57) is that *all* geometrical operators are represented as matrix operations. The vectors require to be 4-vectors and the **R, T, S** matrices are square (4 × 4) operators.

In order to register two datasets, four or more corresponding points must be identified in both sets. From these may be determined the parameters $\theta_x, \theta_y, \theta_z, s_x, s_y, s_z, b_x, b_y, b_z$ using equation (1.57).

1.2.3. Decoupling translation from rotation and scaling using centre-of-mass coordinates

The way forward to solving for the geometric operators is to uncouple the translation from the rotation and scaling. This is done by working in centre-of-mass coordinates. We shall regard the rotation as taking place about the centre-of-mass of the reference points in the p set. Suppose p_{cm} is the centre-of-mass vector of the set of reference points in p and p'_{cm} is the centre-of-mass vector of the set of reference points in p'. This does not mean the centre of mass of the real tissue enclosed by such points; instead imagine unit mass at each of the points with air between. That is

$$p_{cm} = \left(\frac{1}{N}\right) \sum_{i=1}^{N} r_i \tag{1.58}$$

and

$$p'_{cm} = \left(\frac{1}{N}\right) \sum_{i=1}^{N} r'_i. \tag{1.59}$$

where there are N reference points in each of the datasets to be registered. Then

$$r_{cm} = r - p_{cm} \tag{1.60}$$

and

$$r'_{cm} = r' - p'_{cm}. \tag{1.61}$$

For the moment return to the 3-vector description. Putting equations (1.60) and (1.61) into (1.42) we have

$$r'_{cm} + p'_{cm} = \mathbf{S}\,\mathbf{R}\,(r_{cm} + p_{cm}) + b$$

from which since

$$b = p'_{cm} - S R p_{cm} \tag{1.62}$$

$$r'_{cm} = S R r_{cm} + S R p_{cm} - S R p_{cm}. \tag{1.63}$$

The last two terms in equation (1.63) cancel leaving

$$r'_{cm} = S R r_{cm}. \tag{1.64}$$

The image datasets in the centre-of-mass coordinates are now linked only by the rotation and scaling operators **S R**.

1.2.4. *Methods of solution*

There is no universally acceptable method of solving the registration problem. Kessler (1989) has classified the available techniques into four groups. These are point-matching, line-matching, surface matching and interactive matching. The first three of these attempt to force one dataset to coincide with another by minimizing some merit or cost function representing the mismatch. The fourth method instead relies on a user attempting to register anatomical (e.g. bony) landmarks. All methods have the advantage that the imaging volumes do not have to both encapsulate the entire volume of interest, unlike other methods (e.g. Gamboa Aldeco *et al* 1987) which use moments of the distributions and can only assume isotropic scaling.

1.2.5. *Solving by point-matching method*

The point-matching method, as its name suggests, relies on the use of some external fiducials. These may be small spots of olive oil or $CuSO_4$-doped water for MRI studies, solder for x-ray CT or $^{99}Tc^m$ for SPECT or ^{68}Ga for PET. Alternatively, the points may be observer-determined anatomic point landmarks (e.g. Hill *et al* 1991a).

1.2.5.1. Solving for the rotation operator. Equation (1.64) describes the relationship between any of the reference points, expressed in the centre-of-mass frame, from the two datasets. The operator **S R** is determined by minimizing the sum of the squares

$$\chi^2 = \sum_{i=1}^{N} \|r'_i - S R r_i\|^2. \tag{1.65}$$

where the sum is taken over all the N reference points and implicitly over all the three indices of the vectors. (*Important note:* the $_{cm}$ subscript has been omitted for clarity in the following mathematical development. In section 1.2.5.1 all vectors are in the centre-of-mass frame.) This method has been used by Hawkes *et al* (1990) and Hill *et al* (1990) for aligning MRI and SPECT data. Their solution generally regards the scaling as isotropic so only four quantities are calculated $\theta_x, \theta_y, \theta_z$ and s. Similar 'point matching' has been performed by Kessler (1989) using from 5 to 12 points to improve the sensitivity of the fit. Chen *et al* (1985) used point structures visible in both modalities to register MR and CT images for radiotherapy planning. Evans *et al* (1989) registered PET images with MR images the same way. Phillips *et al* (1991) registered MRI and CT this way for treatment planning and Hill *et al* (1991a) registered MRI and CT data by a point-feature based technique. Rizzo *et al* (1990) registered 2D slices from PET and SPECT with MRI and CT after these had been identified from 3D datasets. Hill *et al* (1990) registered 2D gamma-camera images and x-radiographs this way and also used the method for 3D registration of MRI and SPECT data.

1.2.5.2.   The method of finding the rotation matrix **R**.   First regard the centre-of-mass dataset r_i' as also being de-scaled by the operator **S**, in which case

$$\chi^2 = \sum_{i=1}^{N} \| r_i' - \mathbf{R} \, r_i \|^2. \tag{1.66}$$

The method of minimizing χ^2 in equation (1.66) has been described by Arun *et al* (1987). This is sometimes known as a 'Procrustes problem' (Schönemann 1966). Expand the right-hand side to obtain, ^{t} representing matrix transpose,

$$\chi^2 = \sum_{i=1}^{N} \left(r_i' - \mathbf{R} \, r_i \right)^t \left(r_i' - \mathbf{R} \, r_i \right). \tag{1.67}$$

So

$$\chi^2 = \sum_{i=1}^{N} \left(r_i'^t r_i' + r_i^t \mathbf{R}^t \mathbf{R} \, r_i - r_i'^t \mathbf{R} \, r_i - r_i^t \mathbf{R}^t r_i' \right). \tag{1.68}$$

Because **R** is a rotation matrix $\mathbf{R}^t \mathbf{R} = \mathbf{I}$. The last two terms in the above expression are transposes of each other and identical because these two terms are also the (sum over i of) the inner products of two vectors r_i' and $\mathbf{R} \, r_i$ (recall the inner product of two vectors a and b is $a^t b = b^t a$ and is a simple scalar quantity (Korn and Korn 1968)). Hence these two terms are the same and we can write

$$\chi^2 = \sum_{i=1}^{N} \left(r_i'^t r_i' + r_i^t r_i - 2 r_i'^t \mathbf{R} \, r_i \right). \tag{1.69}$$

We require to minimize χ^2, which is achieved by maximizing

$$F = \sum_{i=1}^{N} r_i'^t \mathbf{R} \, r_i. \tag{1.70}$$

This is the trace of the (sum over i of the) inner product of two vectors r_i' and $R r_i$. Using the reversibility property of the trace of two matrices (i.e. Tr $M L =$ Tr $L M$), equation (1.70) can be written

$$F = \text{Tr} \left(\sum_{i=1}^{N} R \, r_i r_i'' \right) = \text{Tr} \; (R H) \tag{1.71}$$

where the matrix H is defined as

$$H = \sum_{i=1}^{N} r_i r_i''. \tag{1.72}$$

The problem has now reduced to finding the R which maximizes F in equation (1.70). A neat way exists to achieve this. First it is necessary to find the singular value decomposition of H. The SVD is given by

$$H = U \Lambda V' \tag{1.73}$$

where U and V are (3×3) *orthonormal* matrices and Λ is a diagonal (3×3) matrix with non-negative elements (recall that an orthonormal matrix L has $L \, L' = L'L = I$). A method for obtaining the singular value decomposition has been described by Press *et al* (1988).

Now we construct the matrix

$$X = V U' \tag{1.74}$$

which itself is orthonormal. We can now show that X is the optimum R rotation required to minimize χ^2 above. To do this construct

$$X H = V U' U \Lambda V' \tag{1.75}$$

from which

$$X H = V \Lambda V'. \tag{1.76}$$

But this is a symmetrical (because it is equal to its transpose) and positive definite matrix. It is a theorem from the Schwartz inequality (Gellert *et al* 1977—see Arun *et al* 1987) that for such matrices and for any (3×3) orthonormal matrix L one cares to choose,

$$\text{Tr} \; (X H) \geqslant \text{Tr} \; (L X H) \tag{1.77}$$

and so of all the orthonormal matrices, X maximizes F in equation (1.70). If the determinant of X is $+1$, then this is a rotation, which is what we want. X is the optimum R. If the determinant of X is -1, then this is a reflection, which is not what we want. The correct X (or R) is formed by changing the sign of the third column of V corresponding to the smallest singular value of H.

So the algorithm is summarized by equations (1.72)–(1.74), from which **R** may be determined. This was the method used by Arun *et al* (1987), Hawkes *et al* (1990) and Hill *et al* (1990,1991a). Similar methods have been developed by Evans *et al* (1989). How successful the solution is depends on the noise in the problem and the degeneracy in the choice of points to register. These problems are discussed by Arun *et al* (1987) and by Hill *et al* (1991a).

1.2.5.3. Solving for the translation operator.   The translation is simply given by

$$\boldsymbol{b} = \boldsymbol{p}'_{\text{cm}} - \boldsymbol{S}\,\boldsymbol{R}\,\boldsymbol{p}_{\text{cm}}. \tag{1.78}$$

Now we have the six parameters required to describe all the solid geometrical registration, namely $\theta_x, \theta_y, \theta_z, b_x, b_y$ and b_z. Whilst these have been determined using the 3-vector description, it is to the 4-vector description that we return to *implement* the results and register datasets by the simple matrix operations in equation (1.52).

1.2.6. Solving by line-matching method

Kessler (1989) and Kessler *et al* (1991) have shown how to use line-matching methods to obtain the transformation parameters under circumstances where it is difficult to ensure that single points are in the field of view. A custom mask for the head was constructed to which four glass rods were attached; the orientations of the rods were chosen to maximize the sensitivity of the method. The rods were filled with the same materials as for point-matching. A merit function involving the centres of the rods was minimized.

1.2.7. Solving by surface-matching method

This is perhaps the most difficult method to engineer. It requires that two surfaces can be constructed from each dataset and matched. The obvious application is for the head, where the skull surface has been used. The jargon 'hat-head' fitting or 'computer docking' has come into use to describe this technique (Kessler *et al* 1991b, Kessler 1987, 1989, Pelizzari *et al* 1987,1989, Levin 1990, Levin *et al* 1988, 1989,1991, Chen *et al* 1990, Salehpour *et al* 1991, Neiw *et al* 1991, Bidaut 1991).

The technique first requires that from each dataset a tiled surface is constructed. Many methods for doing this have been proposed (e.g. Christiansen and Stephenson 1985, Fuchs *et al* 1990). The vertices of triangular tiles in the *rotated* 'hat' r_i are joined to a common point below the 'head'. The points r'_i are defined as the intersections of these lines with the 'head' (figure 1.21). The same merit function as in equation (1.65) is then minimized. This oversimple description masks a wealth of computational arrangements.

For example, Neiw *et al* (1991) have recently automated the technique with the following improvements. The contours from each tomographic slice were

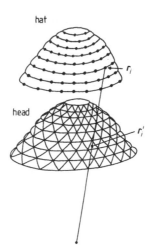

Figure 1.21. *The 'hat-head' method of image registration. One tomographic dataset defines the contours of the 'hat'. Selected points r_i are located along the hat contours. A second tomographic dataset (generally a different modality) defines the 'head'. All the points r_i are joined to a common point below the 'head' intersecting the head surface at r_i'. The 'hat' is fitted to the 'head' by a minimization process.*

extracted automatically (by thresholding CT and MR data and by the Marr–Hildreth (1980) edge detector for PET data). This automation of the feature extraction phase overcomes one of the major difficulties, the need for human interaction. The merit or cost function was defined using data which could be pre-stored. The minimization used simulated annealing to avoid the optimization becoming trapped in a local minimum.

Levin *et al* (1991) have used this technology for merging images for a variety of applications, not all connected with radiotherapy. These include:

- displaying the surface of the brain in 3D to show distortions of the sulci, gyri and convolutions resulting from tumours beneath the surface. This can assist the surgeon operating without damaging the important functions of the residual brain.
- displaying internal tumours through the semi-transparently rendered brain surface.
- developing the technique of computer-simulated craniotomy (using 3D MR data).
- displaying integrated PET and MRI 3D data to show abnormal brain function correlated with anatomical structures (figure 1.22). When abnormal function is matched to apparently normal structural details, these merged images can be used in radiotherapy treatment planning. Levin *et al* (1991) have a 'plane picker' tool. Planes are identified on the 3D displays and the corresponding 2D tomographic sections are displayed. Provided the correct registration has

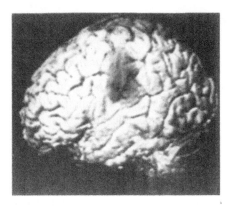

Figure 1.22. *Integrated* 3D *display of the brain of a child with intractable seizures due to extensive encephalitis. There is a mild abnormality of the lower segments of the motor and sensory strips just posterior to the speech area. This abnormality, which was confirmed at surgery, was not clearly shown by the original cross-sectional images (from Levin et al (1991)) (reprinted with permission from '3D imaging in medicine', ©CRC Press Inc, Boca Raton, Florida, USA).*

been made, these will be exactly corresponding planes from CT, MRI or PET. Salehpour *et al* (1991) extended the method to make use of multiple matched surfaces including, for example, the inner table and the frontal sinuses.

Chen *et al* (1991) incorporated the structural information from MRI and CT images into PET image reconstruction. To do this they located the important anatomical boundaries from the MRI and CT data and then showed that when these were used in a Bayesian image reconstruction scheme (rather than the convolution and back-projection method), the reconstructed PET images were sharper, as well as less noisy. This finding was from simulated data and only stated in qualitative terms.

Bidaut (1991) implemented the method to assist interpretation of brain images from PET as follows. MRI and PET data were contoured by a Laplacian of Gaussian filter, followed by interactive editing. The MRI data formed the 'head' and the PET data the 'hat'. The matching algorithm minimized the volume between the two. After registration, MRI data (being the modality with the higher spatial resolution) were re-sliced to match PET reference slices. Regions of interest were transferred between the two. Side-by-side display and colour-wash were available. The implementation was set up in a local framework which allowed for expansion to include other modalities in the future (e.g. SQUID magnetoencephalography (Reichenberger 1990)).

Hill *et al* (1991a) point out that this method is only suitable when the skin surface is itself not distorted by the method of supporting the head and when there is sufficient area of skin to provide a good matching. Because of the above constraints they did not use the method for registering 3D images of the skull base, instead preferring landmark-based correlation.

1.2.8.   *Interactive and optimized iterative matching*

Guesses for the matching are tried until obvious anatomical landmarks from one dataset match those in the other. As well as being fast (Evans 1992a,b) this can also provide the first estimate for the other three techniques. Interactive image matching is further discussed in chapter 6. This 'forward' method has to be used if the scaling factors are unknown because in those circumstances the analytic solution of Arun *et al* (1987) cannot be constructed.

Ende *et al* (1992) reported a method of anatomical landmark-based correlation with the following philosophy. At least three matching anatomical 'points' were selected from each of two unregistered 3D datasets. The points were expressed as 4-vectors. The two matrix operations taking each set of three points into a common coordinate system with one point at the origin and a second lying along a z-axis was determined. From this the single matrix operation taking one set of three points into the other was determined. By definition, one of the points from each dataset then has to coincide with its matching relative in the other dataset. The other two points will in general not exactly coincide. The error in the transformation was characterized in terms of the mismatch of these other two points. Two sets of three points can be so matched in six separate ways. By using more (n) points the number of possible ways increases to $6n!/3!(n-3)!$. The optimum transformation was selected as that which gave the minimum mismatch between the transformed sets of (n-1) points. Ende *et al* (1992) found the error could be reduced to within the spatial resolution of the datasets. They matched PET, MRI and CT data.

1.2.9.   *Lock-and-key matching*

The registration methods discussed so far have relied on identifying corresponding points in images from two different modalities, be they anatomical landmarks identified *a posteriori* or *a priori* placed external markers. Hill *et al* (1991b) addressed the problem of how to register two datasets when corresponding features were *not* able to be imaged by the two modalities. They cited the example of trying to discover the geometric relationship between major blood vessels on the surface of the brain and the surface fissures of the brain. The best way to image the blood vessels is by digital subtraction angiography. The Guys group have developed methods for reconstructing the 3D blood-vessel structure from biplane angiograms (called SARA for 'system for angiographic reconstruction and analysis'). Such vessels do not image well with MRI, which conversely can easily produce images from which the 3D brain surface can be reconstructed. Hill *et al* (1991b) used T_1- weighted spin-echo images and developed a manual method of reconstructing surfaces. The task was then to discover the best solid-body transformation to locate specific major vessels into the convolutions of the brain surface. Clearly the problem was not to register corresponding structures but to arrange the correct relationship between two structures (the vessels and the brain surface) known to be adjacent. To do this, Hill *et al* (1991b) defined and minimized appropriate cost functions, assumed geometric scaling factors were known and solved for the three translation and three rotation variables. Several optimization methods were tried including the simplex and simulated annealing (see also chapter 2). The aim was to fit two structures together by analogy with fitting a key into a lock. They tested the method with images of a brain from a cadaver.

The clinical objective was to improve the success of stereotactic procedures such as biopsy, implantation of radiotherapy sources and placement of sub-dural EEG electrodes. There is otherwise significant morbidity and mortality associated with neurosurgical procedures resulting from the accidental damage to blood vessels. Another application was to combine blood vessel images with tissue-perfusion images to assess the clinical significance of blood-vessel disease.

1.2.10. Structure mapping

Structure mapping may be performed in two ways. Either:

regions of interest (ROI) and volumes of interest (VOI) from one dataset may be transformed and displayed on the other, generally necessitating resampling (e.g. Levin 1990, Chen *et al* 1985)

or:

one whole dataset is transformed to match the other. Then a number of possibilities arise for (i) merging modalities by substituting some of the voxels from one study by those of another. Hill *et al* (1991a) used this method to generate very interesting 'fused' images of the skull base, wherein the bony structure was displayed from CT data and all other structure from MRI data (figure 1.23), (ii) side-by-side display (see e.g. Levin 1990), (iii) 'chequerboard' display (i.e. 'white' squares might receive full colour representing function (SPECT/PET), 'black' squares coding grey-scale anatomy (MRI/CT) (Hawkes et al 1990), (iv) 'modulated display' (i.e. *each* pixel colour coded for function but intensity modulated for anatomy) (Hill *et al* 1990), (v) 'linked cursor' display, whereby two corresponding slices are displayed side by side with a line joining equivalent pixels (Hill *et al* 1990, 1991a), (vi) resampled ROIs/VOIs, whereby ROIs/VOIs created from one modality are resampled by intersecting planes from the other (Chen *et al* 1985, Kessler 1989).

All the techniques described in section 1.2 assume global inelastic transformations. There is a large literature covering elastic transformations. Most is beyond the present scope of this discussion. A good starting place is with Bajcsy and Kovacic (1989), Dann *et al* (1989), Broit (1981), Bajcsy *et al* (1988), Evans *et al* (1991) and Gee *et al* (1991). Holupka and Kooy (1991) and Holupka *et al* (1991) have also described non-marker-based image correlation.

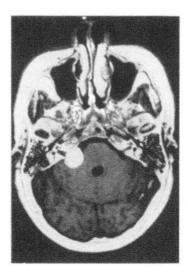

Figure 1.23. *Single fused display of* MR *and* CT *data for the skull base. The bone information is from the* CT *image and the soft-tissue information is from the* MR *image. The two datasets have been registered in* 3D *by a point-feature based technique due to Arun et al (1987). Images like these were used by the group at Guy's Hospital for surgical planning but could equally well form the basis of* 3D *radiotherapy treatment planning (from Hill et al (1991a)).*

1.3. SUMMARY AND THE NCI STUDY OF 3D RADIATION THERAPY PLANNING

Three-dimensional radiation treatment planning is not synonymous with planning conformal therapy but it is an essential component in the process. Starting around 1984, four centres in the US collaborated under contracts from the National Cancer Institute to evaluate the state of the art. The four Institutes and their Principal Investigators were: Massachusetts General Hospital (M Goitein), Memorial Sloan-Kettering Cancer Centre (J S Laughlin), University of Pennsylvania (P Bloch and M Sontag) and Mallinckrodt Institute of Radiation (J Purdy). Their work took many years to complete and the findings occupied a whole issue of the International Journal of Radiation Oncology, Biology and Physics (May 1991) running into several hundred pages. This report is a seminal publication in this area. Many of the technologies discussed have been reviewed in this chapter, but readers should consult the lengthy papers for detailed discussions of the collaborators' findings. The NCI Contractors studied just 16 patients (275 treatment plans) in a common framework in order to make meaningful comparisons. To conclude this chapter some significant statements are extracted from these reports.

Any review of the 'state of the art' has to decide whether this is to be 'that which is commonly available' or 'the best which is currently available' (Sontag and Purdy 1991). The NCI Contractors adopted the latter definition, yet even within these four Centres it was clear that different technological capabilities existed (Laughlin *et al* 1991). In this chapter, reviewing the work at these and other Centres, the same approach was taken. The tables in the publications summarizing the technological capabilities at these four Centres might be regarded as 'wish lists' against which developments elsewhere could be checked.

Certain 'growth areas' may be identified including the increased use of tools for objective scoring of treatment plans, tools for computing uncertainty in dose (Urie *et al* 1991), increased use of dose–volume histograms, computation of TCP and NTCP, correlated imaging applied to treatment planning and monitoring, and real-time visualization of beams and anatomy via the BEV. Despite years of development there is still a feeling that more effort is needed on implementing practical tissue-inhomogeneity corrections, especially to account for perturbations in electron transport. Not surprisingly, non-coplanar field treatments (called 'unconstrained' in the NCI contract) have received higher scores than coplanar field treatments. Perhaps more surprisingly, the use of very high-energy photons was de-emphasized (Laughlin *et al* 1991).

In the next chapter aspects of optimization of conformal radiotherapy will be discussed, starting with some general two-dimensional considerations and working towards three-dimensional conformal radiotherapy.

REFERENCES

Aird E G A 1988 *Basic physics for medical imaging* (Oxford: Heinemann)

Arun K S, Huang T S and Blostein S D 1987 Least squares fitting of two 3D point sets *IEE Trans. on Pattern analysis and machine intelligence* **PAM1-9** (5) 698–700

Bajcsy R and Kovacic S 1989 Multiresolution elastic matching *Comp. Vision. Graphics and Image Processing* **46** 1–21

Bajcsy R, Reivich M and Kovacic S 1988 Evaluation of registration of PET images with CT images *Technical report* MS-CIS-86-89 GRASP LAB 163, Department of Computer and Information Science, University of Pennsylvania

Bidaut L 1991 Composite PET and MRI for accurate localisation and metabolic modeling: a very useful tool for research and clinic *Proc. SPIE* **1445** (Image Processing) 66–77

Bleehan N M, Glatstein E and Haybittle J L 1983 *Radiation therapy planning* (New York: Marcel Dekker)

Boesecke R, Bauer B, Schlegel W, Schad L and Lorenz W J 1987 Medical workstation for radiation therapy planning *The use of computers in radiation therapy* ed I A D Bruinvis *et al* (Proc. 9th ICCRT) pp 181–183

Boesecke R, Doll J, Bauer B, Schlegel W, Pastyr O and Lorenz W J 1988 Treatment planning for conformation therapy using a multi-leaf collimator *Strahlentherapie und Oncologie* **164** 151–154

Brahme A 1992 Recent developments in radiation therapy planning *Proc. ART91 (Munich)* (abstract book p 6) *Advanced Radiation Therapy: Tumour Response Monitoring and Treatment Planning* ed A Breit (Berlin: Springer) pp 379–387

Brahme A, Lind B K and Källman P 1992 Application of radiation—biological data for dose optimisation in radiation therapy *Proc. ART91 (Munich)* (abstract book p 73) *Advanced Radiation Therapy: Tumour Response Monitoring and Treatment Planning* ed A Breit (Berlin: Springer) pp 407–415

Broit C 1981 Optimal registration of deformed images *Doctoral dissertation* University of Pennsylvania

Burman C, Kutcher G J, Emami B and Goitein M 1991 Fitting of normal tissue tolerance data to an analytic function *Int. J. Rad. Oncol. Biol. Phys.* **21** 123–135

Chen C T, Ouyang X, Ordonez C, Hu X, Wong W H and Metz C E 1991 Incorporation of structural CT and MR images in PET image reconstruction *Proc. SPIE* **1445** (Image Processing) 222–225

Chen G T Y, Kessler M and Pitluck S 1985 Structure transfer between sets of three-dimensional medical imaging data *Proc. Comp. Graph. (Dallas)* p 171–177

Chen G T Y, Pelizzari C A and Levin D N 1990 Image correlation in oncology *Important advances in oncology 1990* ed V T DeVita Jr, S Hellman and S A Rosenberg (Philadelphia: J B Lippincott Co) pp 131–141

Chen G T Y, Spelbring D R, Pelizzari C A, Balter J M, Myrianthopoulos L C, Vijayakumar S and Halpern H 1992 The use of beams-eye-view volumetrics in the selection of non-coplanar radiation portals *Int. J. Rad. Oncol. Biol. Phys.* **23** 153–163

Christiansen H and Stephenson M 1985 *MOVIE.BYU* (Provo, Utah: Community Press)

Cunningham J R 1982 Tissue inhomogeneity corrections in photon beam treatment planning *Progress in medical radiation physics 1* ed C G Orton (New York: Plenum) pp 103–131

—— 1983 Physics and Engineering perspective *Computed tomography in radiation therapy* ed C C Ling, C C Rogers and R J Morton (New York: Raven) pp 259–261

—— 1988 Computer applications in radiotherapy *Selected topics in Physics of radiotherapy and imaging* ed U Madhvanath, K S Parthasarathy and T V Venkateswaran (New Delhi: McGraw-Hill) pp 31–47

Dann R, Holford J, Kovacic S, Reivich M and Bajcsy R 1989 Evaluation of elastic matching system for anatomic (CT,MR) and functional (PET) cerebral images *J. Comp. Assist. Tomogr.* **13** (4) 603–611

Dobbs H J 1992 Dose specification: is there a concensus? *Proc. 50th Annual Congress of the British Institute of Radiology (Birmingham, 1992)* (London: BIR) p 4

Drzymala R E, Mohan R, Brewster L, Chu J, Goitein M, Harms W and Urie M 1991 Dose–volume histograms *Int. J. Rad. Oncol. Biol. Phys.* **21** 71–78

Emami B, Lyman J, Brown A, Coia L, Goitein M, Munzenrider J E, Shank B, Solin

L J and Wesson M 1991 Tolerance of normal tissue to therapeutic irradiation *Int. J. Rad. Oncol. Biol. Phys.* **21** 109–122

Ende G, Treuer H and Boesecke R 1992 Optimisation and evaluation of landmark-based image correlation *Phys. Med. Biol.* **37** 261–271

Evans A C, Dai W, Collins L, Neelin P and Marrett S 1991 Warping of a computerized 3D atlas to match brain image volumes for quantitative neuroanatomical and functional analysis *Proc. SPIE* **1445** (Image Processing) 236–246

Evans A C, Marrett S, Collins L and Peters T M 1989 Anatomical-functional correlative analysis of the human brain using three-dimensional imaging systems *Proc. SPIE* **1092** (Medical Imaging 3: Image Processing) 264–274

Evans P M, Gildersleve J Q, Morton E J, Swindell W, Coles R, Ferraro M, Rawlings C, Xiao Z R and Dyer J 1992a Image comparison techniques for use with megavoltage imaging systems *Brit. J. Radiol.* **65** 701–709

Evans P M, Gildersleve J Q, Rawlings C and Swindell W 1992b The implementation of patient position correction using a megavoltage imaging device on a linear accelerator *Brit. J. Radiol.* in press

Evans P M, Swindell W, Morton E J, Gildersleve J and Lewis D G 1990 A megavoltage imaging system for patient position reproducibility studies *The use of computers in radiation therapy: Proceedings of the 10th Conference* ed S Hukku and P S Iyer (Lucknow) pp 99–102

Fraass B A 1986 Practical implications of three-dimensional radiation therapy treatment planning *A categorical course in radiation therapy treatment planning* ed B R Paliwal and M L Griem (Oak Brook, IL: Radiological Society of North America) pp 13–21

—— 1990 Clinical application of 3D treatment planning *AAPM Summer School 1990*

Fraass B A and McShan D L 1987 3D treatment planning: 1 overview of a clinical planning system *The use of computers in radiation therapy* ed I A D Bruinvis *et al* (Proc. 9th ICCRT) pp 273–276

Fraass B A, McShan D L, Ten Haken R K and Lichter A S 1990 A practical 3D treatment planning system: overview and clinical use for external beam planning *Int. J. Rad. Oncol. Biol. Phys.* in press

Fraass B A, McShan D L and Weeks K J 1987 3D treatment planning: 3. Complete beams-eye-view planning capabilities *The use of computers in radiation therapy* ed I A D Bruinvis *et al* (Proc. 9th ICCRT) pp 193–196

Fuchs H, Levoy M and Pizer S M 1990 Interactive visualisation of 3D medical data *Visualisation in scientific computing* ed G M Nielson, B Shriver and L J Rosenblum (Los Alamitos, California: IEEE Computer Society Press) pp 140–146

Gamboa-Aldeco A, Fellingham L L and Chen G T Y 1987 Correlation of 3D surfaces from multiple modalities in medical imaging *SPIE Medical Image Processing and Display and Picture Archiving and Communications Systems for Medical Applications* pp 1897–1908

Gee J C, Reivich M, Bilaniuk L, Hackney D, Zimmermann R, Kovacic S and Bajcsy R 1991 Evaluation of multiresolution elastic matching using MRI data *Proc. SPIE* **1445** (Image Processing) 226–234

Geiss R A and McCullough E C 1977 The use of CT scanners in megavoltage photon beam therapy planning *Radiology* **124** 133–141

Gellert W, Kustner H, Hellwich M and Kastner H 1977 *The VNR Concise encyclopedia of mathematics* (New York: Van Nostrand) p 708

Gifford D 1984 *Handbook of physics for radiologists and radiographers* (Chichester: Wiley)

Goitein M 1982 Applications of computed tomography in radiotherapy treatment planning *Progress in medical radiation physics 1* ed C G Orton (New York: Plenum) pp 195–293

—— 1983 Patient position during CT scanning *Computed tomography in radiation therapy* ed C C Ling, C C Rogers and R J Morton (New York: Raven) pp 147–153

—— 1985 Calculation of the uncertainty in the dose delivered during radiation therapy *Med. Phys.* **12** 608–612

—— 1990 Oblique sections need 3D names also *Int. J. Rad. Oncol. Biol. Phys.* **19** 821

—— 1991 The practical incorporation of biological models into treatment planning *Proc. ART91 (Munich)* (abstract book p 72)

Goitein M, Abrams M, Rowell D, Pollari H and Wiles J 1983a Multidimensional treatment planning. 2: Beams eye view, back projection and projection through CT sections *Int. J. Rad. Oncol. Biol. Phys.* **9** 789–797

—— 1983b A multidimensional treatment planning system *Computed tomography in radiation therapy* ed C C Ling, C C Rogers and R J Morton (New York: Raven) pp 175–176

Gildersleve J 1991 private communication

Guzzardi R 1987 *Physics and engineering of medical imaging* (Dordrecht: Martinus Nijhoff (in collaboration with NATO Scientific Affairs Division))

Hahn P, Shalev S, Viggars D and Therrien P 1990 Treatment planning for protocol-based radiation therapy *Int. J. Rad. Oncol. Biol. Phys.* **18** 937–939

Hamilton B (ed) 1982 *Medical diagnostic imaging systems: technology and applications* (New York: Frost and Sullivan)

Hawkes D J, Hill D L G, Lehmann E D, Robinson G P, Maisey M N and Colchester A C F 1990 Preliminary work on the interpretation of SPECT images with the aid of registered MR images and an MR derived 3D neuro-anatomical atlas *NATO ASI Series* Vol F 60 *(3D imaging in medicine)* ed K H Hohne *et al* (Heidelberg: Springer) pp 241–252

Hill D L G, Hawkes D J, Crossman J E, Gleeson M J, Cox T C S, Bracey E E C M L, Strong A J and Graves P 1991a Registration of MR and CT images for skull base surgery using point-like anatomical features *Brit. J. Radiol.* **64** 1030–1035

Hill D L G, Hawkes D J and Hardingham C R 1991b The use of anatomical knowledge to register 3D blood vessel data derived from DSA with MR images *Proc. SPIE* **1445** (Image Processing) 248–357

Hill D L G, Hawkes D J, Lehmann E D, Crossman J E, Robinson L, Bird C F and Maisey M N 1990 Registered high resolution images in the interpretation of radionuclide scans *Proc. Annual Conf. of the IEEE Engineering in Medicine and Biology Society* **12** (1) 143–144

Holupka E J, Makrigiorgos G M and Kooy H M 1991 Verification of an image correlation algorithm using a multi-modality phantom *Med. Phys.* **18** 606

Holupka E J and Kooy H M 1991 A geometric algorithm for multi-modality image correlation *Med. Phys.* **18** 601

Hopewell J W, Morris A D, Dixon-Brown A 1987 The influence of field size on the late tolerance of the rat spinal chord to single doses of x-rays *Brit. J. Radiol.* **60** 1099–1108

Horwich A 1990 The future of radiotherapy *Radiother. Oncol.* **19** 353–356

ICRU 1993 Prescribing, recording and reporting photon beam therapy *ICRU report* 50 (Bethesda, MD: International Commission on Radiation Units and Measurement) in press

Ishigaki T, Itoh Y, Horikawa Y, Kobayashi H, Obata Y and Sakuma S 1990 Computer-assisted conformation radiotherapy *Medical Physics Bulletin of the Association of Medical Physicists of India* **15** (3,4) 185–189

Jacky J and White C P Testing a 3D radiation therapy planning program *Int. J. Rad. Oncol. Biol. Phys.* **18** 253–261

Källman P, Ågren A and Brahme A 1992b Tumour and normal tissue responses to fractionated non-uniform dose delivery (submitted to Int. J. Rad. Oncol. Biol. Phys.) and also in Källman *PhD thesis* Optimization of radiation therapy planning using physical and biological objective functions (Stockholm University, Department of Radiation Physics)

Källman P, Lind B K and Brahme A 1992a An algorithm for maximising the probability of complication-free tumour control in radiation therapy *Phys. Med. Biol.* **37** 871–890

Kessler M L 1987 Computer techniques for correlating NMR and X Ray CT imaging for radiotherapy treatment planning *The use of computers in radiation therapy* ed I A D Bruinvis *et al* (Proc. 9th ICCRT) pp 441–444

Kessler M L 1989 Integration of multimodality imaging data for radiotherapy treatment planning *PhD thesis* University of California at Berkeley

Kessler M L, ten Haken R, Fraass B and McShan D 1991a Expanding the use and effectiveness of dose-volume-histograms for 3D treatment planning *Med. Phys.* **18** 611

Kessler M L, Pitluck S, Petti P and Castro J R 1991b Integration of multimodality imaging data for radiotherapy treatment planning *Int. J. Rad. Oncol. Biol. Phys.* **21** 1653–1667

Kijewski P 1987 Data dependencies in a three-dimensional treatment planning system *The use of computers in radiation therapy* ed I A D Bruinvis *et al* (Proc. 9th ICCRT) pp 53–56

Korn G A and Korn T M 1968 *Mathematical handbook* (New York: McGraw-Hill)

Kutcher G J and Burman C 1989 Calculation of complication probability factors for non-uniform normal tissue irradiation: the effective volume method *Int. J. Rad. Oncol. Biol. Phys.* **16** 1623–1630

Kutcher G J, Burman C, Brewster L, Goitein M and Mohan R 1991 Histogram reduction method for calculating complication probabilities for three-dimensional treatment planning evaluations *Int. J. Rad. Oncol. Biol. Phys.* **21** 137–146

Kutcher G J, Burman C and Lyman J 1990 Response to letter by Dr Yaes *Int. J. Rad. Oncol. Biol. Phys.* **18** 975–6

Laughlin J S, Goitein M, Purdy J A and Sontag M 1991 (Writing Chairs) Evaluation of high energy photon external beam treatment planning *Int. J. Rad. Oncol. Biol. Phys.* **21** 3–8

Levin D 1990 MR and PET data merge in 3D images of brain *Diagnostic Imaging* (Jan/Feb) pp 28–33

D N Levin, X Hu, K K Tan, S Galhotra, C A Pelizzari, G T Y Chen, R N Beck, C T Chen, M D Cooper, J F Mullan, J Hekmatpanah and J P Spire 1989 The brain: integrated three-dimensional display of MR and PET images *Radiology* **172** 783–789

D N Levin, X Hu, K K Tan, S Galhotra, A Hermann, C A Pelizzari, G T Y Chen, R N Beck, C T Chen and M D Cooper 1991 Integrated three-dimensional display of MR, CT and PET images of the brain *3D imaging in medicine* ed J K Udupa and G T Herman (Boca Raton, FL: Chemical Rubber Company) pp 271–283

D N Levin, C A Pelizzari, G T Y Chen, C T Chen and M D Cooper 1988 Retrospective geometric correlation of MR, CT and PET images *Radiology* **169** 817–823

Lichter A S 1986 Clinical practice of modern radiation therapy treatment planning *A categorical course in Radiation therapy treatment planning* ed B R Paliwal and M L Griem (Oak Brook, IL: Radiological Society of North America) pp 7–12

—— 1991 Three-dimensional conformal radiation therapy: a testable hypothesis *Int. J. Rad. Oncol. Biol. Phys.* **21** 853–855

Lichter A S, Fraass B A, van de Geijn J, Fredrickson H A and Glatstein E 1983 An overview of clinical requirements and clinical utility of computed tomography based radiotherapy treatment planning *Computed tomography in radiation therapy* ed C C Ling, C C Rogers and R J Morton (New York: Raven) pp 1–21

Lyman J T 1985 Complication probability as assessed from dose volume histograms *Rad. Res.* **104** S-13–S-19

Lyman J T and Wolbarst A B 1987 Optimisation of radiation therapy 3. A method of assessing complication probabilities from dose-volume histograms *Int. J. Rad. Oncol. Biol. Phys.* **13** 103–109

—— 1989 Optimisation of radiation therapy 4. A dose-volume histogram reduction algorithm *Int. J. Rad. Oncol. Biol. Phys.* **17** 433–436

Margulis A R 1992 Diagnostic imaging in oncology-present and future *Proc. ART91 (Munich)* (abstract book p 4,5) *Advanced Radiation Therapy: Tumour Response Monitoring and Treatment Planning* ed A Breit (Berlin: Springer) pp 3–7

Marr D and Hildreth E C 1980 Theory of edge detection *Proc. R. Soc. (London)* **207** 187–217

McCullough E C 1983 Specifying a CT scanner for use in radiation therapy planning *Computed tomography in radiation therapy* ed C C Ling, C C Rogers and R J Morton (New York: Raven) pp 143–146

McShan D L 1986 Treatment plan evaluation and optimisation *A categorical course in Radiation therapy treatment planning* ed B R Paliwal and M L Griem (Oak Brook, IL: Radiological Society of North America) pp 33–39

—— 1990a Conformal treatment planning at the University of Michigan *The use of computers in radiation therapy: Proceedings of the 10th Conference* ed S Hukku and P S Iyer (Lucknow) p 80

—— 1990b Conformal treatment planning *Medical Physics Bulletin of the Association of Medical Physicists of India* **15** (3,4) 190–199

McShan D L and Fraass B A 1987 3D treatment planning: 2. Integration of gray scale images and solid surface graphics *The use of computers in radiation therapy* ed I A D Bruinvis *et al* (Proc. 9th ICCRT) pp 41–44

—— 1990 A practical 3D treatment planning system: anatomy and display *Int. J. Rad. Oncol. Biol. Phys.* in press

McShan D L, Fraass B A and Lichter A S 1990 Full integration of the beam's eye view concept into computerized treatment planning *Int. J. Rad. Oncol. Biol. Phys.* **18** 1485–1494

McShan D L, Matrone G M, Fraass B A and Lichter A S 1987 A large screen digitiser system for radiation therapy planning *Med. Phys.* **14** (3) 459

McShan D L, Silverman A, Lanza D M, Reinstein L E and Glicksman A S 1979 A computerized three-dimensional treatment planning system utilizing colour graphics *Brit. J. Radiol.* **52** 478–481

Moerland M A, Bhagwandien R, Beerama R and Bakker C J G 1991 Determination of geometric distortions in MR imaging; implications for radiotherapy treatment planning *Proc. 1st biennial ESTRO meeting on physics in clinical radiotherapy (Budapest 1991)* p 25

Moores B M, Parker R P and Pullan B R (eds) 1981 *Physical aspects of medical imaging* (Chichester: Wiley)

Nahum A E and Tait 1992 Maximising local control by customized dose prescription for pelvic tumours *Proc. ART91 (Munich)* (abstract book p 84) *Advanced Radiation Therapy: Tumour Response Monitoring and Treatment Planning* ed A Breit (Berlin: Springer) pp 425–431

Neiw H M, Chen C T, Lin W C and Pelizzari C A 1991 Automated three-dimensional registration of medical images *Proc. SPIE* **1445** (Image Processing) 259–264

Niemierko A and Goitein M 1991 Calculation of normal tissue complication probability and dose–volume histogram reduction schemes for tissues with a critical element architecture *Radiother. Oncol.* **20** 166–176

Orton C G 1988 A unified approach to dose–effect relationships in fractionated radiotherapy *Selected topics in Physics of radiotherapy and imaging* ed U Madhvanath, K S Parthasarathy and T V Venkateswaran (New Delhi: McGraw-Hill) pp 25–30

Parker R P and Hobday P A 1981 CT scanning in radiotherapy treatment planning: its strengths and weaknesses *Computerised axial tomography in oncology* ed J E Husband and P A Hobday (Edinburgh: Churchill Livingstone) pp 90–100

Parker R P, Hobday P A and Cassell K J 1979 The direct use of CT numbers in radiotherapy dosage calculations for inhomogeneous media *Phys. Med. Biol.* **24** 802–809

C A Pelizzari and G T Y Chen 1987 Registration of multiple diagnostic imaging scans using surface fitting *The use of computers in radiation therapy* ed I A D Bruinvis *et al* (Proc. 9th ICCRT) pp 437–440

C A Pelizzari, G T Y Chen, D R Spelbring, R R Weichselbaum and C T Chen 1989 Accurate three-dimensional registration of CT, PET, and/or MR images of the brain *J. Comp. Assist. Tomog.* **13** 20–26

Phillips M H, Kessler M, Chuang F Y S, Frankel K A, Lyman J T, Fabrikant J I and Levy R P 1991 Image correlation of MRI and CT in treatment planning for radiosurgery of intracranial vascular malformations *Int. J. Rad. Oncol. Biol. Phys.* **20** 881–889

Press W H, Flannery B P, Teukolsky S A and Vatterling W T 1988 *Numerical recipes: The Art of Scientific Computing* (Cambridge: Cambridge Univ Press) p 52 et seq

Purdy J A, Wong J W, Harms W B, Drzymala R E, Emami B, Matthews J W, Krippner K and Ramchandar P K 1987 Three dimensional radiation treatment planning system *The use of computers in radiation therapy* ed I A D Bruinvis *et al* (Proc. 9th ICCRT) pp 277–279

Reichenberger H 1990 Biomagnetic diagnostics *Imaging systems for medical diagnostics* ed E Krestel (Berlin and Munich: Siemens AG)

Reinstein L E, McShan D L, Land R E and Glicksman A S 1983 Three-dimensional reconstruction of CT images for treatment planning in carcinoma of the lung *Computed tomography in radiation therapy* ed C C Ling, C C Rogers and R J Morton (New York: Raven) pp 155–165

Reinstein L R, McShan D L, Webber B M and Glicksman A S 1978 A computer-assisted three-dimensional treatment planning system *Radiology* **127** 259–264

Rizzo G, Gilardi M C, Bettinardi V, Carutti S and Fazio F 1990 Integration of multimodal medical images *Proc. Annual Conf. of the IEEE Engineering in Medicine and Biology Society* **12** (1) 145–146

Roeske J C, Pelizzari C A, Spelbring D, Blend M, Belcaster G and Chen G T Y 1991 Registration of SPECT and CT images for radiolabelled antibody biodistribution analysis *Med. Phys.* **18** 649

Rosenman J 1991 3D imaging in radiotherapy treatment planning *3D imaging in medicine* ed J K Udupa and G T Herman (Boca Raton, FL: Chemical Rubber Company) pp 313–329

Sailer S L, Bourland J D, Rosenman J G, Sherouse G W, Chaney E L and Tepper J E 1990 3D beams need 3D names *Int. J. Rad. Oncol. Biol. Phys.* **19** 797–798

Salehpour M R, Pelizzari C A, Balter J M and Chen G T Y 1991 A multi-structure approach to image correlation *Med. Phys.* **18** 600

Schlegel W 1987 Progress and trends in computer assisted radiotherapy *Computer Assisted Radiology* ed H U Lemke, M L Rhodes, C C Jaffee and R Felix (Berlin: Springer) pp 295–299

Schlegel W, Boesecke R, Bauer B, Alandt K and Lorenz W J 1987 Dynamic therapy planning *The use of computers in radiation therapy* ed I A D Bruinvis *et al* (Proc. 9th ICCRT) pp 361–365

Schönemann P H 1966 A generalized solution of the orthogonal Procrustes problem *Psychometrika* **31** (1) 1–10

Schultheiss T E, Dixon D O, Peters L J and Thames H D 1987 Calculation of late complication rates for use in optimisation algorithms *The use of computers in radiation therapy* ed I A D Bruinvis *et al* (Proc. 9th ICCRT) pp 127–129

Schultheiss T E, Orton C G and Peck R A 1983 Models in radiotherapy: volume effects *Med. Phys.* **10** 410–415

Shalev S, Hahn P, Bartel L, Therrien P and Carey M 1987 The quantitative evaluation of alternative treatment plans *The use of computers in radiation therapy* ed I A D Bruinvis *et al* (Proc. 9th ICCRT) pp 115–118

Shalev S, Viggars D, Carey M and Hahn P 1991a The objective evaluation of alternative treatment plans. 2: score functions *Int. J. Rad. Oncol. Biol. Phys.* **20** 1067–1073

Shalev S, Viggars D, McCracken D and Gatschuff J 1991b The limitations imposed on conformal treatment planning by patient movement and set up errors *Proc. ART91 (Munich)* (abstract book p 67)

Sherouse G W, Bourland J D, Reynolds K, McMurry H L, Mitchell T P and Chaney E L 1990 Virtual simulation in the clinical setting: some practical considerations *Int. J. Rad. Oncol. Biol. Phys.* **19** 1059–1065

Sherouse G W and Chaney E L 1991 The portable virtual simulator *Int. J. Rad. Oncol. Biol. Phys.* **21** 475–482

Smith A R 1990 Evaluation of new radiation oncology technology *Int. J. Rad. Oncol. Biol. Phys.* **18** 701–703

Sontag M R, Altschuler M D, Bloch P, Reynolds R A, Wallace R E and Waxler G K 1987 Design and clinical implementation of a second generation three-dimensional treatment planning system *The use of computers in radiation therapy* ed I A D Bruinvis *et al* (Proc. 9th ICCRT) pp 285–288

Sontag M and Purdy J A 1991 (Writing Chairs) State-of-the-art of external photon beam radiation treatment planning *Int. J. Rad. Oncol. Biol. Phys.* **21** 9–23

Swindell W and Gildersleve J Q 1991 Megavoltage imaging in radiotherapy *Rad. Magazine* **17** (196) 18–20

Suit H D and Verhey L J 1988 Precision in megavoltage radiotherapy *Megavoltage*

Radiotherapy 1937–1987 (Brit. J. Radiol. Supplement 22 pp 17–24)

Suit H D and Westgate S J 1986 Impact of improved local control on survival *Int. J. Rad. Oncol. Biol. Phys.* **12** 453–458

Tait D M and Nahum A E 1990 Conformal therapy *Eur. J. Cancer* **26** 750–753

Ten Haken R K, Kessler M L, Stern R L, Ellis J H and Niklason L T 1991 Quality assurance of CT and MRI for radiation therapy treatment planning *Quality assurance in radiotherapy physics* ed G Starkschall and J Horton (Madison, WI: Medical Physics Publishing) pp 73–103

Tepper J E and Shank B 1991 (Writing Chairs) Three-dimensional display in planning radiation therapy: a clinical perspective *Int. J. Rad. Oncol. Biol. Phys.* **21** 79–89

Urie M M, Goitein M, Doppke K, Kutcher K G, LoSasso T, Mohan R, Munzenrider J E, Sontag M and Wong J W 1991 The role of uncertainty analysis in treatment planning *Int. J. Rad. Oncol. Biol. Phys.* **21** 91–107

Viggars D A and Shalev S 1990a Dose optimisation in 3D treatment planning *The use of computers in radiation therapy: Proceedings of the 10th Conference* ed S Hukku and P S Iyer (Lucknow) pp 308–311

—— 1990b Colour graphics for 3D treatment planning *The use of computers in radiation therapy: Proceedings of the 10th Conference* ed S Hukku and P S Iyer (Lucknow) pp 111–114

Wambersie A, Hanks G and Van Dam J 1988 Quality assurance and accuracy required in radiation therapy: biological and medical considerations *Selected topics in Physics of radiotherapy and imaging* ed U Madhvanath, K S Parthasarathy and T V Venkateswaran (New Delhi: McGraw-Hill) pp 1–24

Webb S (ed) 1988 *The physics of medical imaging* (Bristol: Adam Hilger)

—— 1990 Non standard CT scanners: their role in radiotherapy *Int. J. Rad. Oncol. Biol. Phys.* **19** 1589–1607

—— 1991 Optimisation by simulated annealing of three-dimensional, conformal treatment planning for radiation fields defined by a multileaf collimator *Phys. Med. Biol.* **36** 1201–1226

Wells P N T (ed) 1982 *Scientific basis of medical imaging* (Edinburgh: Churchill Livingstone)

Wheldon T 1988 *Mathematical models in cancer research* (Bristol: Adam Hilger)

Wong J W 1991 Imaging in radiotherapy *Proc. 1st biennial ESTRO meeting on physics in clinical radiotherapy (Budapest, 1991)* p 17

Wong J W and Purdy J A 1990 Advances in radiotherapy treatment planning and verification *The use of computers in radiation therapy: Proceedings of the 10th Conference* ed S Hukku and P S Iyer (Lucknow) pp 72-75

Yaes R J 1990 Volume effects *Int. J. Rad. Oncol. Biol. Phys.* **18** 975–6

Yu X and Chen G T Y 1991 Applying correlated CT and MR images of the brain in compensation for attenuation in PET image reconstruction *Med. Phys.* **18** 649

CHAPTER 2

TREATMENT PLAN OPTIMIZATION

2.1. GENERAL CONSIDERATIONS

There are many aspects of delivering external beam radiotherapy which should be addressed when attempting treatment-plan optimization. For the moment, however, and for the purposes of introducing the problem, let us suppose that a target volume has been identified by an imaging method and consider a simple transaxial slice of the patient in which this target volume reduces to a simple target area, possibly with some simple, for example circular, outline. The ideal aim in this reduced version of the 3D planning problem would be to deliver a specified uniform dose to that area and to deliver zero dose elsewhere, clearly impossible since to reach the target area, photons must travel across surrounding tissue depositing dose *en route*. Hence the aim is generally modified to that of delivering a specified high dose to the target area and as low a dose as possible elsewhere. Let us look at how this could be done by rotation therapy †. Instead of the conventional small number of fixed flat or wedged fields, imagine that the beam is deliverable from a full range of orientations in $0-2\pi$. Furthermore imagine that the *profile* of the beam may be varied, i.e. the intensity may be non-uniform over the width of the beam port.

Set up a coordinate system as shown in figure 2.1, called dose space, with dose points specified by $d(r, \theta)$. Define the origin of this space as also the axis of rotation of the treatment machine. Let $f(r_\phi, \phi)$ specify the intensity of the beam at angle ϕ and at a distance r_ϕ from the projection of the axis of rotation, along the beam. $f(r_\phi, \phi)$ is non-zero only for postive r_ϕ. The normal from the origin to this projection is the first ray of the beam and the beam is blocked for negative r_ϕ.

---

† Nowadays, rotation therapy, common many years ago (see appendix 2A), has fallen rather out of favour. This is possibly because using a small number of fixed fields allows sensitive organs to be out of the shadow of the primary beam completely, whereas, unless synchronous protection is allowed (see section 2.4), all tissue receives some dose in rotation therapy. However, consideration of rotation therapy provides a good insight into the dose-planning problem and is a natural way to begin to consider solving the inverse problem of determining the beam intensities *from* a dose prescription.

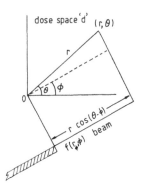

Figure 2.1. *Showing the dose space and the relationship between dose at a point (r, θ) and the beam intensity $f\left(r_\phi, \phi\right)$ at a particular position r_ϕ in the treatment beam port at angle ϕ.*

Now imagine for the moment that the beam is unattenuated on passage through the patient (ie is of sufficiently high energy). Then we observe that the dose $d(r, \theta)$ is proportional to the intensity $f(r \cos(\theta - \phi), \phi)$ of the beam at angle ϕ. Again for simplicity we shall ignore proportionality (and conflicts of units and dimensions) and say the dose *is* this intensity. After a full rotation through 2π the dose will be

$$d(r, \theta) = \left(\frac{1}{\pi}\right) \int_{\theta - \left(\frac{\pi}{2}\right)}^{\theta + \left(\frac{\pi}{2}\right)} f(r \cos(\theta - \phi), \phi)\, d\phi \qquad (2.1a)$$

because there will only be contributions to the dose $d(r, \theta)$ for the half of the 2π rotation when this dose point is in the unblocked beam. Because $d(-r, \theta + \pi) = d(r, \theta)$, we may write more simply

$$d(r, \theta) = \frac{1}{(2\pi)} \int_0^{2\pi} f(r \cos(\theta - \phi), \phi)\, d\phi \qquad (2.1b)$$

where the extra factor $\frac{1}{2}$ is included because this range evaluates the dose to each point twice. (We shall return later to the apparent paradox of ignoring attenuation, in which case there would be no photon collisions in the material at which to deposit dose.) Equation (2.1b) is the familiar Radon Transform, the relationship which, in x-ray computed tomography (CT), expresses a 2D quantity in terms of its 1D projections (Herman 1980). Once again, as in CT, the unknown quantity is under the integral sign whereas the known quantity (here taken to be known in the sense of a specified dose prescription) is to the left of the equality, i.e. *given* a dose *prescription* $d(r, \theta)$ we require to compute the distribution of *profiles* $f\left(r_\phi, \phi\right)$. We need an inverse Radon Transform. This similarity between reconstruction tomography and optimized dose planning has been pointed out by a number of authors including Cormack (1987), Cormack and Cormack (1987), Cormack and Quinto (1989), Brahme (1988), Webb (1989, 1990a), Bortfeld *et al* (1990a, 1992),

Jones (1990) and Lind and Källman (1990). It is appropriate to call optimized dose planning 'inverse computed tomography'.

The degree of approximation in equation (2.1b) depends on the energy of the beam. If a very fine beam of monoenergetic radiation falls collinearly on opposite sides of a slab of material of attenuation coefficient μ, and r is the distance from the centre of the slab, the intensity of primary radiation at r is proportional to $\cosh \mu r$. For 10 MeV radiation, $\mu \simeq 0.02$ cm^{-1}, so for a slab 30 cm thick, the largest value of μr is 0.3. As $\cosh 0.3 = 1.045$, the zero-order approximation ($\mu \simeq 0$) is correct at this energy to better than 5%. For ^{60}Co radiation, however, the error would be 50%. The zero-order approximation also ignores electronic disequilibrium near the surface and also scatter.

Cormack (1987) provides an argument that, for parallel-opposed pencil beams, the isodose curves *including scatter* will be cylinders concentric with the line of the beams, in which case equation (2.1b) holds provided f inside the integral is replaced by a function f_s where

$$f_s\left(r_\phi, \phi\right) = \int_{-\infty}^{+\infty} f\left(t, \phi\right) g\left(t - r_\phi\right) \mathrm{d}t \qquad (2.1c)$$

and $g\left(s\right)$ is the dose at a distance s from the primary beam line.

The $\mu = 0$ approximation is a simplification developed to allow a discussion of the impossibility of true inverse computed tomography, even under this unphysical assumption. In section 2.3 we shall return to what can be analytically deduced when the $\mu \neq 0$ condition is restored.

2.2. THE IMPOSSIBILITY OF TRUE INVERSE COMPUTED TOMOGRAPHY

Some might feel that it is obvious that equation (2.1b) cannot be inverted for the most general real dose prescription. Indeed this is true. However, a nice intuitive demonstration of the difficulties has been provided by Cormack (1987) and is summarized here. We can draw the 'f-space' $f\left(r_\phi, \phi\right)$ in two dimensions (figure 2.2). This might also be called the 'beam element space'. Draw a line from the origin in f-space to a point T a distance r away at an angle θ. Then construct a circle on this diameter 0T and any triangle 0TQ where Q is also on the circumference. It is now clear that the right-hand side of equation (2.1b) is simply proportional to the mean of the values of f around the circumference of this circle (for a full rotation of the beam implemented via equation (2.1b) the point Q actually goes twice round the circle). An ideal arrangement for delivering a high-dose *point* would be for f to be unity on this circumference and zero elsewhere.

Now consider (figure 2.3) trying to deliver a uniform dose to a small circular area in dose space, a distance h from the origin and of radius a. If we place this area in beam space (figure 2.4), then we can construct circles on all the diameters

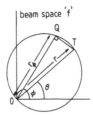

Figure 2.2. *Showing f-space or beam-element space. Integration of f around the circumference of the circle produces the dose at (r, θ). This corresponds to rotation of the beam shown in figure 2.1 around a half circle.*

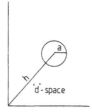

Figure 2.3. *A target region of high dose in dose space. The region is shown as a circular area of radius 'a' and centre a distance 'h' from the origin.*

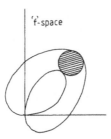

Figure 2.4. *The corresponding area in f-space contributing to the high-dose region shown in figure 2.3 (shown hatched). The area is bounded by a limaçon of Pascal.*

having ends at the origin, within, and on, the circumference of this circular area. It follows that the beam intensity requires to be non-zero around all these circles and zero elsewhere. The envelope of these circles is the limaçon of Pascal

$$r_\phi = |h \cos \theta - a| \tag{2.2}$$

in a coordinate system where the centre of the target region of radius a is at distance h from the origin along the x-axis. In principle, too, this is no problem.

Aside: the envelope is constructed as follows (refer to figure 2.5). Let C be any point on the circumference of the circle. The envelope is then the pedal curve, with respect to the origin, swept out by the intersection of the normal to the tangent at

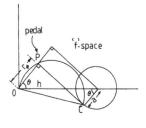

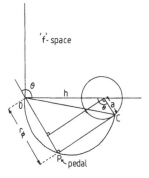

Figure 2.5. *Showing the construction of the envelope of the circles passing through the circular dose region (see text for symbols for figures 2.5 to 2.7).*

C, i.e. the locus of the point which is the intersection of the tangent to C and the perpendicular to this tangent which passes through the origin. In figure 2.5(a) the pedal is on the inner loop of the limaçon and in figure 2.5(b) it is on the outer loop. The pedal is at the origin when the tangent to C also passes through the origin; this is the transition from inner to outer loop. This gives the polar equation $r_\phi = |h\cos\theta - a|$ *(Cormack 1987)†.*

Now consider, as happens in practice, that there is another region (an 'organ at

† It may not be immediately obvious why the pedal curve is the envelope. Korn and Korn (1968) show that the envelope of any one-parameter family of plane curves described by $\psi(x, y, \lambda) = 0$ is obtained by eliminating the parameter λ between this equation and its partial derivative $\partial\psi(x, y, \lambda)/\partial\lambda = 0$. Consider figure 2.6 in which (r_ϕ, α) is a general point on the circumference of the circle whose diameter is 0C and the line from C to the origin of the circular area of high dose makes an angle β with the line joining the origin to the centre of that circular area. The equation of the circle is then

$$r_\phi + a\sin\alpha\sin\beta - \cos\alpha\,(h - a\cos\beta) = 0 \qquad (2.3)$$

The 'one parameter' to be eliminated is then β. Differentiating this equation and setting the result to zero gives

$$a\sin\alpha\cos\beta - a\cos\alpha\sin\beta = 0 \qquad (2.4)$$

from which $\alpha = \beta$, i.e. the line joining C to the point (r_ϕ, α) is a *tangent*. Putting $\alpha = \beta$ in the general equation (2.3) gives the limaçon $r_\phi = h\cos\alpha - a$.

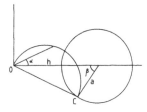

Figure 2.6.   *Showing the construction for the limaçon.*

Figure 2.7.   *High- and low-dose regions in dose space.*

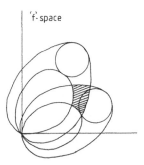

Figure 2.8.   *The corresponding area in f-space contributing to the high- and low-dose regions shown in figure 2.7. The hatched region is the area of overlap.*

risk') which it is required to shield, geometrically separated from the first region in space. Suppose for argument this is also a circle (figure 2.7). Now imagine placing this circle also in beam-space and making the same construction as before (figure 2.8). The difficulty is that the two limaçons *intersect*. A conflict arises. The region inside the limaçon embracing the high-dose target region requires non-zero intensity f, whereas the region inside the limaçon embracing the sensitive region (organ at risk) requires zero f. This is clearly impossible.

Thus Cormack argued that in general the inverse tomography problem is insoluble and confirms intuition. *There is in general no distribution of beam intensities which can give a prescribed dose distribution exactly*, even when the assumptions are as simple as those going to make up equation (2.1b). Cormack calls this the zero-order approximation: it does not reflect reality but it is a good start allowing us to see that even then the problem is insoluble. Sometimes it is stated that the problem would be soluble if there could be negative beam intensities, but this is obviously unphysical. Cormack (1987) gives a solution to inverting equation (2.1b) involving negative beam weights. He then suggests simply adding a constant so as to make the smallest beam weight zero. This has a certain elegance, but in a sense the real dose-planning problem has not been solved. The treatment-planning physicist requires a solution with non-zero beams.

Thus we appreciate that in speaking of 'solving' the inverse problem we mean obtaining an approximation to the solution. Often this is done by iterative methods (see section 2.5.4).

If the patient is considered to have a circular outline of radius R and one cannot make the assumption that $\cosh \mu r$ is 1, then the following equation (2.5) representing a first-order approximation replaces equation (2.1b). The dose space is considered to be irradiated with parallel-opposed fields on one side of the normal through the origin (other side blocked to irradiation), each with the same (but times half the magnitude) profile of intensity and a full 2π rotation (Cormack and Cormack 1987), i.e. the irradiation is as in figure 2.1 with non-zero $f\left(r_\phi, \phi\right)$ for positive r_ϕ only:

$$
\begin{aligned}
d\left(r, \theta\right) = \left(\frac{1}{\pi}\right) \int_{\theta-\left(\frac{\pi}{2}\right)}^{\theta+\left(\frac{\pi}{2}\right)} & f\left(r \cos(\theta - \phi), \phi\right) \\
& \times \exp\left(-\mu\left(R^2 - r_\phi^2\right)^{\frac{1}{2}}\right) \cosh\left(\mu r \sin\left(\theta - \phi\right)\right) d\phi
\end{aligned}
$$

(2.5)

where $r_\phi = r \cos\left(\theta - \phi\right)$. Notice the range of integration, which at first sight might appear to be only a half-rotation of the parallel-opposed beams. This range does represent a *full* 2π rotation. For any particular point $d\left(r, \theta\right)$ there is a contribution only from those profiles at orientations $\theta - \left(\frac{\pi}{2}\right) \leqslant \phi \leqslant \theta + \left(\frac{\pi}{2}\right)$ because for other orientations the dose point is in the line of sight of the block. Putting $\mu = 0$ reduces equation (2.5) to equation (2.1a). The same argument can be made regarding scatter. The problem is now to solve the attenuated Radon Transform problem. This cannot in general be done unless very restricting further (and unphysical) approximations are made (Brahme et al 1982, Lax and Brahme 1982, Cormack and Cormack 1987). This author's view is that, elegant though the mathematics become, it soon loses touch with reality and if it is desired to find a solution for a real physical problem, it is necessary to resort to numerical methods and in particular iterative solutions. These also of course do not *solve* the problem, but they deliver an approximate solution for the beam weights corresponding to a real dose distribution and they can constrain these beamweights to be positive.

2.3.   THE CASE OF A CIRCULARLY-SYMMETRIC DOSE DISTRIBUTION

Now assume the beam profiles at each orientation of the beam are all the same (but still spatially varying along their length), giving as a result a circularly-symmetric dose distribution. Imagine the situation (figure 2.9(a)) where a complete rotation of the single shaped beam shown through 2π takes place. The isocentre is the centre of the circular object. As in all discussions so far, the beam is non-zero for only positive r_ϕ and so must complete a full rotation to generate a circularly-symmetric dose distribution. The dose at radius r is given by:

$$d\,(r) = \left(\frac{1}{\pi}\right) \int_{\text{around circle radius } r} f\,(x = r\cos\phi)$$
$$\times \exp\left(-\mu\left(R^2 - x^2\right)^{\frac{1}{2}}\right) \exp\left(-\mu z\right) d\phi \tag{2.6}$$

where the x-axis is defined so that $x = r\cos\phi$ and the scattered radiation is ignored. Equation (2.6) does not result from equation (2.5) because the latter was for a rotation with parallel-opposed fields whereas equation (2.6) is for a rotation with a single field (Brahme *et al* 1982). Now noting that

$$dx = -r\sin\phi\,d\phi \tag{2.7a}$$

i.e.

$$d\phi = -dx\left(r^2 - x^2\right)^{-1/2} \tag{2.7b}$$

and writing the beam profile $f\,(x)$ in terms of a corresponding dose profile $m\,(x)$ through the origin of dose space, i.e.

$$f\,(x)\exp\left(-\mu\left(R^2 - x^2\right)^{\frac{1}{2}}\right) = m\,(x) \tag{2.8}$$

and also breaking up the circular integral into parts†, and changing to an integral over x, we obtain

$$d\,(r) = \int_0^r \frac{m\,(x)\,2\cosh\mu\left(\left(r^2 - x^2\right)^{1/2}\right) dx}{\pi\left(r^2 - x^2\right)^{1/2}}. \tag{2.9}$$

† The simplest way to obtain equation (2.9) is to note that the dose $d\,(r)$ from a full 2π rotation of the shaped beam is exactly the same as keeping the beam fixed and integrating the dose profile around the circle on radius r. From figure 2.9(a) it is then clear that the integration only needs to be performed over the upper semicircle because the beam is blocked for negative x. The integral is split into two parts for $-\pi/2 \leqslant \phi \leqslant 0$ (negative z) and for $0 \leqslant \phi \leqslant \pi/2$ (positive z). When combining these two parts, and changing from ϕ limits to x limits, to obtain equation (2.9) care must be taken with the sign of the root $z = \left(r^2 - x^2\right)^{\frac{1}{2}}$.

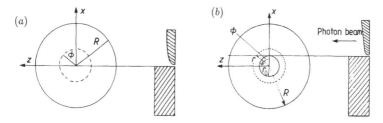

Figure 2.9. *(a) The symmetric rotation therapy case showing the field to be non-zero one side of the projection of the rotation axis and blocked the other side (adapted from Brahme et al (1982)). (b) The experimental arrangement when it is required to shield the region $r \leqslant r_0$, in which case the block extends across the projection of the rotation axis (from Brahme et al (1982)) (see text for discussion).*

Equation (2.9) was derived by Brahme *et al* (1982) (their equation (3))—refer to figure 2.9(a) adapted from that paper for further clarification. Notice the interesting way of relating the dose distribution at a radius r in terms of the distribution $m(x)$ from each projection, along a *central* axis rather than along the beam port. This is reminiscent of the way of handling projections in x-ray CT when projections are often transformed to axes through the origin of rotation in working out the reconstruction mathematics (e.g. Lakshminarayanan 1975). Of course, when equation (2.9) is solved for $m(x)$ for some particular $d(r)$, it is still necessary to correct back to the beam profile $f(x)$ using equation (2.8) above.

Solving equation (2.9) is of interest despite the limiting assumptions of a circular patient outline, ignoring build-up and scatter. Brahme posed the question: what is the solution for the general case where $m(x)$ is zero inside a radius r_0? i.e. $d(r)$ is quite zero inside the circle of radius r_0. The problem is then to invert

$$d(r) = \int_{r_0}^{r} \frac{m(x) \, 2 \cosh \mu \left(\left(r^2 - x^2 \right)^{1/2} \right) \mathrm{d}x}{\pi \left(r^2 - x^2 \right)^{1/2}} \tag{2.10}$$

i.e. $d(r)$ is a given function and $m(x)$ is to be determined.

First make the change of variables

$$x^2 = t + r_0^2 \qquad r^2 = y + r_0^2 \tag{2.11}$$

which transforms equation (2.10) into a convolution equation

$$g(y) = \int_{0}^{y} e(t) \, f(y - t) \, \mathrm{d}t \tag{2.12}$$

where

$$e(t) = \frac{m \left(\left(t + r_0^2 \right)^{1/2} \right)}{\pi \left(t + r_0^2 \right)^{1/2}} \tag{2.13}$$

$$g(y) = d\left((y + r_0^2)^{1/2}\right) \tag{2.14}$$

and

$$f(y - t) = \frac{\cosh\left(\mu(y - t)^{1/2}\right)}{(y - t)^{1/2}}. \tag{2.15}$$

It is now possible to Laplace transform the convolution equation. The Laplace transform is defined by

$$H(s) = \int_0^\infty \exp(-sy)\, h(y)\, dy \tag{2.16}$$

where an upper-case letter represents the Laplace transform of the corresponding lower-case parameter. So

$$E(s) = \frac{G(s)}{F(s)}. \tag{2.17}$$

Now make the assumption that dose is to be a constant D outside $r = r_0$, then the Laplace transform of dose is

$$G(s) = D/s. \tag{2.18}$$

By direct integration of $f(y)$ from equation (2.15) via equation (2.16)

$$F(s) = (\pi/s)^{1/2} \exp\left(\mu^2/4s\right) \tag{2.19}$$

and so

$$E(s) = G(s)\,(s/\pi)^{1/2} \exp\left(-\mu^2/4s\right). \tag{2.20}$$

Inverting the Laplace transform and putting back the original variables we come to the final result

$$m(x) = \frac{Dx \cos\left(\mu\left(x^2 - r_0^2\right)^{1/2}\right)}{\left(x^2 - r_0^2\right)^{1/2}}. \tag{2.21}$$

Some interesting results immediately follow:

1. Suppose we want to have a uniform dose from the origin to a distance r_1 and as small a dose as possible for larger r. Simply put $r_0 = 0$ in equation (2.21) to obtain

$$m(x) = D \cos(\mu x) \qquad \text{for } x \leqslant r_1 \tag{2.22a}$$

and

$$m(x) = 0 \qquad \text{for } x > r_1 \tag{2.22b}$$

i.e. the lateral dose distribution should decrease quite slowly as the cosine of the distance. Beyond this the dose profile should be set to zero. The delivered dose is then given by putting equations (2.22a), (2.22b) into equation (2.9), i.e. the dose at a radius r, $d(r)$, is given by

$$d(r) = \int_0^r \frac{m(x)\, 2\cosh\mu\left((r^2 - x^2)^{1/2}\right) dx}{\pi\left(r^2 - x^2\right)^{1/2}} \tag{2.9}$$

with

$$m(x) = D \cos(\mu x) \qquad \text{for } x \leqslant r_1 \qquad (2.22a)$$

and

$$m(x) = 0 \qquad \text{for } x > r_1. \qquad (2.22b)$$

The integral can be numerically computed for $d(r)$. This would show a uniform dose D up to the radius r_1, and for $r > r_1$ the dose falls off with increasing radius due to the fact that radiation has to pass through the outer annulus $r > r_1$ to get to the circular high-dose area $r \leqslant r_1$. The curve shows that the dose falls to half the value of the dose in the uniform region at about $1.4r_1$. For very large (and unphysical) values of r, like $r \geqslant 50$ cm, the dose profile begins to rise again because of the behaviour of the cosh term. This need not concern us.

Now we may imagine making r_1 very small. Equation (2.9) then defines the *point irradiation distribution* for a full 2π rotation irradiation by a pencil beam. This is a very important fundamental distribution and can form the basis of a method of dose planning by techniques of 'inverse computed tomography' (Brahme 1988, Lind and Källman 1990). We return to this in detail in section 2.5.

2. To shield a circular region inside $r = r_0$, simply use equation (2.21). Figure 2.9(b) shows the field arrangement with the block extending beyond $x = 0$ to $x = r_0$. Equation (2.21) shows the linear dose profile theoretically rising to infinity as x approaches r_0; Brahme *et al* (1982) state that in practice this infinity can be simply 'large'. The equation (2.21) can be realized in practice by having a central absorbing pin and symmetrically-placed wedge filters (with their narrow ends touching the pin) to correspond to the profile in equation (2.21) corrected back to the beam space by equation (2.8). Recently this has been engineered by Casebow (1990a) (see section 2.4). If the patient were circular, a half-rotation with such a wedge-plus-pin filter would suffice; in practice a full rotation would be needed if the field were half-blocked as in figure 2.9(b).

Thus it is possible to arrange for a zero-dose region to lie inside a region of uniform dose (primary beams only being considered) in this special rotationally-symmetric geometry. Cormack's argument is not violated because this stated that *in general* the problem is not always soluble for arbitrary dose regions, one of which does not contain the other. It can easily be seen that for the two concentric regions here (zero dose within $r = r_0$ and uniform dose within $r = r_1$, but outside $r = r_0$) the limaçons *become the circles themselves* in f-space ($h = 0$ in the general equation (2.2)). All that is required then is to arrange zero f-values within the circle $r = r_0$ and non-zero f-values between the annulus $r = r_0$ and $r = r_1$. Then any circle, such as that shown in figure 2.2 built on a diameter connecting the origin to any point on the circumference of the $r = r_1$ circle, partially lies in the inner circular region and partially lies in the outer circular annulus. Interestingly, for all such circles created for values of r between $r = r_0$ and $r = r_1$, the f-values in the non-zero region, with Brahme's solution, sum to the same value (because the dose is uniform in this region).

Brahme *et al* (1982) were the first to state the similarity between this problem and the problem of reconstruction in x-ray CT. The problems are mirrors of each other. The mathematics is a nice insight into the problem even though the assumptions are rather limited (no scatter, circular patient, circular treatment area (or shielded area)). It gives directly the point irradiation distribution.

When the dose distribution is not constant with r an analytical expression can still be obtained for $m(x)$ (Brahme *et al* 1982) and is

$$m(x) = \frac{d}{dx} \int_{r_0}^{x} \frac{\cos\left(\mu\left(x^2 - r^2\right)^{1/2}\right) D(r) r \, dr}{\left(x^2 - r^2\right)^{1/2}}. \tag{2.21a}$$

In section 2.5.1 we describe how Lind and Källman (1990) used this expression to compare with the result from experiment and also a method of inverse computed tomography which is actually not restricted to cylindrically-symmetric dose distributions.

Returning to the result in equation (2.22) we saw that the dose will not suddenly reduce from the uniform value within r_1 to zero outside because of the passage of photons through the outer region beyond r_1. If we really did want this to happen the intensity function $m(x)$ would have to be *negative* for $r > r_1$. Recently Barth (1990) has given the appropriate formalism. For the purposes of the argument to follow, just for the moment assume that these negative intensities are possible— i.e. suspend disbelief. In that case Barth (1990) shows how the formalism in equation (2.21a) can be adjusted so that the dose distribution produced is radially-symmetric about some point other than the rotational axis of the beam. Essentially this involves a translation and a term to account for the different path lengths of the radiation. Then he argues that because everything is linear one could create say two such circular distributions of uniform dose with zero dose elsewhere. This simply involves adding two intensity profiles varying with angle. Furthermore it is then possible to imagine that *any* concave area, within which uniform dose is required, can be decomposed as a set of (possibly a very large number of) circular areas with different centres and different radii. So in principle by superposition of the computations for each of these, a uniform dose can be constructed in any arbitrary, and if needed concave, area (figure 2.10). The mathematics are elegant.

All so well and good except for the problem of needing negative beam-weights. When these are set to zero, the dose distribution ceases to be zero outside the required area and linearity breaks down. Studies such as that of Barth (1990) are very interesting philosophically but cannot really address the real-world problems. Analytic mathematics has also been limited to scatter-free calculations. It is these limitations which iterative, and admittedly computationally expensive, dose-planning algorithms can effectively remove. As we have seen, it is apparent that, without unphysical negative intensities, perfect dose distributions can never be achieved to exactly match prescription. It is interest in seeing how close one can get which fuels the problem solving.

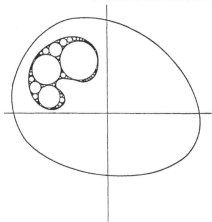

Figure 2.10. *A convex phantom of arbitrary cross section with a concave dose distribution. The arbitrary concave dose distribution is built up out of N small radially-symmetric non-intersecting dose distributions of different radii, each centred about a different point. By invoking negative beam intensities zero dose can be obtained outside the concave area (from Barth (1990)). (Reprinted with permission from Pergamon Press Ltd, Oxford, UK.)*

Goitein (1990) provides a useful critique of the problem posed by negative beam intensities. When the inverse problem is solved to give spatially non-uniform beams, followed by setting negative intensities to zero, Goitein asks in what sense can the resulting dose distribution be considered optimum? It is certainly not the true solution. Indeed the trade-off between dose delivered outside the target volume and dose uniformity inside the target has not been addressed and may not be optimal. Methods of solving the inverse problem which constrain the beam intensities to be positive (e.g. Webb 1989) avoid this difficulty altogether (see section 2.5.4) by seeking practical solutions which minimize some cost function based on the difference between dose prescription and dose delivered to both target volume and organs at risk. Cormack (1990) argues that such methods should at least be given a chance to prove themselves, given work on the inverse problem is relatively new in the near-100 year history of radiotherapy dose planning.

2.4. PRIMITIVE BLOCKED ROTATION THERAPY

For many years treatment planning ignored the possibility of calculating $f(r_\phi, \phi)$†, but recognized that continuous rotation of an open field about a suitable isocentre would give high (if non-uniform) dose distributions sparing normal tissue somewhat. Rotation therapy was improved by a technique of shielding a particular and localized circular structure by suspending a highly absorbing cylinder in the radiation field upon a plate which pivoted under gravity. This

---

† The impetus for such calculations really comes from the 'post-CT era' from those who understood well the mathematics of CT.

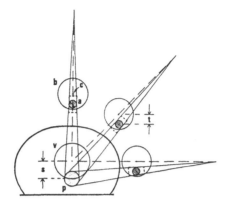

Figure 2.11. *Synchronous protection in rotation therapy using an absorbing rod (shown hatched). The shaded region 'p' is always in the shadow of the gravity-supported circular block in the field. The dose to the crescent shaped region 'v' will be non-uniform. The rod hangs at a distance 't' below the line from the source to the isocentre. This projects to a distance 's' at the isocentre (see text for details). (From Casebow (1990a).)*

is illustrated in figure 2.11. The shielded area 'p' lies below the isocentre by a distance 's'. The area treated is 'v'. The absorbing pin is 'a', pivoting at 'c'. The distance 't' of the pin below the pivot is adjusted so the magnified distance measured from the isocentre becomes 's'. The plate 'b' can rotate freely under gravity. The point 'c' lies on the axis of the treatment beam.

It can be appreciated easily that this arrangement will shield the circular area shown hatched. By changing the size and offset of the absorbing pin, the size and offset of the shielded region changes. The treated volume is crescent shaped. This idea of gravity-assisted blocking was first suggested by Proimos (1961,1963) and Proimos *et al* (1966). Takahashi *et al* (1961) developed similar rotation techniques to those of Proimos. Possibly these were inspired by Takahashi's pioneering work in radiological tomography in the pre-CT era (see Webb 1990c). Two main problems arise:

1. The shielding is limited by the finite size and the attenuation coefficient of the pin. To obtain the same shielding (for a relatively smaller region) from a relatively smaller pin, the material would have to change to one of a higher linear attenuation coefficient. Proimos tried using gold, lead and platinum absorbers.

2. The rod absorber perturbs the dose uniformity in the exposed region.

To some extent problem 1 can be overcome by using a thicker block and gearing arrangements whereby, as well as gravity positioning, the block is turned synchronously with gantry rotation to present a profile of fixed absorbing depth whatever the gantry orientation (Proimos 1961), i.e. the lead blocks remain parallel to the direction of the rays whatever the gantry position. For lead blocks with thickness in excess of some 7 cm the direct primary radiation leakage is very small (depending of course on beam energy—Proimos was using 2 MeV radiation

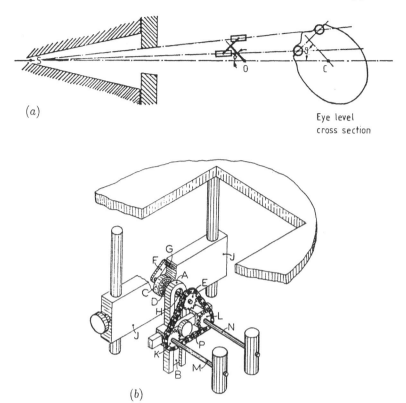

(a)

Eye level
cross section

(b)

Figure 2.12. *(a) Synchronous protection of the eyes using two lead blocks. The eyes are always in the shadow of the blocks whatever the beam orientation. In this drawing the source is stationary and the patient rotates about a vertical axis through C and the blocks are on a co-rotating plate in the field with an additional arrangement (shown in (b) on the alternative gravity-support mechanism for stationary patient and rotating source) to keep them parallel to the rays at all rotational positions of the plate. (From Proimos (1961, 1963).)*

from a Van de Graff generator in 1961). The limit to the protection afforded to the shielded region in 'total eclipse' is determined by the scattered radiation. Figure 2.12 shows an example of how this technique was used to shield the eyes.

Proimos described two possible arrangements for treatment. Either the source remained stationary, the patient rotated about a vertical axis and the shields rotated on a plate with an arrangement to keep the lead blocks parallel to the rays (as in the example shown in figure 2.12(a) and as orginally proposed by Trump *et al* (1961), or alternatively the same result was achieved by rotating the gantry about a horizontal stationary patient with the shielding suspended by gravity with synchronous block turning (figure 2.12(b)). A form of this second arrangement is also illustrated in figure 2.13, where an absorber shaped to the geometry of the

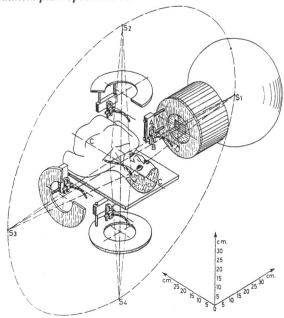

Figure 2.13. *Synchronous protection of the spinal cord. The absorber, shaped to the cord, hangs simply under gravity. As the absorber has a circular cross section, it does not need to be rotated. As the source rotates, the spinal chord is always in the shadow of the absorber. (From Proimos (1961).)*

spinal cord is being used to shield it.

Proimos (1961) also used this rotational technique for shaping fields in an attempt to conform the high-dose region to tumours with irregular outlines. Figure 2.14 illustrates this. As the patient rotates anticlockwise about the axis C, a pair of absorbing ellipses *separately* co-rotate to present a beam whose cross section depends on the orientation. The high-dose area is the ellipse shown in the patient section. The figure diagrammatically shows the patient rotating and the field-shaping ellipses co-rotating relative to a stationary x-ray beam; in practice the beam would rotate about a stationary patient and once again the appropriate arrangements would be needed for the field-shapers, i.e. the two elliptical absorbers would require to be rotated in the opposite sense to the gantry. As for protectors, this can be arranged by letting them hang under gravity, as shown in figure 2.15, providing a variable beam width. By arranging that the absorbers create an aperture matched to the projection of the target region in the direction of the beam, more complicated high-dose regions could be achieved. (In a sense this is not unlike the modern shaping of the leaves of a multileaf collimator to the beam's-eye-view of the target region—see chapter 5.) The shield A in figure 2.14 would hang also under gravity as described above. If, instead of a circular rod, this were a block, there would also need to be a separate mechanism to rotate the block, aligning it with the rays. Proimos presented mathematical arguments to

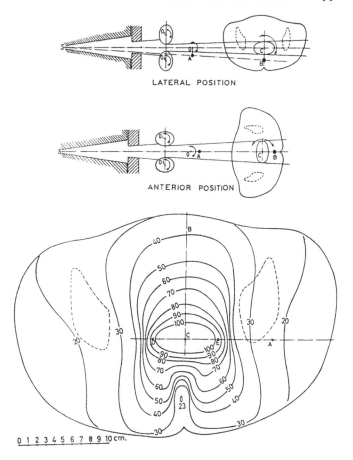

Figure 2.14. *Treatment of an elliptical area in the pelvis by synchronous field shaping. Note that the circular absorber at A protects the spinal cord at B. The dose distribution obtained is shown below the irradiation arrangement. (From Proimos (1961).)*

show that the dosimetry was independent of the position of the isocentre relative to the high-dose area.

In his earliest paper on synchronous megavoltage therapy, Proimos (1960) had proved geometrically that if the pair of field-shaping absorbers were to co-rotate about an axis, not at their centre of gravity but offset by some fixed distance, then the corresponding region in the patient which would always receive radiation would be offset from the isocentre by the corresponding magnified offset. He described this as like a searchlight beam following the course of a circling aeroplane. Then it followed immediately that if disks of different size and offset were all set up on a single axis so they could be simultaneously rotated, the high-dose volume would develop a corresponding shape which could be significantly

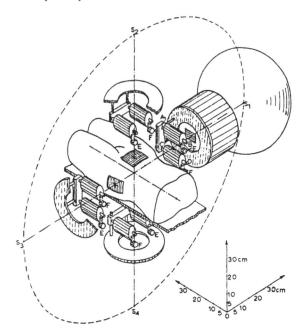

Figure 2.15. *Showing how the use of a pair of shaped blocks rotating under gravity can be used to shape a rotating x-ray field and thus conform the high-dose region to an irregularly shaped treatment region. The absorbers rotate on axes AB, CD and are shown in four orthogonal positions. (From Proimos (1963).)*

different from a right cylinder of constant radius. This is illustrated in figures 2.16 and 2.17. The method was further exploited by Ilfeld *et al* (1971).

Later Proimos (1969) designed more complicated gravity-assisted absorbers, cylinders containing a largely unattenuating demagnified tumour model surrounded by lead shot to tailor the high-dose volume to the irregularly shaped target volume.

Recently a novel arrangement has been engineered by Casebow (1990a) whereby this gravity assist plus geared rotation is combined with the use of a circular absorbing pin and two wedge filters whose thinnest ends abut the pin. This is analogous to the inverse of the filtered backprojection used in computed tomography imaging (figure 2.18) and was shown by 'end-on' film dosimetry to give a better dose uniformity in the unshielded region than the use of the pin alone.

Casebow (1990b) has suggested that one might logically consider extending this treatment technique to make use of a more complex absorber with an irregular shape and perhaps constructed of materials of differing x-ray absorption. This could introduce a beam intensity function $f\left(r_\phi, \phi\right)$ with the properties required to *approximate* the prescribed dose $d\left(r, \theta\right)$ via equation (2.1b). Suppose the required functions $f\left(r_\phi, \phi\right)$ could be obtained by some method (see e.g. section 2.5); then the absorber would be specified by the attenuation matrix $\mu\left(x, y\right)$ which is the CT

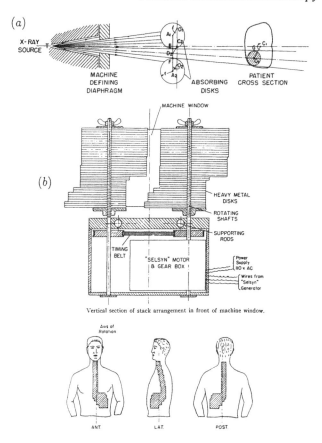

Figure 2.16. *(a) How a region not centred on the rotation isocentre can be irradiated by using a pair of offset corotating field-shaping cylinders. The patient is assumed to rotate in the fixed beam. The patient rotates about C_1 and the field shapers corotate about O_1 and O_2. As they do so the beam is swept like a search-light following the course of the target volume. (b) By stacking sets of these field-shaping disks on two corotating shafts, the high-dose volume can be conformed to a highly asymmetric volume, not centred on the axis of rotation. (From Proimos (1960).)*

reconstruction from $f\left(r_\phi, \phi\right)$. In principle this would provide a very neat solution, but there may be practical difficulties such as:

- there may be no consistency in $f\left(r_\phi, \phi\right)$ to give a solution,
- $f\left(r_\phi, \phi\right)$ may not be properly normalized (recall that in x-ray CT scanning the area under all projections must be the same),
- and the solution, if it exists, may not be well sampled (CT typically uses hundreds of projections).

Most of the papers describing synchronous field shaping and synchronous blocking were somewhat qualitative and did not directly address the dosimetry.

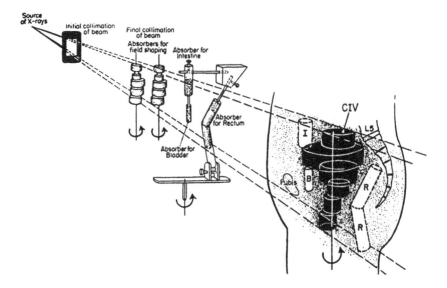

Figure 2.17. *How synchronous absorbers for field shaping and synchronous protection can be arranged to provide a high-dose volume in the continuously irradiated volume (CIV) and protect the intestine (I), bladder (B) and rectum (R). (From Ilfeld et al (1971).) (Courtesy of the American Röntgen Ray Society.)*

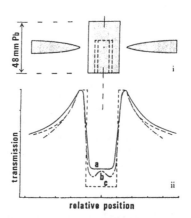

Figure 2.18. *Design of wedge + pin absorbing filter. The filter is mounted on a plate which hangs under gravity and the absorbers synchronously rotate by an arrangement of gears. The curves show the transmission profiles, normalized to block width (a,b, measured at 10 cm depth in water for 1.85 cm and 2.8 cm width blocks and c is a design calculation). (From Casebow (1990a).)*

This was put right by Rawlinson and Cunningham (1972). They noted that the dose within the shielded region was *not* negligible and that the dose in the

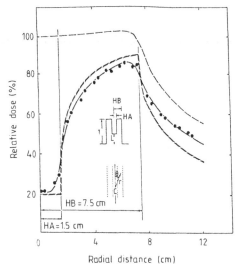

Figure 2.19. *The radial dependence of the dose profile in rotation therapy with synchronous protection. The absorber is of width 2 HA and the field of width 2 HB (shown as an inset). The data points are experimental measurements. The heavy broken curve is the result from the simple theory given by equations (2.26) to (2.28) showing good agreement. The dashed curve is the result without the block (open beam). A full 360° rotation was used. The phantom was 30 cm in diameter. The values for HA and HB are shown. (From Rawlinson and Cunningham (1972).)*

continuously irradiated region was *not* uniform. They made measurements using a full 360^0 degree rotation of a uniform field with width $2HB$ with a central absorber of width $2HA$ (figure 2.19) irradiating a cylindrical phantom of 30 cm diameter with the isocentre in the centre of the phantom. Note the approximately 20% dose to the shielded region, the scatter into the shielded region (on the experimental curve) and the non-uniformity in the continuously irradiated region (because the cosine dependence needed (equation (2.21)) was not included).

For this circularly-symmetric situation they computed the expected radial dose variation by simple trigonometry. From equation (2.1b) , we note that we may set θ to any value we choose. The θ-dependence of d and the ϕ-dependence of f disappear and setting $\theta = \pi/2$ equation (2.1b) becomes

$$d(r) = \frac{1}{2\pi} \int_0^{2\pi} f(r \sin \phi) \, d\phi. \tag{2.23}$$

Recall that the approximation in equation (2.1b), and hence also in equation (2.23), is that the dose to any point in the dose space is given by backprojecting the beam element 'connecting to' that point. Put another way, the dose beneath the absorber is a constant value for each position in the field. Set this to unity for the open field and to L for below the absorber, i.e.

$$f(r \sin \phi) = L \qquad \text{for } (r \sin \phi) < HA \tag{2.24}$$

and

$$f(r \sin \phi) = 1 \qquad \text{for } HA < (r \sin \phi) < HB. \tag{2.25}$$

It follows immediately that

$$d(r) = L \qquad \text{for} \qquad r < HA. \tag{2.26}$$

For $HA < r < HB$, $f = L$ for a total rotation angle of $4 \sin^{-1}(HA/r)$ and unity for the remaining part of the 2π rotation. Hence

$$d(r) = 1 - \frac{(1-L) \sin^{-1}(HA/r)}{\pi/2} \qquad \text{for } HA < r < HB. \tag{2.27}$$

For $r > HB$, $f = L$ for a total rotation angle of $4 \sin^{-1}(HA/r)$ and unity for a total rotation angle of $4 \sin^{-1}(HB/r) - 4 \sin^{-1}(HA/r)$. Hence

$$d(r) = \frac{\sin^{-1}(HB/r)}{\pi/2} - \frac{(1-L) \sin^{-1}(HA/r)}{\pi/2} \qquad \text{for } r > HB. \tag{2.28}$$

Rawlinson and Cunningham plotted equations (2.26) to (2.28) (see figure (2.19)) and found that despite the very simple approximations (such as ignoring photon attenuation and scatter) in this model there was good agreement with the experimental data points. The poor uniformity in the continuously irradiated zone is the result of rotation wherein for part of the time this region 'sees' the absorber. They suggested using a wedge with its thinnest end abutting the absorber as also suggested by Casebow (1990a) and discussed above.

By a similar geometrical construction as used to get equations (2.26) to (2.28) it can easily be shown (see e.g. Brahme 1988 and Boyer *et al* 1991) that when a uniform beam of width $2a$ rotates about 360^0 the primary dose for any radius $r > a$ is given by

$$d(r) = \frac{2}{\pi} \sin^{-1}\left(\frac{a}{r}\right) \tag{2.29}$$

within the $\mu = 0$ approximation. This non-zero dose is unwanted (Boyer *et al* 1991 call it the 'transit dose skirt') but unavoidable if the region $r < a$ is to be approximately uniform. The radius r_{50} at which the dose becomes 50% of the dose in the continuously irradiated area is $a\sqrt{2}$, which is very similar to the value found from the theory in section 2.3 when attenuation (but not scatter) was included and the beam profile had a spatially varying intensity (see equations (2.9) and (2.22a, b)) to obtain an exactly uniform dose inside the radius r_1.

In this section an account of synchronous field shaping and protection in rotation therapy has highlighted the contemporary interest in those techniques. In passing we should also note that alternative techniques called 'tracking therapy' for conformation therapy were being developed in London (Davy 1985, Brace 1985). Work started as early as 1959 (Green 1959) at the Royal Northern Hospital, progressed in several phases (Green 1965) and the Mark 2 machine was eventually sited at the Royal Free Hospital. One of the prime movers, W A Jennings, has recently provided a historical overview of these developments (Jennings 1985). Green (1965) wrote '...the basis of attaining the principle of restricting the radiation to the diseased area while sparing the normal tissues around appears to be well within sight.....'.

2.5. METHODS FOR 2D AND 3D OPTIMIZATION

2.5.1. *Analytic and iterative deconvolution theory of 2D inverse computed tomography*

In section 2.3 we showed how it was possible to compute a point-irradiation distribution corresponding to the rotation of a very thin pencil beam (of width $2r_1$) around a full 2π rotation. The dose distribution was uniform inside a circle of radius r_1 and thereafter fell off monotonically. It is this monotonic decrease which rotation therapy attempts to exploit. To emphasize this we now show that using a parallel-opposed pair of fields only would give much more dose to the region traversed by the fields. Brahme (1988) described the central-axis depth dose distribution in a single photon beam by the function

$$d_s(z) = D_{\text{surface}} (\exp(-\mu z) - \nu \exp(-\mu_e z)) \tag{2.30}$$

where the first exponential describes the fall-off in absorbed primary dose and the second exponential describes the dose build-up due to secondary electrons, μ_e represents the attenuation coefficient of the secondary electrons and ν is a measure of the electron contamination in the beam. Now if two such beams are parallel opposed, directed to a cylinder of radius R, the dose normalized to D_0 at the centre of the cylinder and expressed in terms of the distance r from the centre is $d_{\text{PO}}(r)$, where

$$d_{\text{PO}}(r) = D_0[\exp(-\mu(R+r)) - \nu \exp(-\mu_e(R+r)) + \exp(-\mu(R-r))$$
$$- \nu \exp(-\mu_e(R-r))]/[2(\exp(-\mu R) - \nu \exp(-\mu_e R))] \tag{2.31}$$

and PO refers to 'parallel opposed'. This expression reduces, using hyperbolic functions, to

$$d_{\text{PO}}(r) = D_0 \left(\cosh \mu r - \frac{\cosh \mu_e r - \cosh \mu r}{(1/\nu) \exp((\mu_e - \mu)R) - 1} \right). \tag{2.32}$$

(If electron build-up is ignored, then one may set $\nu = 0$ and only the first term remains namely

$$d_{\text{PO}}(r) = D_0 \cosh \mu r \tag{2.33}$$

which is of course the simple expression from Cormack (1987) (see section 2.1 just above equation (2.1c)). Brahme (1988) has plotted the distribution given by equation (2.33) on the same diagram as that given by equation (2.9) with (2.22a), (2.22b) for the point irradiation distribution corresponding to the rotation of a very thin pencil beam (of width $2r_1$) (where one reads D_0 for D in equation (2.22a)). These two curves are shown in figure 2.20. The trade-off is now clear:

- *rotation therapy* delivers dose everywhere in the slice but concentrates it at the centre of rotation;
- *parallel-opposed pairs* give dose only in the beams' path (ignoring scatter), thus sparing tissue not in the 'line of sight' but giving a much less 'concentrated' dose.

Let us now return to considering the point-irradiation distribution. As shown, this can be calculated with certain approximations. Alternatively it could be measured experimentally. It could also be deduced from the point-spread function as follows, using the notation of Brahme (1987, 1988). Define the *point-spread function h (r)* as the ratio of the mean energy imparted per unit volume at the point r by the photon interacting at the origin $r = 0$ (this could be computed by Monte Carlo methods for example). For a pencil beam directed down a z-axis, the *pencil-beam dose distribution p (r)* is given by

$$p(r) = \int \exp(-\mu z) h(r - z) \, dz. \qquad (2.34)$$

From the pencil-beam distribution the cylindrical convergent point-irradiation distribution may be computed by

$$d_\phi(r) = \oint p(r) \, dl / 2\pi r \qquad (2.35)$$

where l refers to distance around the contour.

One can also imagine a point-irradiation distribution d_{spoke} from a finite number of pencil beams, equispaced in 2π, which would have the form of a spoke pattern centred on the rotation axis (Lind and Källman 1990). From here on we shall use the symbol $d(r)$ to represent the rotational point-irradiation distribution *howsoever formed*, with it implicit that this is appropriately selected by one of the methods given above. It is the 'fingerprint of the irradiation' (Brahme 1992).

The question arises of the spatial invariance of the function $d(r)$. Brahme (1988) argues that the functions are only weakly spatially variant and, at least to a first approximation, may be considered spatially invariant. True spatial invariance would require an infinite source-to-skin distance (SSD). As the SSD is reduced, the divergence of the beam increases and the approximation becomes less appropriate. In the experimental verification of the inverse tomography algorithm of Lind and Källman (1990) (see later) 2 m SSDs were used to minimize the problem of the positional sensitivity of $d(r)$.

Brahme then argues that from the superposition principle the dose $D(r)$ at any vector r in a cross section of the patient may be obtained from the *density* $\Phi(r)$

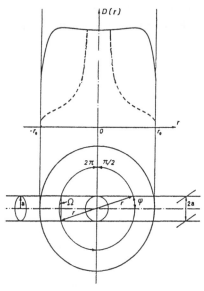

Figure 2.20. *Showing the difference between the point irradiation distribu-*
tion from a full 2π *rotation (dotted) and the central axis distribution (solid)*
for a parallel-opposed pair of pencil beams. The dotted curve is for a slit
beam of width 2a, *as shown in the right-hand side of the lower part of the*
figure. (Adapted from Brahme (1988).)

of point irradiations (sometimes known as the *kernel density* (Lind 1990)) via the
convolution integral

$$D(r) = \int \int \Phi(r') d (|r - r'|) d^2 r' \qquad (2.36)$$

where r' is the centre coordinate of each point irradiation. Equation (2.36) is
a straightforward convolution equation. It is the space-invariant form of the
Fredholm equation of the first kind:

$$D(r) = \int \int \Phi(r') d (r, r') d^2 r' \qquad (2.37)$$

where $d(r, r')$ is an elementary dose kernel representing the dose at r from an
elementary irradiation convergent at r'. From this we see that the irradiation
density function $\Phi(r)$ is given implicitly by an integral equation containing $D(r)$
and $d(r)$. Equation (2.36) is the *forward planning problem*, i.e. knowing the two
quantities on the right-hand side, the left-hand side follows. The *inverse problem*
is alternatively that of inverting equation (2.36) to obtain this irradiation density
function. In principle this can be achieved by a deconvolution, since, if we cast
equation (2.36) in Fourier space, this becomes

$$D(s) = \Phi(s) d(s) \qquad (2.38)$$

where the Roman font represents the Fourier transform of the corresponding italic-font character and s is the Fourier-space variable inverse to r. Equation (2.38) can then be inverted to give

$$\Phi(r) = F^{-1} \left[\frac{D(s)\, q(s)}{d(s)} \right]. \tag{2.39}$$

The term $q(s)$ is a low-pass filter function which should be designed to ensure that the deconvolution does not become ill-conditioned when the function $d(s)$ has zeros.

The final step in calculating beam profiles is now to make 'inverse backprojections' (or forward projections) of the density function $\Phi(r)$ weighted by the *positive* exponential factor representing attenuation of the beam, i.e.

$$f(r_\phi, \phi) = \int_{\text{line}} \Phi(r) \exp(\mu z(r))\, dz \tag{2.40}$$

where the lower case ϕ represents the angle of the projection, as in section 2.1, and r_ϕ represents the position within that beam; the vector r labels the position in the slice where the intensity function is $\Phi(r)$ and $z(r)$ is the depth of the point r in the direction of the projection of the line integral. If the slice is very inhomogeneous, the single exponential should be replaced by a ray-trace through the tissues of differing attenuation (see also Lind 1990). If the point-irradiation distribution were a simple spoke pattern, then the projections should only be made in the same directions as the angles of the spokes. Of course one cannot have negative beam weights, so where these arise the beam-weights must be set to zero. In this sense this calculation is only an approximation. The beam profile in equation (2.40) must be computed by ray-tracing along fan-lines connecting the point r to the x-ray source point.

There is an alternative way (Lind and Källman 1990) to compute the resultant dose distribution $D(r)$ from the beam profiles $f(r_\phi, \phi)$. Instead of using equation (2.36), which uses the density of point irradiations and the point-irradiation distribution, one can backproject a function $f^*(r_\phi, \phi)$ derived from the beam profiles via

$$D(r) = \sum_\phi \int f^*(r'_\phi, \phi) \exp(-\mu z(r))\, dr'_\phi \tag{2.41}$$

where f^* is f convolved with the line spread function (which could be obtained from a Monte Carlo calculation). Equation (2.41) provides a direct method of dose calculation completely analogous to the usual way of computing dose distributions from superposed open or wedged fields. The only difference here is that the beam profiles are highly non-uniform.

At least in principle, then, the spatial profile of beams at differing orientations may be computed and then delivered either using a multileaf collimator or sets of fixed compensators or scanning photon pencil beams. Perhaps this is easier to imagine than to achieve. Convery and Rosenbloom (1992) have recently provided a potential solution in terms of moving jaws (or multileaf collimator leaves) with a time-varying velocity. The above theory is due to Brahme (1988). Similar arguments have been advanced by Mackie *et al* (1985).

Instead of using the deconvolution equation (2.39) to invert the integral in equation (2.36), an iterative method may alternatively be used. Lind and Källman (1990) have used the iterative scheme:

$$\Phi_0(r) = aD(r) \tag{2.42}$$

$$\Phi_{k+1}(r) = C\left(\Phi_k(r) + a\left(D(r) - \Phi_k(r) \bigotimes d(r)\right)\right) \tag{2.43}$$

where the subscript k refers to the kth iteration and $\bigotimes$ means convolution. They use typically 20–30 iterations. C is a positivity constraint operator and a controls the speed of convergence and deals with the units and dimensions. Lind (1990) writes the same equation in matrix notation as

$$\Phi_{k+1} = C\left(\Phi_k + a\left(D - d\Phi_k\right)\right) \tag{2.44}$$

where the symbols are now read as vectors (Φ, D) and matrix operator (d). It turns out that this solution never gives underdosage. Lind (1990) also give the iterative scheme

$$\Phi_{k+1} = C\left(\Phi_k + a\left(d^t D - d^t d\Phi_k\right)\right) \tag{2.45}$$

which is the result of minimizing in a least-squares sense. This is formally identical to iterative methods to remove blurring from medical images (see e.g. Webb 1988).

Lind and Källman (1990) investigated the first iterative method for two model clinical problems. The first attempted to obtain a uniform dose inside a 12 cm square within a 25 cm diameter cylinder using just three beams at 120° to each other. For this case the point-irradiation distribution (dose kernel) was of the d_{spoke} type described earlier with three spokes and was generated by Monte Carlo methods. Figure 2.21(*a*) shows the desired dose distribution, figure 2.21(*b*) the point-irradiation density and figure 2.21(*c*) the result of applying equations (2.42) and (2.43). Figure 2.21(*d*) shows the resulting beam profiles. The delivered dose distributions were then *computed* by equation (2.41) and *measured* by constructing compensators to give the required beam profiles (thickness related to the log of the beam profile divided by the linear attenuation coefficient of the compensator material). There was good agreement, but, needless to say, a perfect square cannot be achieved with this small number of fields.

The second dose distribution was a cylindrically-symmetric high-dose volume with two dose levels (figure 2.22(*a*)). The point irradiation density corresponded to a full 360° rotation (figure 2.22(*b*)). The density of point irradiations resulting from iterative equations (2.42) and (2.43) is shown in figure 2.22(*c*). In this case all profiles are of course the same. Again the delivered dose distribution was

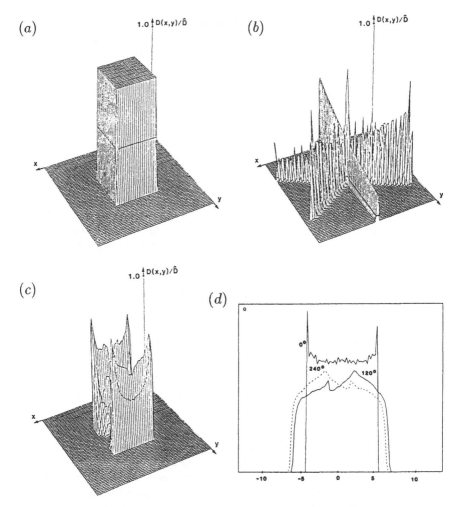

Figure 2.21. *(a) The desired dose distribution, a uniform dose inside a square of side 12 cm within a 25 cm diameter circle (not shown). (b) The point irradiation distribution from just three equally spaced 2 MeV beams. (c) The density of point irradiation from the iterative calculation. (d) The three beam profiles obtained by ray-tracing throught the irradiation densities in c. (From Lind and Källman (1990).)*

computed by equation (2.41) and *measured* experimentally with film dosimetry with good agreement.

Remembering the theory developed in section (2.3) for this circularly-symmetric problem there exists an analytic solution (equation (2.21a)). Figure 2.23 shows a comparison of the analytic solution obtained from this equation with the absorbed dose profile through the centre of the target volume calculated from the iterative algorithm (2.42) and (2.43) and equation (2.41).

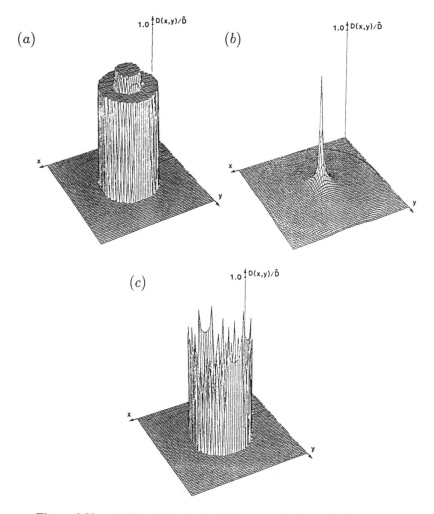

Figure 2.22. *(a) The desired circularly-symmetric dose distribution, a two-level uniform dose within a 12 cm circle. (b) The point irradiation distribution from a continuous 360⁰ 2 MeV irradiation. (c) The density of point irradiation from the iterative calculation. (From Lind and Källman (1990).)*

Note the characteristic 'ears' to the density of point irradiations near the sharp boundaries of the dose distributions (figures 2.21(c), 2.22(c)). This is to compensate for the absence of in-scatter from surrounding volumes.

Lind and Brahme (1992) have discussed a number of different geometrical dose kernels $d(r)$ appropriate to other treatment geometries which may also be used in the same way with the convolution equation (2.36) as the basis of treatment optimization.

The theory of iterative deconvolution has recently been extended by Källman *et*

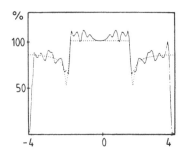

Figure 2.23. *Comparison of the absorbed radial dose profile computed by the iterative method (solid line) and the analytic method (dotted line). (From Lind and Källman (1990).)*

al (1992) to include modelling the biological response of normal and pathological tissues. The equations were reformulated to maximize the probability of uncomplicated tumour control taking into account the distribution of clonogenic cell density in the target and the 'organization' of the tissues in terms of the serial and parallel nature of the functional subunits. Källman *et al* (1992) applied both this biologically based optimization and the 'dose-only-based' iterative optimization to the same model problem, with the result that the former gave a small increase in the probability of uncomplicated tumour control. It was also shown that the resulting 1D beam profiles were smoother with the former method. The biological optimization does *not* require the target volume to have homogeneous dose, but tends to reduce the dose to parts of the target close to organs-at-risk and increase the dose to compensate elsewhere in the target volume.

Bortfeld *et al* (1990a,1992) also develop the theme of computing beam profiles by the techniques of inverse computed tomography for situations with only a small (and odd, such as 7 or 9) number of beams. They describe the function $f\left(r_\phi, \phi\right)$ as an *intensity modulation function* or IMF. The method is similar to that of Brahme (1988), but not the same. For instance they do not refer to forming projections of the density of point irradiations, as in equation (2.40) but instead refer to a first step of forward projecting the dose prescription itself, i.e. although they do not explicitly write the equation, they perform

$$f\left(r_\phi, \phi\right) = \int_{\text{line}} D\left(r\right) \exp\left(\mu z\left(r\right)\right) \mathrm{d}z. \qquad (2.46)$$

This is reasonable given the first-order approximation from Lind and Källman's method that $\Phi_0\left(r\right) = aD\left(r\right)$ (equation (2.42)). However when they compute the backprojection from these first-order profiles

$$D\left(r\right) = \sum_\phi f\left(r_\phi, \phi\right) \exp\left(-\mu z\left(r\right)\right) \qquad (2.47)$$

the resultant dose distributions were not as tightly conforming as required. To overcome this, they high-pass filtered $f\left(r_\phi, \phi\right)$ *before* the backprojection (setting

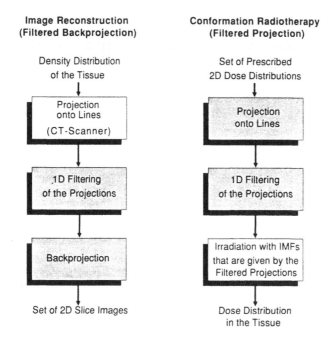

Figure 2.24. *Illustrates the formal similarity between image reconstruction by the* CT *technique of filtered backprojection and the filtered-projection method of Bortfeld for conformation radiotherapy. The actions in the lighter boxes are the practical tasks. The actions in the darker boxes are the backprojection or formation of, and the actions on, the projections respectively. (From Bortfeld et al (1990).)*

any negative values to zero). This improved the degree of conformation. It would appear that this filtering is needed simply because Bortfeld *et al* (1990a,1992) did not choose to forward project (as in equation (2.40)) the *density function* Φ which would have arisen from a *deconvolution* of the point-irradiation distribution (equation (2.36)). They provide a diagram (figure 2.24) which clarifies this point whilst discussing the analogy with computed tomography.

Bortfeld *et al* (1990a,1992) then adjust the beam elements by an iterative method, starting from the filtered projections, which minimizes a 'cost function' based on the RMS difference between the prescription and the running estimate of the dose distribution. This is similar to the simultaneous iterative reconstruction technique in medical imaging tomography. After some 10 iterations for a test problem the result is a substantial improvement.

We have seen that the point-spread function (PSF) is central to these computations. So now we introduce a lengthy interlude on convolution techniques for dosimetry. Central to these is the use of the PSF, usually computed by Monte Carlo methods. Readers wanting to stay with the central theme of optimization should proceed now to section 2.5.4.

2.5.2. *A note on convolution dosimetry based on terma*

Provided the pattern of energy spread around a point of primary photon interaction is spatially invariant, convolution methods may be used in dosimetry. Such methods are very modern and have not yet found their way to all planning computers used for routine patient dosimetry, which at present still use some of the methods reviewed over ten years ago by Cunningham (1982). Equation (2.34) was a specific example of a convolution for calculating a pencil-beam dose distribution. Strictly, spatial invariance will only hold for uniform homogeneous phantoms. Ahnesjø *et al* (1987a) defined the *point-spread function* (sometimes called an *energy deposition kernel* (Ahnesjø 1989)) as the fractional mean energy imparted in a small volume d^3r at vector position r when a photon interacts at the origin of coordinates, i.e.

$$h(E, r) = \frac{dE(r)}{E\, d^3r}. \tag{2.48}$$

The dose is the convolution of this function with the *terma* at energy E, the total energy released by primary photon interactions per unit mass $T_E(r)$. Terma has dimensions energy per unit mass. The terma at energy E, $T_E(r)$, is given in terms of the *fluence* at energy E, $\Gamma_E(r)$ (dimensions photons per unit area), via:

$$T_E(r) = \mu/\rho(E, r)\, E\Gamma_E(r) \tag{2.49}$$

where $\mu/\rho(E, r)$ is the mass attenuation coefficient of the primary photons of energy E at point r (dimensions area per unit mass). The *attenuation* rather than the *absorption* coefficient is used because the *terma* includes energy carried by both photons and electrons (compare with section 2.5.3, equation (2.62)). The energy carried by electrons alone would be the traditional *kerma*. (In other papers (e.g. Ahnesjø 1989, Ahnesjø and Trepp 1991) energy is combined with fluence to give 'energy fluence'.) From equation (2.49) the dose at energy E is

$$D_E(r) = \int\!\!\int\!\!\int T_E(s)\, h(E, r - s)\, d^3s \tag{2.50}$$

where the specific dependence on energy of the point-spread function has been inserted. The fluence at energy E is given by (figure 2.25)

$$\Gamma_E(r) = \Gamma_E(r_0) \left(\frac{r_0}{r}\right)^2 \exp - \int_{r_0}^{r} \frac{\mu}{\rho}(E, l)\, \rho(l)\, dl \tag{2.51}$$

where $\rho(l)$ is the density. Equation (2.51) is simply the inverse-square law with exponential attenuation applied to the photon fluence at the surface $\Gamma_E(r_0)$. Equation (2.51) is rather oversimplified and Mackie *et al* (1987) explain that, for example, at the field edges the penumbral fluence must be developed by convolving with a Gaussian point-spread function, which accounts for the finite x-ray source size. The small penetration through the collimation can also

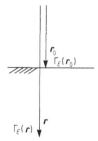

Figure 2.25. *Showing the relationship between the fluence at distance r and r₀ from the source (see equation (2.51)).*

be included. Blocks and wedges can be similarly handled by building in a modification to $\Gamma_E(r_0)$.

The total (rather than differential) terma is simply (from equation (2.49))

$$T(r) = \int \mu/\rho(E,r) E\Gamma_E(r)\,\mathrm{d}E \qquad (2.52)$$

and the corresponding total dose from photons of all energies in the incident spectrum is (from equation (2.50))

$$D(r) = \int\int\int\int T_E(s)\,h(E,r-s)\,\mathrm{d}^3s\,\mathrm{d}E. \qquad (2.53)$$

Looking at how Brahme's equation (2.34) fits into this scheme we see that Brahme combined the energy into the point-spread function (dimensions energy per unit volume; Ahnesjø *et al* (1987a) had simply per unit volume) and took the energy *out* of the other term under the integral.

Ahnesjø *et al* (1987a) computed the point spread functions by Monte Carlo methods, properly accounding for all leptons, and their paper indicated they will supply these distributions to other workers. These distributions can also be represented by analytic expressions (see below). They used them to compute depth-dose curves with enormous accuracy. For example, typically 500 000 histories were used per point-spread function, taking about 80 h on a VAX 750. Using 200 000 dose points in equation (2.53) is thus equivalent to a Monte Carlo run taking 1800 years of VAX 750 time!

The above convolution equations strictly only hold for a space-invariant point-spread function. Ahnesjø (1987) considered this further. Firstly, the point-spread function can only be spatially invariant for an infinite medium and an error would arise if the kernel was used near to the boundary of a medium. Ahnesjø developed a 'worst-case' kernel, the spatial average of a point-spread function corresponding to the primary interaction site forced to the centre of the entry and exit surfaces of a large phantom. This was compared with the point-spread function developed from an interaction point in the centre of the phantom by using both to compute depth-dose curves. The mean difference between the two was 0.2% and the maximum difference was 1.6% (at small depths for large fields). Thus spatial invariance was considered a reasonable approximation.

Secondly, the convolution equation assumes a kernel aligned to the main axis of the incident field. Ahnesjø (1987) worked out that for diverging beams the error in this approximation increased with depth and with decreasing source-to-skin distance. For example, the error was 3% at 25 cm depth for $SSD = 100$ cm. Mackie *et al* (1985, 1987) explain that if the beam is tilted relative to the patient, the dose kernels must tilt correspondingly.

Ahnesjø (1987) found almost no change in kernel due to beam hardening. If a kernel $h(r)$ is defined from the kernel with energy dependence E, $h(E, r)$ via a weighting by the fluence

$$h(r) = \sum \Gamma_{E_i} h(E_i, r) / \sum \Gamma_{E_i} \qquad (2.54)$$

then the fourth integral in equation (2.53) can be removed to give

$$D(r) = \int \int \int T(s) h(r - s) \, \mathrm{d}^3 s. \qquad (2.55)$$

This considerably shortens computation times. Ahnesjø (1989) included a multiplicative 'fudge factor' in equation (2.55) as the ratio of dose calculated for homogeneous water by equation (2.53) to that from equation (2.55).

Mackie *et al* (1987) explain that it is simply too time consuming to work out dosimetry for spectral beams by convolution with monoenergetic kernels and subsequently weighting the results by the amplitude of the spectral components. Computing a dose kernel weighted by the spectral components is likely to be the only practical solution to the problem. Additionally, a potential problem arises because the mass attenuation coefficient for a spectral beam depends on depth into the scattering medium because of beam hardening. In principle, a spectral-averaged mass attenuation coefficient can be computed by a formula given by Mackie *et al* (1987). However, this is implicit in equation (2.52) above, where the differential terma (which requires the same spectral weighting) is integrated to get the total terma.

Ahnesjø and Mackie (1987b) found that the monoenergetic point-spread function $h(E, r)$ could be represented accurately in the scatter zone by the analytic function $B \exp(-b \mid r \mid) / \mid r \mid^2$, where B and b depend on energy and the angle θ of r to the direction of the impinging photons. Ahnesjø (1989) found that the polyenergetic kernels $h(r)$ (equation (2.54)) could be represented accurately by the analytic function in water:

$$h(r, \theta) = (A_\theta \exp(-a_\theta r) + B_\theta \exp(-b_\theta r)) / r^2 \qquad (2.56)$$

where A_θ, a_θ, B_θ and b_θ are functions of the accelerating potential and of the angle θ from the measurement point r to the direction of the impinging photons (figures 2.26 and 2.27). Large tables were provided by Ahnesjø (1989) for a range of θ at

Figure 2.26. *A gamma ray interacts at a point. The point-spread function is determined by the distance **r** and the angle θ.*

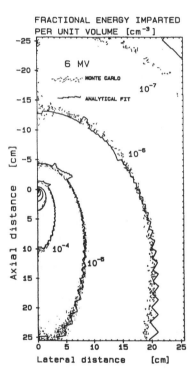

Figure 2.27. *Showing the energy deposition kernel for 6 MV photons in water. The thin hatched lines are calculated with Monte Carlo methods, whereas the thick solid lines are the corresponding analytical fits using equation (2.56). (From Ahnesjø (1989).)*

different accelerating potentials. The first term is mainly primary dose; the second is mainly scatter.

By deconvolving the point-spread function (convolution kernel) from a *measurement* of dose at depth (using for example equation (2.55)) the energy fluence may be computed (Ahnesjø 1990). Ahnesjø and Trepp (1991) amplified this theme.

Clearly, lack of homogeneity in the tissues must render the point-spread function space variant. Some workers have overcome this problem by scaling the kernel in terms of radiological distance from an interaction site to the site of dose measurement (e.g. Ahnesjø 1987, 1989, Mackie *et al* 1985, 1990a,b, 1991, Nilsson and Knöös 1991). Equation (2.50) can be cast into the form

$$D_E(r) = \int \int \int T_E(r-s)\, h(E,s)\, \mathrm{d}^3 s \qquad (2.57)$$

which becomes

$$D_E(r) = \int \int \int \mu/\rho\,(E, r-s)\, E\Gamma_E(r-s)\, h(E, s, \omega)\, |s|^2\, \mathrm{d}|s|\, \mathrm{d}\Omega. \quad (2.58)$$

When the 3D integral is explicitly broken into parts, the terma is written in terms of the fluence and ω is introduced into the argument of the point-spread function to indicate the direction s from the interaction site. Equation (2.58) holds for homogeneous media. The form in an inhomogeneous medium follows by simply scaling the vector s into the radiological length $\rho l(s)$ namely

$$D_E(r) = \int \int \int \mu/\rho\,(E, r-s)\, E\Gamma_E(r-s)\, h(E, \rho l(s), \omega)\, |s|^2\, \mathrm{d}|s|\, \mathrm{d}\Omega.$$
$$(2.59)$$

This is the form of the equation given by Mackie *et al* (1990a).

The use of scaled kernels removes the opportunity to perform convolutions in Fourier space. Zhu and Boyer (1990) have presented an alternative way to compute the dose in 3D for inhomogeneous materials. The kernels remain spatially invariant but the *fluence* is computed by ray tracing through the 3D CT matrix of inhomogeneous densities. This does not jeopardize the possibility of using Fourier space with its attendant speed advantages. Zhu and Boyer (1990) show the method compares well with measurements for fairly simple inhomogeneous phantoms. Obviously it fails in the region of large interfaces, where there is no electron equilibrium.

When the kernel is space invariant the convolution equations given above can be implemented in Fourier space by more time-efficient *multiplications* (Boyer and Mok 1985, Boyer *et al* 1988, Field and Battista 1987, Mackie *et al* 1985, 1990a). Typical speed-up factors are given in section 2.5.3.1. Alternatively, Bortfeld *et al* (1990b) describe how the 3D convolutions can be separated into 2D and 1D convolutions. When, on the other hand, the kernel is not space invariant, these methods are unavailable (Ahnesjø 1989). The real-space calculation can be very time consuming and Mackie *et al* (1990a,b) discuss some methods of speeding up the computation and report the method in use in the AB Helax treatment planning system. Zhu *et al* (1992) have used the method of convolution dosimetry for irregular fields defined by a multileaf collimator and showed that the results matched experimental determination to an accuracy of some 3 mm. These

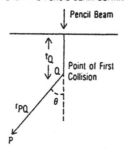

DIFFERENTIAL PENCIL BEAM DEFINITION

Figure 2.28. *The definition of the differential pencil beam. The differential pencil-beam dose distribution is the dose distribution relative to the point of first collision of the primary photons. It is so-called because its line integral gives the dose due to a pencil beam. (From Mohan et al (1986).)*

calculations were performed with a voxel size of 5 mm and took some 30 min per field (on a VAXstation 3500) for a $64 \times 64 \times 128$ dose matrix.

2.5.2.1. Convolution dosimetry by the differential pencil-beam method. Mohan *et al* (1986) introduced the differential pencil beam method of convolution photon dose calculation. The formalism is given here and it can be seen that it has similarities with the method based on the concept of terma.

The following definitions were introduced. A *pencil beam* is a parallel monoenergetic photon beam with infinitessimal cross-sectional area (figure 2.28). A *differential pencil beam (DPB)* at a point (such as Q) is the fraction of the pencil-beam photons which have their first interaction in a small volume surrounding that point. The *collision density* is the number of first collisions per unit volume in a small volume surrounding the point of first collision. The differential pencil-beam dose distribution is the dose distribution relative to the position of the first collision, per unit collision density, for such a monoenergetic pencil beam in an infinite homogeneous medium of unit density. The dose distribution at any point is then computed as the convolution of the collision density and the differential pencil-beam dose distribution.

Following Mohan *et al* (1986), the collision density at point Q is $\Gamma_Q(E)\,\mu(E)\exp\left(-\mu(E)\,t_Q\right)$, where $\Gamma_Q(E)$ is the fluence at Q of the unattenuated photons of energy E, μ is the linear attenuation coefficient for these photons and t_Q is the depth of point Q in water. The differential pencil beam (for dose deposited at point P from photons having their first interaction at point Q) is written $DPB\left(r_{PQ}, \theta, E\right)$. Hence the dose at point P for photons of energy E is given by

$$D_{P_E} = \int \Gamma_Q(E)\,\mu(E)\exp\left(-\mu(E)\,t_Q\right) DPB\left(r_{PQ}, \theta, E\right)\mathrm{d}t_Q. \qquad (2.60)$$

Note the collision density has dimensions $\left[L^{-3}\right]$ and the DPB has dimensions $\left[\text{dose} \times L^3\right]$.

If we compare equations (2.60) and (2.50) we see that the first quantity under the integral is terma multiplied by (ρ/E) and the second quantity, the DPB, is the dimensionless point-spread function h multiplied by (E/ρ). So we see that the formalism introduced by Ahnesjø *et al* (1987a) can be related to that of Mohan *et al* (1986) when the beam is considered to be a pencil.

Equation (2.60) shows that the dose can be specified by a convolution of the collision density with the DPB dose distribution. The latter is obtained by Monte Carlo calculations. Mohan *et al* (1986) show a large number of examples. These DPB dose distributions include the effects of both the transport of electrons and photons and include all the phenomena contributing to the dose for a homogeneous medium.

DPB dose distributions were computed for 'water-like' material with relative electron density 0.4, and when these were overlaid on to the corresponding water DPB dose distributions with distance scaled by 0.4 they were found to almost coincide. From this it was concluded that it is the material along the path from the first collision to the point of dose computation which mainly determines the effect of different material on dose transport. This suggested the means to account for tissue inhomogeneities:

- first compute the collision density taking into account the non-uniform photon attenuation between the source point and the point of first collision (using the average linear attenuation coefficient along this path in the exponential term in equation (2.60));
- weight the photon fluence by the linear attenuation coefficient in the voxel at which the first scatter occurs to account for secondary electron and photon production, which is in direct proportion (because attenuation coefficient scales with electron density if Compton processes dominate);
- scale the path between the point of first scatter and the dose computation point by an effective density.

This recipe requires modification for materials whose composition differs from water, such as bone. Mohan *et al* (1986) suggest the first step should instead use a ratio of *total* linear attenuation coefficient of bone and water rather than a ratio of electron density to scale the attenuation. The third step should also use an effective density when scaling the argument of the DPB kernel. The effects of different composition will matter more for higher-energy photons when pair production is a significant interaction. At ^{60}Co energies the simple scaling by electron density suffices.

This method is not expected to work in the build-up region since this is not part of the DPB model. However, it can compute the dose in the presence of wedges, beam blocks and collimators, provided the effects of these can be modelled into the fluence.

In section 2.5.3 we shall see that inhomogeneities can be accounted for in a convolution model based on kerma. As here, the fluence is scaled accurately and weighted by electron density to account for changed production. The change in the kerma kernels is accounted for analytically.

2.5.3. *Convolution dosimetry by analytic methods based on kerma*

We have seen how the dose to a point can be related to the terma convolved with the point-spread function or kernel which includes the transport of *all* energy including that carried away by secondary photons. Since dose and terma have the same dimensions, this convolution kernel is dimensionless. The kernel was evaluated by Monte Carlo methods by Ahnesjø. Provided the kernel is space invariant, the convolution could be evaluated by Fourier techniques.

There is an alternative way of looking at convolution dosimetry without the concept of terma and instead using the more familiar concept of *kerma*. *Kerma* is the kinetic energy released by primary photon interactions per unit mass and therefore includes only that energy imparted to charged particles such as electrons. It specifically excludes the energy carried away by Compton-scattered secondary photons and other higher-order scattering. It is not enough to calculate therefore the kerma deposited by the primary fluence alone, since the total dose is the result of kerma deposited by primary, secondary and higher-order multiply scattered photon fluences. All these must be included in the dose calculation. With this in mind it might be thought that it would not be possible to reduce the calculation to a convolution of primary fluence and kerma, but Boyer and Mok (1985) elegantly showed how this could be done and the following analysis stems from their work.

Some assumptions need to be introduced for this to be possible. Assume an infinite homogeneous medium into which the photons enter. Assume all incident photons have the same energy. Energy will not appear as a specific dependence in what follows, but is implicitly present. If the incident beam has a spectrum of energies, the calculations must be performed for each energy and the result subsequently weighted by the spectral intensities. For example, Boyer *et al* (1989) show how a linac spectrum can be represented as five discrete energies; the fluence calculated from Monte Carlo (EGS4) and five separate convolutions can be summed to give the total dose distribution.

As in section 2.5.2, the primary photon fluence is given by

$$\Gamma_p(r)\left[\text{photon cm}^{-2}\right] = \Gamma_p(r_0)\left[\text{photon cm}^{-2}\right]\left(\frac{r_0}{r}\right)^2 \\ \times \exp\left(-\int_{r_0}^{r}\frac{\mu}{\rho}(l)\,\rho(l)\,dl\right) \tag{2.61}$$

where $\rho(l)$ is the density. Equation (2.61) is simply the inverse-square law with exponential attenuation applied to the photon fluence at the surface $\Gamma_p(r_0)$. r_0 is the SSD and r is $SSD + d$, where d is the distance into the phantom. The subscript p is to indicate the primary fluence.

The kerma deposited by this primary fluence is

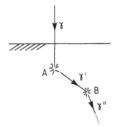

Figure 2.29. *Shows a gamma ray entering tissue (surface shown hatched) and having a first interaction at a point A. Electron transport ensures the dose is delivered to a region surrounding A. The first-scattered photon travels to point B where it undergoes a second scatter. In the analytic model it is assumed the dose from the second scatter is deposited locally at B. In practice, of course, there is a further small component of electron transport. Multiple scatter follows thereafter.*

$$K_p(r)\left[cGy\right] = \Gamma_p(r)\left[\text{photon cm}^{-2}\right]\frac{\mu_{en}}{\rho}(E_0)\left[\text{cm}^2\text{ g}^{-1}\right]$$
$$\times E_0\left[\text{MeV}\right]c\left[cGy(\text{MeV g}^{-1})^{-1}\right] \tag{2.62}$$

where $\mu_{en}/\rho\,(E_0)$ is the mass energy *absorption* (N.B. not attenuation) coefficient at energy E_0. The subscript $_p$ on this kerma also serves to indicate that it is the kerma from the primary fluence. Later we shall meet secondary and multiple-scatter kerma. The factor $c = 1.602 \times 10^{-8}$ converts from MeV g^{-1} to cGy. Now this energy is not of course delivered solely at the interaction site but is carried out to surrounding points by the electron transport (figure 2.29). This energy transport may be represented by a dimensionless function $f(r)$ which may be computed by Monte Carlo methods. Including this transport, the primary dose at r, $D_p(r)$ becomes

$$D_p(r) = \int\int\int \Gamma_p(r')\frac{\mu_{en}}{\rho}(E_0)E_0cf(r-r')\,dV'. \tag{2.63}$$

This is a convolution equation of the primary fluence with what we may call a kernel for electron transport $k_e(r)$, .i.e

$$D_p(r) = \int\int\int \Gamma_p(r')k_e(r-r')\,dV' \tag{2.64}$$

with

$$k_e(r) = c\frac{\mu_{en}}{\rho}(E_0)E_0f(r). \tag{2.65}$$

Notice that by the stage of writing down equation (2.64) the concept of kerma cannot be directly identified under the integral. It is shared between the two functions Γ_p and k_e. This differs from the simpler corresponding equation involving terma instead of kerma (see section 2.5.2; equation (2.50)) where terma appears explicitly under the integral and is convolved with a dimensionless spread function including *all* energy transport (electrons and secondary photons). Note that k_e involves analytic factors but also a function f which cannot easily be expressed analytically and must be calculated once again by Monte Carlo methods.

This kernel has the dimensions $[cGy\ cm^2]$. We may observe that equation (2.64) with (2.65) is another convolution (this time fluence with a kerma kernel) giving dose, analogous to equations (2.50) (based on terma convolved with a point spread function (PSF)) and (2.60) (based on DPB convolved with collision density). Both equation (2.64) and the latter involve convolution with either the primary fluence directly or a quantity proportional to it but (because of the factor μ in equation (2.60) operating on the fluence to get the collision density), the DPB kernel has dimensions $[dose \times L^3]$, whereas the kerma kernel has dimensions $[dose \times L^2]$. The relationship between the DPB model and the method with the PSF operating on terma has already been discussed in section 2.5.2.1.

Now we consider the contribution to dose from the first-scattered photons. These give rise to a scattered fluence $\Gamma_s(r)$ at a lower energy E_s. In turn each part of the scattered fluence with energy E_s produces a monoenergetic kerma K_s due to the first-scattered monoenergetic fluence given by identical formalism to that for the primary kerma by

$$K_s(r)\,[cGy] = \Gamma_s(r)\,[photon\ cm^{-2}]$$
$$\times \frac{\mu_{en}}{\rho}(E_s)\,[cm^2\ g^{-1}]\,E_s\,[MeV]\,c\,[cGy(MeV\ g^{-1})^{-1}]$$
(2.66)

where $\mu_{en}/\rho\,(E_s)$ is the mass energy *absorption* (NB not attenuation) coefficient at energy E_s. The total scattered photon fluence can be written down in terms of the primary fluence as

$$\Gamma_s(r) = \int\int\int \frac{\Gamma_p(r')}{(r-r')^2} \rho_e \frac{d\sigma}{d\Omega}(r-r')\exp(-\mu'|r-r'|)\,dV'. \quad (2.67)$$

Equation (2.67) considers the fluence arriving at r from the primary fluence at r', which depends on the Compton differential-scattering cross section, the inverse-square law for the distance between these two points and the exponential attenuation (coefficient μ' characteristic of energy E_s). ρ_e is the electron density. The point r receives scatter from a summation over all points r'; so the scattered fluence from equation (2.67) is no longer monoenergetic. Defining a first-scatter kerma kernel $k_s(r)$, the total kerma due to the first-scatter fluence may be written

$$K_s(r) = \int\int\int \Gamma_p(r')\,k_s(r-r')\,dV' \quad (2.68)$$

with

$$k_s(r') = \frac{1}{(r')^2}\rho_e\frac{d\sigma}{d\Omega}(r')\exp(-\mu'|r'|)c\frac{\mu'_{en}}{\rho}(E_s(r'))\,E_s(r'). \quad (2.69)$$

The single prime on μ'_{en}/ρ indicates the coefficient is for the first-scattered photons of energy E_s. Inside the integral, the fluence at r from r' is monoenergetic and the monoenergetic relation (2.69) applies. After integration over all space, the first-scatter kerma on the LHS of equation (2.69) includes photons scattered to a range of energies. Now Boyer and Mok (1985) introduce a big assumption, that the electron transport from the first-scatter kerma can be neglected completely, i.e. that the dose is delivered exactly where the secondary kerma is released. In this case the first-scatter dose is exactly equal to the first-scatter kerma and we write

$$D_s\,(r) = K_s\,(r)\,. \tag{2.70}$$

For reference we therefore repeat that

$$D_s\,(r) = \int\int\int \Gamma_p\,(r')\,k_s\,(r - r')\,dV'\,. \tag{2.71}$$

Thus the first-scatter dose also derives from a convolution of the primary photon fluence with a kernel.

Now we must turn to the dose which is deposited by higher-order multiple scattering. Boyer and Mok (1985) argue that after many scatters the remaining photon fluence (which we call the multiple-scatter photon fluence) $\Gamma_m\,(r)$ will be more or less isotropic. Photons will have 'forgotten' their original direction of incidence. In this case the scattered fluence can be represented by a steady-state diffusion process described by a Helmholtz equation. It turns out that, if we make one further assumption (not strictly true, but a simplification) that all the multiply scattered photons end up with the same effective energy E_{avg}, a closed solution exists for the Helmholtz equation. If once again we also assume that the multiply scattered fluence leads to a multiply scattered kerma K_m which is deposited locally where it is produced (i.e. electron transport is again ignored for these greatly energy-degraded photon interactions), then dose D_m equals kerma K_m. We thus arrive at

$$D_m\,(r) = \int\int\int \Gamma_m\,(r')\,k_m\,(r - r')\,dV' \tag{2.72}$$

where

$$k_m\,(r - r') = \frac{\exp\left(-|r - r'|/L\,(E_{avg})\right)}{4\pi D\,(E_{avg})\,|r - r'|}\,\frac{\mu_{en}{''}}{\rho}\,(E_{avg})\,E_{avg}c. \tag{2.73}$$

L is a diffusion length and D is a diffusion coefficient. The double primes indicate the coefficient is for the multiply scattered photons of energy E_{avg}.

Unfortunately we see that the formula for the multiply scattered dose, although being a convolution, does not in this form involve the primary fluence. However, some straightforward but tedious algebra can reduce the expression to this wanted form, namely

$$D_m\,(r) = \int\int\int \Gamma_p\,(r'')\,k_{s,m}\,(r - r'')\,dV'' \tag{2.74}$$

where $k_{s,m}$ is a composite kernel evaluated from k_s and k_m (see Boyer and Mok (1985) for details).

Summarizing, equations (2.64),(2.71) and (2.74) give the convolution equations for the primary dose, first-scattered dose and multiply scattered dose in terms of the primary photon fluence and three kerma kernels, which can largely be written down in analytic form. Be aware the kerma kernels do *not* have the dimensions of kerma because the fluence remains as a separate term under the integral. The kernels simply give the 'extraction factors' which, operating on the primary fluence, give dose. The three separate contributions can be added to give the total dose

$$D_t(r) = \int \int \int \Gamma_p(r') k_t(r - r') \, dV' \qquad (2.75)$$

where

$$k_t(r) = k_e(r) + k_s(r) + k_{s,m}(r). \qquad (2.76)$$

This kernel has the dimensions $[cGy\ cm^2]$.

This expression for k_t is the three-dimensional impulse response of the radiation dose transport process operating on the primary fluence. If we accept the limitations which have led to the convolution equation, we may evaluate the dose by making *multiplications* in Fourier space rather than the time-consuming three-dimensional *convolutions* in real space. Implementing the fast Fourier transform (FFT) (Cooley–Tukey algorithm) leads to relatively fast speeds. The Fourier transform of k_t can be thought of as a *modulation transfer function*. Indeed all the ideas from medical imaging carry across on a formal basis when considering the implementation of equation (2.75). Boyer and Mok present figures showing the first-scatter kernel k_s (forward peaked) and the isotropic multiple-scatter kernel k_m for ^{60}Co radiation. They use these, together with the kernel k_e for primary kerma to compute some simple dose distributions which were checked against published BJR Supplement 11 data with good agreement.

When they come to implement equation (2.75) in practice, Boyer *et al* (1989) use kernels, with these dimensions, computed (by Mackie) by Monte Carlo methods. Additionally, for spectral beams they use five such discrete convolutions at five different energies and add the results. The method is inherently fast because the five convolutions are completely independent and so can be executed in parallel if appropriate hardware is available. Boyer *et al* (1989) show how this method can reproduce measurable depth-dose distributions for regular field sizes to an accuracy of about 2 mm. However, the power of the method is that in principle it can be applied to irregular fields, fields with wedges and blocks etc. All that is required is to be able to obtain an accurate model of the fluence and the dose follows from convolution with the total kerma kernel.

This analytic theory is undoubtedly very elegant. It carries over to radiation dosimetry much of the formalism which is well understood in medical imaging. However one must remember that there were some very limiting assumptions which went into the arguments. The results are best applied to single-energy radiation entering infinitely large uniform water phantoms. Any departure from this must be considered with care. The analytic forms have the important merit of giving a deep insight into the dosimetry of photon interactions. As yet, analytic convolution methods have not been implemented into many treatment-planning systems. It remains to be seen whether, elegant though they are, they will replace existing routine treatment planning algorithms.

Photon dose calculation by convolution methods provide:

- a clear description of the underlying physics,
- the possibility of a very fast method of 3D dose calculation, taking advantage of the fast Fourier transform. When the conditions are particularly simple, it is expected that the result will have a high accuracy.

Unfortunately the theory already includes some approximations and is already of moderate complexity when the medium is homogeneous. When similar formalisms are attempted for dose computation in heterogeneous tissues, further assumptions have to be made which may prove to be limitations on accuracy. The aim is to arrive at a formalism by which the dose can be calculated without resort to space-variant kernels, so that the advantages of calculating in Fourier space are not lost. These advantages are very great when the dose matrix is large (Boyer *et al* 1988) and have spurred on efforts towards an analytic theory. Boyer and Mok (1986) have provided the formalism for kerma-based convolution dose calculations in inhomogeneous media. We shall not repeat the equations here but give the major assumptions and limitations as specified.

Firstly, the *primary* fluence involves ray tracing through the heterogeneous matrix of CT numbers, taking into account the exact fraction of each voxel intercepted and using the voxel-dependent linear attenuation coefficient. This can be done exactly with no loss of accuracy. The electron density of each voxel relative to water $\rho(r)$ then modifies the intensity of the electron and secondary photon scattering in all directions in direct proportion. The primary dose $D_p(r)$ is then a convolution of the electron-density-weighted fluence with an electron transport kernel, but now the kernel does not take account of electron scattering nor density-dependent attenuation. If it were intended to include these effects, the kernel would have to be calculated by Monte Carlo methods, since Boyer and Mok (1986) state this is too complicated to formulate analytically.

In heterogeneous media, the *first scatter* fluence, analogous to equation (2.67) again has the primary fluence weighted by the relative electron density at r', and the exponential attenuation term becomes space variant. Boyer and Mok (1986) show that by expanding this term in a Taylor series and neglecting high-order terms it is possible to write down the first-scatter dose $D_s(r)$ as *two* convolutions of the electron-density-weighted primary fluence with (i) the same first-scatter kerma kernel $k_s(r')$ as in equation (2.69) and (ii) a correction kernel, which is $|r'| \mu'(r') k_s(r')$, acting on fraction $(1 - \rho(r'))$ of the weighted primary fluence. Implicit in these convolutions is (once again) the assumption that the first-scatter kerma is untransported and set equal to the first-scatter dose.

The same assumptions are made for the multiple-scatter dose for heterogeneous and homomogeneous tissues, namely: the photons are randomly distributed and can be modelled with a diffusion equation; the multiple-scatter component is given an average energy and this is set to be the same as for the homogeneous calculation; the multiple-scattered kerma is not transported and is set equal to the multiple-scatter dose. It turns out that (unlike for homogeneous tissue) the multiple-scatter dose can *not* be specified in terms of a convolution with the *primary* fluence but is instead a convolution of the multiple-scattered fluence with a multiple-scatter kerma kernel. The assumption that this is spatially invariant is less worrying because of the smaller fraction of dose which is from multiple scatter.

Boyer and Mok (1986) apply their theory to inhomogeneous phantoms with 2% accuracy compared with measurements. They do not claim more for the theory and point out that it is yet to be tested for irregularly shaped fields, for fields with blocks and wedges and compensating filters.

2.5.3.1. Limitations and gains from the convolution approach. We have seen that the three-dimensional dose distribution can be written as the convolution integral

$$D_t(r) = \int \int \int \Gamma_p(r') \, k_t(r - r') \, dV'. \tag{2.75}$$

This is only possible when the kernel k_t (sometimes also described as a Green's function) is spatially invariant. The more general form for the dose to a point r is

$$D_t(r) = \int dE \int d\Omega \int \int \int \Gamma_p(r', E, \Omega, Z, \rho) \, k_t(E, \Omega, Z, \rho, r, r') \, dV' \tag{2.77}$$

where E is the incident photon energy, Ω is the incident photon direction, Z is the atomic number distribution in the patient and ρ is the density distribution in the patient. Equation (2.77) is a superposition integral requiring the computation of a Green's function for every point in the patient. Clearly this is impractical since it takes typically hundreds of VAX 780 hours to determine the Green's function with sufficient accuracy for *just one point*. Field and Battista (1987) list the assumptions required for replacing equation (2.77) by the more manageable equation (2.75):

1. the medium is considered of infinite extent so the kernels do not sense boundaries;
2. the patient/phantom is considered homogeneous and water-equivalent;
3. there is no beam divergence (infinite SSD);
4. atomic number dependence is considered unimportant in the energy range of interest;
5. photons are monodirectional;
6. photons are monoenergetic.

Mackie has computed the Green's function for many sets of monoenergetic and monodirectional photons using the electron–gamma shower transport code EGS4. The kernels were stored separately for electrons set in motion from primary photons, first-scattered photons, second-scattered photons, multiply scattered photons and annihilation and bremsstrahlung photons. Since the kernels associated with these vary on different spatial scales, this method of storage can assist in the choice of digitization used for convolution.

Indeed Mackie *et al* (1985) were possibly the first to write down expressions whereby the dose can be computed from a convolution of fluence with what they called dose-spread arrays. In the early work the dose-spread arrays were separated into the dose from primary photons, the truncated first-scatter (TFS) dose-spread array and the residual first- and multiple-scatter (RFMS) dose-spread array. The TFS specifically excluded first-scatter dose at large distances from the interaction site and this was incorporated into the RFMS array. Provided the modification to the photon fluence of shielding, beam modifying devices etc, could be modelled, the dose distribution immediately followed. Mackie *et al* (1985) invoked O'Connor's theorem to scale dose kernels in inhomogeneous material or homogeneous non-water-like material (with the same atomic number Z). This stated that the dose in two media with different densities (but the same Z) will be the same provided all distances are scaled inversely with the density of the material. By this means dose-spread arrays were stored in pixel steps labelled by the product of density and length. The size of the pixel steps was different for the different orders of scatter.

Mackie *et al* (1985) also discuss an interesting reciprocity. The dose-spread arrays describe the transport and absorption of dose throughout a phantom due to primary interactions in a single voxel. They also describe the absorption of dose in a voxel due to an equal magnitude of primary interactions throughout the volume. This leads to two different ways to compute dose from the same starting relation depending on whether the dose is needed at just a few or many points in the medium.

If the photon fluence is dependent only on depth into the phantom (e.g. for a square uniform field incident normal to a plane surface), the 3D Green's function can be *collapsed* via

$$k_t(k) = \sum_{i=1}^{N} \sum_{j=1}^{N} k_t(i, j, k) \tag{2.78}$$

to a one-dimensional Green's function. This reduces the convolution to a one-dimensional convolution

$$D_t(z) = \int \Gamma(z') k_t(z - z') \, dz' \tag{2.79}$$

which is consequently speedier.

Boyer *et al* (1988) compared dose calculation by FFT techniques with similar computations made by ray tracing and with look-up tables. The relative speeds depended obviously on the size of the dose matrix and on the size restrictions applied to the convolution kernels, but to give a flavour we state:

- ray tracing with look-up tables was some five times faster than without;

- a $64 \times 16 \times 128$ dose calculation was some 80 times faster than by direct ray tracing;
- all FFT calculations obeyed the logarithmic proportionality.

On the subject of speed, Field and Battista (1987) compute the time advantage of working in Fourier space. A one-dimensional convolution of two vectors of length N requires N^2 real multiplications. The corresponding operation in Fourier space requires that each function be padded to length $2N$. Three FFTs are required, each needing $2N \log_2 2N/2$ complex multiplications. $2N$ further complex multiplications are needed to multiply the two transforms before back-transformation. They assume each complex multiplication takes four times the time for a real multiplication, so the ratio $(R/F)_1$ of 'time-for-real' to 'time-for-Fourier' operation becomes

$$\left(\frac{R}{F}\right)_1 = \frac{N^2}{4\left(\frac{3}{2}(2N)\log_2(2N) + 2N\right)} \tag{2.80}$$

which simplifies to

$$\left(\frac{R}{F}\right)_1 = \frac{N}{\left(12\log_2(N) + 20\right)}. \tag{2.81}$$

The corresponding ratios for two-dimensional and three-dimensional operations are (deduced by squaring and then cubing N and $2N$ in the expression (2.80))

$$\left(\frac{R}{F}\right)_2 = \frac{N^2}{\left(48\log_2(N) + 64\right)}. \tag{2.82}$$

$$\left(\frac{R}{F}\right)_3 = \frac{N^3}{\left(144\log_2(N) + 176\right)}. \tag{2.83}$$

From equations (2.80)–(2.83) the advantage of using FFTs over real-space convolution may be calculated. For $N = 64$, $(R/F)_1 = 0.69$, $(R/F)_2 = 11.6$ and $(R/F)_3 = 252$. For $N = 1024$, $(R/F)_1 = 7.3$, $(R/F)_2 = 1927$ and $(R/F)_3 = 664\,444$. Clearly the advantage of using FFTs increases with the dimensionality of the problem. (For example with $N = 4$ the FFT is never quicker even in three-dimensions. In two-dimensions the break-even point exactly falls at $N = 16$. In one-dimension $N = 64$ is below and $N = 128$ is above the break-even point. In three-dimensions $N = 8$ is below and $N = 16$ is above the break-even point.) These speed increases are well known in medical imaging.

2.5.4. *Iterative* 2D *optimization by simulated annealing*

Equation (2.41) shows the computation of the dose distribution $D(r)$ from contributions $f(r_\phi, \phi)$ from a number of orientations ϕ. In section 2.5.1 it was shown how $f(r_\phi, \phi)$ could be determined analytically from the density function of point irradiations (equation (2.40)). Recently there have been successful attempts to determine $f(r_\phi, \phi)$ by numerical inversion of equation (2.41) by the technique of *simulated annealing* (Webb 1989, 1991b). The essence of this method is that $f(r_\phi, \phi)$ is built up in small increments (known as grains) by randomly polling the orientations ϕ and elements r_ϕ. Suppose that at any iteration N the elements are such that use of equation (2.41) gives $D_N(r)$ and suppose the dose prescription is $D_P(r)$. A quadratic cost function χ_N may be defined

$$\chi_N = \sum_r (D_P(r) - D_N(r))^2. \tag{2.84}$$

The aim is to minimize χ_N. Hence, if addition of a grain leads to a new dose distribution $D_N(r)$ nearer to $D_P(r)$ than $D_{N-1}(r)$, then this grain is an acceptable change. If the change $\Delta\chi_N = \chi_N - \chi_{N-1}$ is positive, the change is only accepted with a probability $\exp(-\Delta\chi_N/kT)$, where k is the Boltzmann constant and T is a temperature. This ensures the iterative solution does not become trapped at a local minimum in the cost function (figure 2.30). After a very large number of iterations, starting with high temperatures and cooling slowly, the dose distribution gradually converges to a good approximation to the prescription. The similarity between this process and crystal growth gives the method its name (Radcliffe and Wilson 1990). There are numerous computational arrangements which must be made for the method to be practicable and the reader is referred to the papers for these details. Figure 2.31 shows how, by combining fields from different directions with the beam profile shaped in intensity, a high-dose volume can be made to conform to a target with an irregular and concave border. Figure 2.32 shows a transaxial slice with dose represented by a grey-scale. The high-dose area has been shaped to an irregular target.

Webb (1989) applied the method to tailoring beam profiles to give dose distributions in a plane to treatment areas with concave outlines. One advantage of the iterative method is that the beam profiles can automatically be constrained to be positive, leading to a practical solution. As Goitein (1990) has pointed out, this is not an optimum solution in the sense that the dose distribution produced does not match the prescription exactly. It could only do that if negative beam intensities were allowed. Since these are specifically excluded in the iterative method, the best description of the resulting profiles is that, whilst remaining physically realizable, they minimize the difference between the prescription and the resultant dose distribution. Different cost functions may be used to equation (2.84) if required and one could for example build in smoothing constraints. Morrill *et al* (1991a,b) applied a similar method to optimizing the selection of wedged beams by maximizing a cost function representing the probability of complication-free treatment, building biological response into the calculation. Different cooling schedules were used. The main disadvantage of the simulated annealing method is the heavy computational burden.

We have now seen how non-uniform one-dimensional beam profiles may be determined by either analytic and iterative deconvolution techniques or by

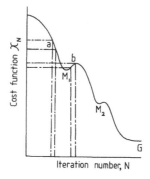

Figure 2.30. *The curve shows a possible behaviour of a cost function defined by equation (2.84) as the iterations progress. Ideally this would be a monotonic decline to the global minimum 'G'. However there are likely to be local minima; two are shown at M_1 and M_2. In order not to be trapped in the local minima some changes must be accepted which increase the cost function (as, for example, at 'b'). More usually, increasing iterations decrease the cost function (as, for example, at 'a'). Provided some positive changes are accepted, according to the distribution function described in the text, the global minimum will be reached. The situation is analogous to a skier coming downhill. The overall aim is to reach the bottom of the hill, but occasionally the skier must ascend in order not to be trapped in local troughs.*

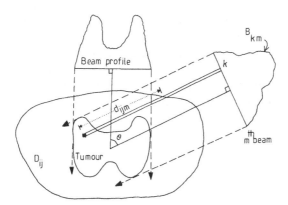

Figure 2.31. *An illustration of how multiple fields which are shaped in one dimension can be combined to give a dose area in a slice which has an irregular concave border. Two positions of the field are shown, labelled by 'm'. 'k' labels the beam element in the mth field at angle θ to the anterior direction. The dose points are labelled by (i, j) and $d_{i,j,m}$ is the depth of the (i, j)th dose point from the surface in the direction of the mth beam. This form of conformal therapy assumes the target is cylindrically symmetric (has constant cross section).*

simulated annealing. These methods may be based on dose optimization alone or on optimizing the probability of uncomplicated tumour control. Söderström

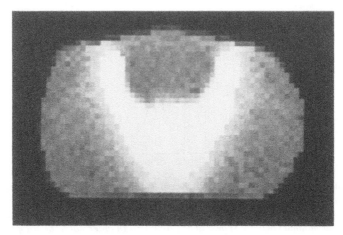

Figure 2.32. *Dose represented as a grey-scale. The brightest region corresponds to the highest dose. Here the dose envelope has been shaped to an irregular concave outline by the superposition of many one-dimensionally modulated beams of the type shown in figure 2.31.*

and Brahme (1992) have computed two quantities for each of the members of the set of 1D profiles. The first is the entropy and the second is an integral of the low-frequency part of the Fourier transform of each profile. They argue that a beam with a large amount of structure will have low entropy and have a large influence on the shape of the resulting dose distribution. Also, profiles with a large value of the above integral in Fourier space will give a large contribution to the gross structure of the dose distribution. Söderström and Brahme (1992) thus argue that beams with a low entropy and large value of this integral should be selected from the set of shaped fields determining 'preferred directions of irradiation'.

2.5.5. Iterative 3D optimization techniques

With modern multislice tomographic imaging techniques it is possible to create a planning problem in three-dimensions with full 3D specification of the location of tumour volumes and sensitive structures (see chapter 1). Treatment fields may be shaped to the beam's-eye-view of a target volume based on such structures using multileaf collimators (see chapter 5). The default in this situation is to simply weight the fields equally (Bauer 1992). However the need arises to optimize the placement of such beams and their intensities. There is no doubt this cannot be done by *ad hoc* trial and error and that computer optimization is essential. Several workers have begun recently to tackle this problem. Mohan (1990) used the simulated annealing technique to optimize a cost function constructed from the dose distribution combined with tumour control probabilities and normal-tissue complication probability data. The geometrically shaped fields were equispaced in a coplanar arrangement together with non-coplanar positions oriented at 15^0 to 30^0 from the main plane of rotation. Furthermore, the possibility to modulate the intensity of the radiation within fields was considered. No results were presented. Simultaneously and independently, Webb (1990b, 1991a, 1992c) constructed a simulated annealing technique for optimizing the intensities of geometrically shaped multileaf-collimated treatment fields with the possibility of non-coplanar fields. A limited spatial modulation was included but the cost function did not include the biological response. The results of applying this method to a number of model problems were presented. It was shown that collimating the fields to exclude the view of sensitive structures (organs at risk) 'in line' with the target volume gave too poor a dose homogeneity to the target volume; an observation also made by Mohan (1990). On the other hand, if a set of optimized weights were determined, one per port, in achieving a uniform dose to the target volume, the dose to sensitive tissue remained unacceptably high. Two solutions were determined. In the first (figure 2.33), each field received two weights, one for that part of the field seeing only target volume and the other for that part of the field seeing both target volume *and* sensitive tissue. The two weights per field were different for each BEV at each orientation, determined by the optimization. The second solution, almost certainly impractical in the early 1990's, was to allow the intensity of the beam to vary at will across each BEV. Both these solutions have been studied theoretically and gave good separation between the DVHs of target and sensitive volumes for model problems, even when the former entwined the latter (Webb 1992a,b,c). Bortfeld *et al* (1992), Galvin *et al* (1991), Mohan *et al* (1991) and Smith *et al* (1991) have also provided a solution for the inverse problem in 3D with full modulation across the geometrically shaped beam.

The problem of determining the optimum distribution of intensity across the multileaf collimator, given prescribed dose constraints, is formally identical to that of determining optimum 2D radiation compensators, as tackled by Djordjevich *et al* (1990) (see chapter 7).

At present very few treatment machines can deliver spatially modulated beams, but one can envisage how this could be arranged by scanning two pairs of jaws, one in each orthogonal direction behind the leaves of the collimator (the jaws having a non-uniform velocity profile). This method has already been examined for producing one-dimensional spatial modulation (Convery and Rosenbloom 1992). Figure 2.34 shows a possible way to deliver each field when there are two weights per field. Mohan *et al* (1991) have very recently described how arbitrary intensity modulation can be achieved under computer control using a multileaf collimator and Smith *et al* (1991) also describe varying the beam intensity over a 15 × 15 matrix of field elements. Their work was applicable to multileaf collimators, but in practice a computer-manufactured set of compensators was used. Boyer *et al* (1991) have shown how the appropriate distribution of intensities could be arranged by (at each orientation of the gantry) using several settings of the multileaf collimator with the settings tracking the iso-intensity contours at the face of the collimator (see section 2.5.6). Galvin *et al* (1991) show that for a multileaf collimator pixellated to a 15 × 15 matrix, superposition of as few as 10 fields could give the required intensity modulation for the problem studied. Whether this is a

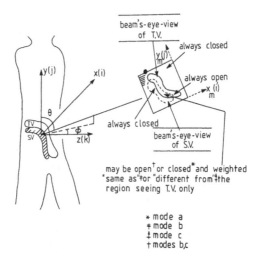

Figure 2.33. *Showing how to optimise the separation between the DVH for target and sensitive volumes when multileaf-collimated BEV ports are used. The multileaf collimator is shown to the right with the projected opening area corresponding to the views of the target volume (TV) and the sensitive volume (SV) (organ at risk (OAR)). These two curves overlap. The question arises of how to handle to overlap region. Several 'modes' of irradiation are identified. In mode a, only the aperture corresponding to a view of the target is irradiated. In mode b, this aperture and the overlap aperture have the same beam-weight. In mode c, a different beam-weight is applied to the aperture viewing target only and the overlap aperture viewing target and organ at risk. (From Webb (1992c).)*

feasible practical tool depends on how fast the fields could be reset.

The group at Galveston, Texas have used the simplex linear-programming algorithm to optimize treatment plans (Rosen *et al* 1991). This finds the best beam weights from a set of possibilities, called the solution space, 'driven' by constraints applied to specified points in the treatment volume. Simplex methods become more time consuming as the number of these constraint points increases. Recognizing that it is possibly unrealistic to specify a single dose constraint for an organ at risk, Morrill *et al* (1991c) have introduced the binary dose- volume constraint for organs at risk wherein the fractional volume is specified together with the maximum dose this fraction may receive in addition to the maximum dose which may be delivered to the rest of the organ. This defines a 'two-bin' dose–volume histogram. Provided a large number of beams are used to aim towards conformal therapy, it is then reasonable to assume that the high-dose volume of any organ at risk, proximal to a target volume, will be closer to the target volume than the low-dose region in the organ at risk. Morrill *et al* (1991c) then define dose-constraint points lying around collars of specified widths encircling the target volume. The widths of the collars are determined from the dose–volume constraints. In this way the dose–volume constraint is built into the simplex

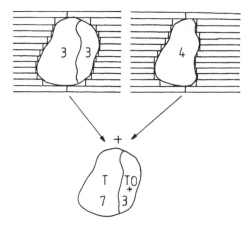

Figure 2.34. *A possible way to deliver a multileaf-collimated field when there are two weights per field. In the lower drawing is shown the shape of the field which encompasses the target volume 'T'. Part of this field 'sees' only target volume and a smaller part 'sees' target plus organ at risk ('T+O'). Assume it is required to deliver seven units of radiation to 'T' and three units of radiation to 'T+O'. This is achieved by delivering the two separate fields shown in the upper part of the figure. The left-hand field gives three units of radiation to the target and organ at risk components; the right-hand field tops up the target-only part to a total of seven units. The idea could be extended to multiple shaped fields at the same gantry orientation, so as to create a spatial variation of beam intensity in two dimensions (see figure 2.36).*

solution. Clinical examples showed the resulting dose–volume histograms after optimization lay entirely to the left of the two-bin prescription, indicating success. Other methods of incorporating dose–volume constraints into linear programming techniques have been developed by Langer and Leong (1989) and Langer *et al* (1990).

A number of authors have developed methods of determining the beam weights for 3D radiation therapy planning using the Cimmino algorithm (Altschuler et al 1985, Censor *et al* 1987, 1988, Powlis *et al* 1985, Starkshall and Eifel 1992). The methods have generally been presented in terms of one weight per field, but there is no inherent limitation precluding working on the general problem of fields with spatially modulated intensities. This is not strictly an optimization so much as a feasibility search. It works as follows. A set of points (labelled by i) spanning the target volume, the organs at risk and other normal structures are chosen. For those points in the target volume, a maximum and minimum dose (D_i^M and D_i^m) are specified. For those points in organs at risk and other normal structure, a maximum dose is specified. All beam-weights are constrained to be positive.

The Cimmino algorithm (Cimmino 1938) then tries to find a vector x of beam-weights (with positive elements x_j, $j = 1, 2.......N$) such that

$$D_i^{\mathrm{m}} \leqslant \sum_j^N A_{i,j} x_j \leqslant D_i^{\mathrm{M}} \tag{2.85}$$

where $A_{i,j}$ is the contribution to the ith dose point from the jth beam. The algorithm is iterative and controlled by a relaxation parameter which governs the speed of convergence.

This method is not like simulated annealing, which is based on minimizing an objective function based on matching dose to a prescription. The goal is not so much to deliver a dose that matches a prescription as to find a set of conditions which deliver a dose within the constraints. It searches for a *feasible* solution. It may be there is no such feasible solution and the strength of the algorithm is that it still provides clinically useful information in the event that the specified dose constraints could not be satisfied by any combination of beam-weights. The planner is presented with the result at that stage (a local minimum of the difference between calculated doses and specified dose bounds) and can interactively make choices whether contraints can be changed to overcome the difficulty.

A second property of the algorithm is that the importance of each dose constraint can be weighted. In their implementation, for example, Starkshall and Eifel (1992) weighted the dose minimum contraint 25 times more than the dose maximum, thus signifying that it was much more important to avoid underdose than overdose to the target volume.

Niemierko (1992) and Niemierko *et al* (1992) have developed a random search algorithm which is able to optimize radiation dose with both physical dose and biological endpoints and constraints. This would appear to be a particularly powerful approach because it allows the specification to be a combination of dose limits, desirable TCP and NTCP values and also for example dose limits specified to fractional volumes of organs-at-risk. The specification of dose–volume constraints is particularly interesting because some hot spots may be allowable provided some large majority of an organ-at-risk is below a specified limit. The approach would seem to be a very useful compromise between the quite natural desire to specify biological endpoints (but with potentially uncertain data and/or model) and the traditional (though surrogate) use of dose. (One might detect here some difference of approach between the USA and the UK. In the latter clinicians are still largely happiest specifying, and thus requesting, optimization of *dose*; in the former it is becoming more common to request quantitative TCP and NTCP for each plan.) The random search algorithm is fast and, although not proven to converge to the global optimum, gave the same results as much more computationally intensive optimization by simulated annealing for identical constraints and end-points. To date this method has only been used to optimize beam-weights and the authors' suggestions may not be applicable to optimizing beam apertures with a multileaf collimator.

Rosen *et al* (1991) have recently reviewed the relative merits of different computer optimization approaches, including linear programming, the method of simulated annealing and the feasibility technique.

2.5.6. *Analytic and iterative deconvolution theory of 3D inverse computed tomography*

Section 2.5.1 discussed an analytic technique from Brahme (1988) to conform a 2D high-dose region to the target area. The method comprised the following steps:

1. Deliver the radiation by a full 2π rotation of a beam whose 1D intensity variation is $f\left(r_\phi, \phi\right)$, ϕ labelling the gantry orientation and r_ϕ the position coordinate within the beam.

2. From a specified 2D dose distribution, $D\left(r\right)$, deconvolve the rotational point-spread dose function (PSF) $d\left(r\right)$ to obtain the density function $\Phi\left(r\right)$ of point irradiations (equations (2.36) and (2.39)). The deconvolution could be performed by an iterative method (equations (2.42) and (2.43)).

3. Forward project the density function, taking account of (spatially variable) photon attenuation and geometry, to obtain the appropriate intensity weighting for the connecting beam element $f\left(r_\phi, \phi\right)$ (equation (2.40)).

We note that the method correctly accounted for photon attenuation in the forward projection stage, for scattered radiation (incorporated into the dose kernel (PSF)) and was constrained to give positive radiation intensities.

Boyer *et al* (1991) have elegantly extended the method to compute 3D dose distributions by a rotational technique. The principles are identical, although the formalism at first sight looks a little more complicated in 3D. The method again assumes a rotational kernel Λ which is spatially invariant. Then the 3D dose distribution $D\left(r\right)$ is related to the density function Ψ by

$$D\left(r\right) = \int \int \int \Psi\left(r'\right) \Lambda \left(\left| r - r' \right|\right) \mathrm{d}^3 r'. \qquad (2.86)$$

Compare this with the corresponding 2D equation (2.36) and note the use of symbols Λ and Ψ instead of d and Φ to distinguish the 3D quantities from their 2D counterparts (and also to follow the notation of Boyer *et al* 1991). Λ is a cylindrically-symmetric kernel representing dose spread in 3D. Now if D and Λ are known, Ψ can be computed from the 3D iterative deconvolution (compare with equations (2.42) and (2.43) in 2D):

$$\Psi_0\left(r\right) = aD\left(r\right) \qquad (2.87)$$

$$\Psi_{k+1}\left(r\right) = C\left(\Psi_k\left(r\right) + a_k\left(D\left(r\right) - \Psi_k\left(r\right) \otimes \Lambda\left(r\right)\right)\right) \qquad (2.88)$$

where the subscript k refers to the kth iteration and $\otimes$ means convolution. C is a positivity constraint operator and a_k controls the speed of convergence. Boyer *et al* (1991) use some 40 iterations and calculate the convolutions using 3D FFT techniques implemented on an array processor. This keeps the computational time to under an hour. The convergence parameter a_k assumes a subscript k indicating that it can be varied as a function of the iteration number. At each stage of iteration

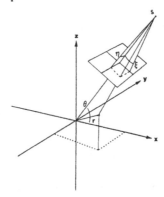

Figure 2.35. *The coordinate system used for 3D conformal planning by optimised rotational therapy. The beam is collimated by a multileaf collimator whose central axis is at an angle θ from an x-axis. The ray from the source at s to the location r in the patient lies at angles ζ and η as shown. The intensity function where this ray intersects the collimator is specified by $f(\zeta(\theta), \eta(\theta))$. (From Boyer et al (1991).)*

the difference $\Delta D_k(r)$ between the dose prescription $D(r)$ and the kth estimate $D_k(r)$ of the dose distribution is

$$\Delta D_k(r) = D(r) - D_k(r).$$ (2.89)

Minimizing

$$\chi^2_{k+1} = \sum_r (\Delta D_{k+1}(r))^2$$ (2.90)

by setting to zero the derivative of this function with respect to a_k gives

$$a_k = \sum_r \Delta D_k(r) \left(\Delta D_k(r) \otimes \Lambda \right) \left[\sum_r \left(\Delta D_k(r) \otimes \Lambda \right)^2 \right]^{-1}.$$ (2.91)

The collimator for the radiation is specified by its angle θ from an x-axis; the ray from the source directed at vector r has angle ζ and η to the central ray directed towards the origin, through which the axis of rotation passes (figure 2.35). The intensity variation at the collimator is then specified by the function $f(\zeta(\theta), \eta(\theta))$ which is given by ray tracing through the attenuating medium to the source position:

$$f(\zeta(\theta), \eta(\theta)) = \int_{\text{line}} \Psi(r_0) \exp\left(\int_{s(\theta)}^{r_0} \mu(r) \, dr \right) d\alpha$$ (2.92)

where $s(\theta)$ represents the position of the source when the gantry is at angle θ and α labels the direction of the ray, along which the line integral is taken.

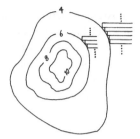

Figure 2.36. *Showing how it might be possible to use multiple fields at any given orientation, each shaped by a multileaf collimator and tailored to conform to the iso-intensity contours across the face of the open field. In this schematic it is implied that the smallest inner part of the field receives '10 units' of radiation intensity, the annulus outside this receives '8 units' and so on. A few leaves are drawn to give an idea of how the multileaf collimator is used.*

Boyer *et al* (1991) thus develop the formalism for computing a spatially varying intensity across the face of the multileaf collimator. They argue that in practice these irradiations could be built up at each gantry orientation as a sequence of small exposures with different multileaf collimator settings defined by the shape of the iso-intensity contours at the plane of the collimator (figure 2.36). This would require a computer-controlled multileaf collimator. At present it is not feasible to implement the method because of the very large number of orientations the multileaf collimator must take up, and also the large number of sub-fields needed at each orientation. Indeed Boyer *et al* (1991) consider this may not even be developed in the near future. Boyer *et al* (1991) use very small (1°) angular increments for the gantry to study the limiting performance of such a method. This is valuable work which helps to establish just how much effort is worth putting into rotational conformal therapy with a multileaf collimator.

They studied several cylindrically-symmetric distributions and came to the somewhat disappointing conclusion that the tails of the skirts of the point-dose kernel led to rather large doses being delivered to normal tissue if this were enclosed by target volume. However, the method coped well with delivering a conformal dose to a volume which had a concave outline. Boyer *et al* (1991) discuss the possibility of using non-coplanar multileaf-collimator-defined fields with spatially varying beam intensities, removing the cylindrical symmetry of the dose kernel Λ. This might lead to better therapeutic ratios. The same conclusion was reached by Webb (1992a,b,c), who computed the spatially varying intensities by a numerical optimization method for non-coplanar fields with a simple dose calculating algorithm.

2.5.7. *The inversion algorithm for proton therapy*

So far this chapter has been concerned with optimizing photon therapy. High-energy protons may also be used for precision conformal therapy. The subject of proton therapy will be discussed in detail in chapter 4. Here we briefly present a theory of optimization which has formal similarities with analytic photon optimization. Figure 2.37 shows a beam of protons emanating from a cyclotron and entering the patient, depositing energy along their track towards a Bragg peak. Suppose the aim is to produce a high-dose volume in the re-entrant volume, shown with a solid line. This can be achieved if the proton beam can make a linear traverse from one side of the region to the other, and at the same time the Bragg peak may be adjusted to different depths z at each position of the beam during its linear traverse. (Alternatively, if the proton energy is fixed (Bragg peaks all at the same depth) to treat the volume requires rotation as well as translation, as for photons). Consider the case with a single, translated, beam portal. Suppose the dose distribution from a narrowly collimated monoenergetic proton beam is $h(r)$ and the desired dose distribution is $D(r)$. If the impulse response h may be considered spatially invariant, then, as for photons, we may write

$$D(r) = \int \int \int \Phi(r') h(r - r') \, d^3 r'. \tag{2.93}$$

i.e. the dose distribution is a convolution of the elementary distribution h with the density Φ of such elementary distributions. (Compare with equation (5.6)

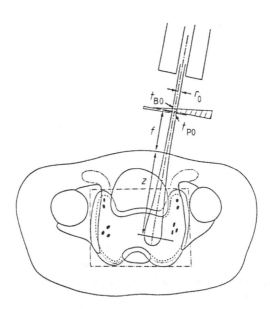

Figure 2.37. *Arrangement for proton therapy with scanned elementary beam. The solid butterfly shaped curve is the treatment area. The elementary beam range is modulated by double-wedge filters. The beam is swept through the target. (From Brahme et al 1989).)*

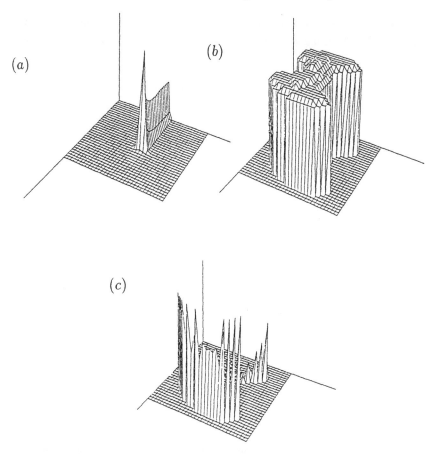

Figure 2.38. *Isometric plot of (a) the elementary proton beam, (b) the desired dose in the target area, and (c) the density of Bragg peaks. (From Brahme et al (1989).)*

(chapter 5) for single-port photon therapy with a multileaf collimator.) The irradiation density function Φ may be obtained in exactly the same way as for photons by equations (2.42) and (2.43), where again positivity is easy to invoke (see also equations (5.7) and (5.8) of chapter 5).

Brahme *et al* (1989) introduced this method, applying it to a number of model problems. For example figure 2.38 shows the elementary beam profile, the dose prescription and the density function. In this figure the beam enters top right. As one expects, the density function peaks at the distal side of the target volume. It is easy to see why. Proton beams with Bragg peaks in this far-distal region will deposit some energy along the track back to the source. Hence the density of elementary beams with Bragg peaks nearer to the source will be lower because these regions already pick up some radiation from the distal-side Bragg peaks. The disadvantage of this single-port technique is that tissue outside the treatment volume, but through which the radiation passes, will obviously be irradiated. This can be overcome by using rotation methods or multiple ports, as discussed by Brahme *et al* (1989).

The function $\Phi(r)$ gives the density of Bragg peaks inside the target volume. At each translation position of the beam, the range must be modified according to this distribution. The range modulation may be achieved by using beam filters (shown as wedges of lead and beryllium in figure 2.37). At every lateral position of the elementary beam port, a set of different filtrations are required with the beam monitor units given by the appropriate irradiation density for each of the points in- line down that elementary track. In practice this would be quite complicated and it may be simpler to keep the range constant and use rotation instead.

The heart of the technique is the elementary function h. This is given by

$$D(r, z) = \frac{D(z) \exp\left(-r^2/\langle r^2(z)\rangle\right)}{\pi \langle r^2(z)\rangle} \tag{2.94}$$

where r is the radial position, z is distance along the central axis of the beam, $D(z)$ is the central axis depth dose of a broad beam and $\langle r^2(z)\rangle$ is the mean square radius at depth z (see also chapter 4).

2.6. SUMMARY

We began this chapter with a consideration of aspects of two-dimensional radiotherapy planning and investigated the intensity modulation needed for beams to combine to give a high-dose region which might not have a convex boundary. This enabled a deeper insight into what had been achieved in the past with unmodulated but partially blocked fields in gravity-assisted rotation therapy. The concept of the point-dose kernel naturally arose and is at the heart of a number of new developments, generically called convolution dosimetry. Whilst these are not yet widely established in routine treatment planning, they are excellent benchmark tools and are expected to play a wider role as computers become ever faster. There have been several attempts to determine how multiple 2D fields, shaped by a multileaf collimator and not necessarily coplanar, might be combined to give a highly conformal 3D dose distribution. Some of these are analytic approaches and others are iterative. The parallel approach in proton therapy was reviewed here, ahead of the main discussion of this topic, because of its formal similarity with optimizing photon therapy.

In the next chapter conformal stereotactic radiotherapy will be discussed and the following chapter will return to discussing the physics of radiotherapy with protons.

2.7. APPENDIX 2A. A HISTORICAL NOTE ON THE ORIGINS OF ROTATION THERAPY

In this chapter we have seen that at the heart of conformal therapy lies the need to arrange for multiple fields at different orientations to the patient. Some methods can be formalized in terms of a rotation model. Rotation therapy has a long history going back to the germ of an idea not long after the turn of the century (Baese 1915, Portes and Chausse 1921). More serious attempts were made in the 1930's and rotation therapy was well established by the 1940's and had become very elaborate by the early 1950's.

To understand the popularity of what is now a rarely used technique, one has to remember that high-energy therapy implied in those days the use of photons at or below some 2 MV, i.e. the radiation was considerably less penetrating than that in common use today from linear accelerators. Often it was as low as 250 kV. The aim was to increase the ratio between the dose to the target volume and the dose to healthy normal tissue through which the radiation had to pass, by rotating a beam about an axis centred in the target volume.

We should also remember that techniques for establishing the target volume were much poorer than today (no CT or MRI), that computers for dose planning were unheard of, that treatment times per fraction could be large and that techniques for positioning the patient were not particularly accurate. Today they would probably be considered quite unacceptable. Against these difficulties rotation therapy persevered and the technique became widely established, building up a huge literature.

Consider for example dose planning. Hand calculations were tedious enough for combining just a few beams. Rotation therapy presented a considerable problem. The source-to-skin distance was continuously varying. Establishing the patient contour was itself difficult (Friedman *et al* 1955). Generally, tissue inhomogeneities were ignored (Wheatley 1955). As a result there was much resort to experimental dosimetric verification with phantoms. For example films were placed in cylinders of scattering material (Lantzl and Skaggs 1955, Friedman *et al* 1955, Peterson *et al* 1955) or ion-chamber measurements were made (Braestrup and Mooney 1955). These gave a crude indication of the expected dose distributions, but, since the phantoms were not customized to match the individual external patient contour and they were homogeneous, the results were wrong. Transit-dose measurements were used to assess by how much they were wrong. A dosemeter was placed well downstream of the patient to record the primary exit radiation only at each orientation of the patient (Pfalzner 1956) and these measurements, together with knowledge of the unattenuated beam intensity, were used to deduce the 'equivalent water thicknesses' at each angle. This naturally led to plots of the 'equivalent contour' for the patient. Fedoruk and Johns (1957) built a special transit-dose collimator to exclude scattered radiation when making such measurements. In 1953 Smithers at the (then) Royal Cancer Hospital accepted the offer of help from a professional sword swallower (figure 2.39). A dosemeter was thus placed in the thorax of a living subject rotating in an x-ray field to assess the limitations of his models for dosimetry (Wheatley *et al* 1953). However, imaginative though this was as an example of early in-vivo dosimetry, it was hardly a practical possibility for all subjects! Accurate dosimetry remained a real problem.

Localization of the target volume was generally estimated from planar radiographs, classical axial transverse tomography or fluoroscopy. Some manufacturers produced diagnostic equipment which mimicked the treatment

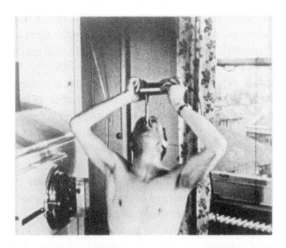

Figure 2.39. *Showing how Smithers used the services of a professional sword swallower in the 1950s to try to estimate dosimetry in the thorax during rotation therapy. (From Smithers' papers.)*

geometry and shared a room with it (see figure 6.3) (Klemm 1957). After target localization, the patient was wheeled across on the couch for treatment. This was appropriate for rotation therapy with the patient in the horizontal position (so-called paraxial conical therapy).

We might be more concerned at the common practice of rotating the patient vertically in a stationary beam. Possibly this was influenced by the then common practice of performing axial transverse tomography in the vertical position, which itself became popular from the need to image fluid levels in the lungs of patients with tuberculosis. Imagine the difficulty of ensuring the target volume was properly located in the beam, of reproducing the patient positioning accurately and of asking the patient to sit quietly for some 10–15 minutes during irradiation (Pfalzner and Inch 1956, Berman, 1957). Errors must have occurred. The difficulties were well known to the workers of the time but they could do little about them. There was much disagreement in the literature over the acceptability of so called circumaxial (i.e. vertical) rotation therapy.

Rotation therapy with a limited arc of travel was sometimes known as pendulum therapy, there being provided a means of reversing the direction of motion of the x-ray source at two end-stops defining the arc of travel (Busse 1956). A form of pendulum therapy was devised by Lupo and Pisani (1956) with apparatus which was the therapy equivalent of classical planigraphy. Some considered the problems would be better solved with the patient recumbant and the necessary rotation arcs obtained by combinations of translations and pivoting of the x-ray source (Peterson *et al* 1955).

For all this the intentions were fine and indeed a quotation from a paper by Friedman *et al* (1955) is still echoed today in introducing the goals of conformal therapy:

'The efficiency of a particular form of radiation therapy is gauged by its ability to produce rather homogeneous dose distribution within the tumour region with minimal irradiation of the surrounding tissues.'

REFERENCES

Ahnesjø A 1987 Invariance of convolution kernels applied to dose calculations for photon beams *The use of computers in radiation therapy* ed I A D Bruinvis *et al* (Proc. 9th ICCRT) pp 99–102
—— 1989 Collapsed cone convolution of radiant energy for photon dose calculation in heterogeneous media *Med. Phys.* **16** 577–592
—— 1990 Photon beam characterization for dose calculations from first principles *The use of computers in radiation therapy: Proceedings of the 10th Conference* ed S Hukku and P S Iyer (Lucknow) pp 64–67
Ahnesjø A, Andreo P and Brahme A 1987a Calculation and application of point spread functions for treatment planning with high energy photon beams *Acta Oncol.* **26** 49–56

Ahnesjø A and Mackie R 1987b Analytical description of Monte Carlo generated photon dose convolution kernels *The use of computers in radiation therapy* ed I A D Bruinvis *et al* (Proc. 9th ICCRT) pp 197–200

Ahnesjø A and Trepp A 1991 Acquisition of the effective lateral energy fluence distribution for photon beam dose calculations by convolution methods *Phys. Med. Biol.* **36** 973–985

Altschuler M D, Powlis W D and Censor Y 1985 Teletherapy treatment planning with physician requirements included in the calculation: 1. Concepts and methodology *Optimisation of cancer radiotherapy* ed B R Paliwal, D E Herbert and C G Orton (New York: AIP (AAPM)) pp 443–452

Baese C 1915 Method of, and apparatus for, the localisation of foreign objects in, and the radiotherapeutic treatment of, the human body by the x-rays *UK Patent no* 100491

Barth N H 1990 An inverse problem in radiation therapy *Int. J. Rad. Oncol. Biol. Phys.* **18** 425–431

Bauer J, Hodapp N and Frommhold H 1992 A comparison of dose distributions with and without multileaf-collimator for the treatment of oesophageal carcinoma *Proc. ART91 (Munich)* (abstract book p 49) *Advanced Radiation Therapy: Tumour Response Monitoring and Treatment Planning* ed A Breit (Berlin: Springer) pp 649–653

Berman H L 1957 Immobilisation of the head during rotational x-ray therapy *Radiology* **68** 579–581

Bortfeld T, Boesecke R, Schlegel W and Bohsung J 1990b 3D dose calculation using 2D convolutions and ray tracing methods *The use of computers in radiation therapy: Proceedings of the 10th Conference* ed S Hukku and P S Iyer (Lucknow) pp 238–241

Bortfeld Th, Burkelbach J, Boesecke R and Schlegel W 1990a Methods of image reconstruction from projections applied to conformation therapy *Phys. Med. Biol.* **35** 1423–1434

Bortfeld T, Burkelbach J and Schlegel W 1992 Solution of the inverse problem in 3D *Proc. ART91 (Munich)* (abstracts book p 46) *Advanced Radiation Therapy: Tumour Response Monitoring and Treatment Planning* ed A Breit (Berlin: Springer) pp 503–508

Boyer A L, Desobry G E and Wells N H 1991 Potential and limitations of invariant kernel conformal therapy *Med. Phys.* **18** (4) 703–712

Boyer A and Mok E 1985 A photon dose distribution model employing convolution calculations *Med. Phys.* **12** 169–177

—— 1986 Calculation of photon dose distributions in an inhomogeneous medium using convolutions *Med. Phys.* **13** 503–509

Boyer A L, Wackwitz R and Mok E C 1988 A comparison of the speeds of three convolution algorithms *Med. Phys.* **15** 224–227

Boyer A L, Zhu Y, Wang L and Francois P 1989 Fast Fourier transform convolution calculations of x-ray isodose distributions in homogeneous media *Med. Phys.* **16** 248–253

Brace J A 1985 Computer systems for the control of teletherapy units *Progress in medical radiation physics:2* ed C G Orton (New York: Plenum) pp 95–111

Braestrup C B and Mooney R T 1955 Physical aspects of rotating telecobalt equipment *Radiology* **64** 17–27

Brahme A 1987 Optimisation of conformation and general moving beam radiation therapy techniques *The use of computers in radiation therapy* ed I A D Bruinvis *et al* (Proc. 9th ICCRT) pp 227–230

—— 1988 Optimisation of stationary and moving beam radiation therapy techniques *Radiother. Oncol.* **12** 129–140

—— 1992 Recent developments in radiation therapy planning *Proc. ART91 (Munich)* (abstracts book p 6) *Advanced Radiation Therapy: Tumour Response Monitoring and Treatment Planning* ed A Breit (Berlin: Springer) pp 379–387

Brahme A, Källman P and Lind B K 1989 Optimisation of proton and heavy ion therapy using an adaptive inversion algorithm *Radiother. Oncol.* **15** 189–197

Brahme A, Roos J E and Lax I 1982 Solution of an integral equation encountered in rotation therapy *Phys. Med. Biol.* **27** 1221–1229

Busse W 1956 Einstellung und Dosisbestimmung bei der Pendelbestrahlung der Parametrien *Strahlentherapie* **101** 400–404

Casebow M 1990a A modified synchronous shielding technique in rotation therapy *Brit. J. Radiol.* **63** 482–7

—— 1990b private communication.

Censor Y, Altschuler M D and Powlis W D 1988 On the use of Cimmino's simultaneous projections method for computing a solution of the inverse problem in radiation therapy treatment planning *Inv. Prob.* **4** 607–623

Censor Y, Powlis W D and Altschuler M D 1987 On the fully discretised model for the inverse problem of radiation therapy treatment planning *Proc. 13th Ann. Northeast Bioengineering Conference* ed K R Foster (IEEE 87-CH2436-4 Library of Congress catalog card number 87-80502) pp 211–214

Cimmino G 1938 Calcolo approssimato per le soluzioni die sistemi di equazioni lineari *La Ricerca Scientifica* Roma 1938-XVI, Ser II, Anno IX, Vol I pp 326–333

Convery D J and Rosenbloom M E 1992 The generation of intensity-modulated fields for conformal radiotherapy by dynamic collimation *Phys. Med. Biol.* **37** 1359–1374

Cormack A M 1987 A problem in rotation therapy with X-rays *Int. J. Rad. Oncol. Biol. Phys.* **13** 623–630

—— A M 1990 Response to Goitein *Int. J. Rad. Oncol. Biol. Phys.* **18** 709–710

Cormack A M and Cormack R A 1987 A problem in rotation therapy with X-rays 2: dose distributions with an axis of symmetry *Int. J. Rad. Oncol. Biol. Phys.* **13** 1921–1925

Cormack A M and Quinto E T 1989 On a problem in radiotherapy: questions of non-negativity *Int. J. Imaging Systems and Technology* **1** 120–124

Cunningham J R 1982 Tissue inhomogeneity corrections in photon-beam treatment planning *Progress in medical radiation physics:1* ed C G Orton (New

York: Plenum) pp 103–131

Davy T J 1985 Physical aspects of conformation therapy using computer-controlled tracking units *Progress in medical radiation physics:2* ed C G Orton (New York: Plenum) pp 45–94

Djordjevich A, Bonham D J, Hussein E M A, Andrew J W and Hale M E 1990 Optimal design of radiation compensators *Med. Phys.* **17** (3) 397–404

Fedoruk S O and Johns H E 1957 Transmission dose measurement for cobalt 60 radiation with special reference to rotation therapy *Brit. J. Radiol.* **30** 190–195

Field C and Battista J J 1987 Photon dose calculations using convolution in real and Fourier space: assumptions and time estimates *The use of computers in radiation therapy* ed I A D Bruinvis *et al* (Proc. 9th ICCRT) pp 103–106

Friedman M, Hine G J and Dresner J 1955 Principles of supervoltage (2 million volts) rotation therapy *Radiology* **64** 1–16

Galvin J M, Smith R M and Chen X-G 1991 Modulation of photon beam intensity with a multileaf collimator *Proc. 1st biennial ESTRO meeting on physics in clinical radiotherapy (Budapest, 1991)* p 5

Goitein 1990 The inverse problem *Int. J. Rad. Oncol. Biol. Phys.* **18** 489–491

Green A 1959 A technical advance in irradiation technique *Proc. R. Soc. Med.* **52** 344–346

—— 1965 Tracking cobalt project *Nature* **207** 1311

Herman G T 1980 *Image reconstruction from projections: The Fundamentals of Computed Tomography* (New York: Academic)

Ilfeld D N, Wright K A and Salzman F A 1971 Synchronous shielding and field shaping for megavolt irradiation of advanced cervical carcinoma *Am. J. Rontgenol.* **112** 792–796

Jennings W A 1985 The tracking cobalt project: from moving beam therapy to three dimensional programmed irradiation *Progress in medical radiation physics:2* ed C G Orton (New York: Plenum) pp 1–44

Jones D 1990 Light engineering *Nature* **343** 516

Källman P, Lind B K and Brahme A 1992 An algorithm for maximising the probability of complication-free tumour control in radiation therapy *Phys. Med. Biol.* **37** 871–890

Klemm T 1957 Localisation and setting techniques in moving-field therapy *SRW News: J. Siemens-Reiniger-Werke AG, Erlangen* **1** 8–12

Korn G A and Korn T M 1968 *Mathematical handbook for scientists and engineers* (New York: McGraw Hill) p 564

Lakshminarayanan A V 1975 Reconstruction from divergent beam data *Technical Report* No 92 (State University of New York at Buffalo, Department of Computer Science)

Langer M, Brown R, Urie M, Leong J, Stracher M and Shapiro J 1990 Large scale optimisation of beam weights under dose–volume restrictions *Int. J. Rad. Oncol. Biol. Phys.* **18** 887–893

Langer M and Leong J 1989 Optimisation of beam weights under dose–volume restrictions *Int. J. Rad. Oncol. Biol. Phys.* **13** 1255–1259

Lantzl L H and Skaggs L S 1955 Revolving beam isodose distributions from a cobalt 60 therapy unit *Semiannual report to the Atomic Energy Commission, Argonne*

Lax I and Brahme A 1982 Rotation therapy using a novel high-gradient filter *Radiology* **145** 473–478

Lind B K 1990 Solutions of inverse problems in radiation therapy *The use of computers in radiation therapy: Proceedings of the 10th Conference* ed S Hukku and P S Iyer (Lucknow) pp 163–166

Lind B K and Källman P 1990 Experimental verification of an algorithm for inverse radiation therapy planning *Radiother. Oncol.* **17** 359–368

Lind B K and Brahme A 1992 Photon field quantities and units for kernel based radiation therapy planning and treatment optimisation *Phys. Med. Biol.* **37** 891–909

Lupo M and Pisani G 1956 Une variante technique de la Stratitherapie pseudo-pendulaire a la palmieri *J. Radiol. et d'Elect.* **37** 39–41

Mackie T R, Ahnesjø A, Dickof P and Snider A 1987 Development of a convolution / superposition method for photon beams *The use of computers in radiation therapy* ed I·A D Bruinvis *et al* (Proc. 9th ICCRT) pp 107–110

Mackie T R, Reckwerdt P J, Gehring M A, Holmes T W, Kubsad S S, Sanders C A, Paliwal B R and Kinsella T J 1991 A fast convolution/ superposition algorithm for radiation therapy treatment planning *Proc. 1st biennial ESTRO meeting on physics in clinical radiotherapy (Budapest, 1991)* p 93

Mackie T R, Reckwerdt P J, Gehring M A, Holmes T W, Kubsad S S, Thomadsen B R, Sanders C A, Paliwal B R and Kinsella T J 1990a Clinical implementation of the convolution/superposition method *The use of computers in radiation therapy: Proceedings of the 10th Conference* ed S Hukku and P S Iyer (Lucknow) pp 322–325

Mackie T R, Reckwerdt P J, Holmes T W and Kubsad S S 1990b Review of convolution/superposition methods for photon beam dose computation *The use of computers in radiation therapy: Proceedings of the 10th Conference* ed S Hukku and P S Iyer (Lucknow) pp 20–23

Mackie T R, Scrimger J W and Battista JJ 1985 A convolution method for calculating dose for 15-MV x rays *Med. Phys.* **12** (2) 188–196

Mohan R 1990 Clinically relevant optimisation of 3D conformal treatments *The use of computers in radiation therapy: Proceedings of the 10th Conference* ed S Hukku and P S Iyer (Lucknow) pp 36–39

Mohan R, Chui C and Lidofsky L 1986 Differential pencil beam dose computation model for photons *Med. Phys.* **13** 64–73

Mohan R, Mageras G and Podmaniczky K C 1991 A model for computer-controlled delivery of 3D conformal treatments *Med. Phys.* **18** 612

Morrill S M, Lane R G, Jacobson G and Rosen I 1991a Treatment planning optimization using constrained simulated annealing *Phys. Med. Biol.* **36** 1341–1361

Morrill S M, Lane R G and Rosen I 1991b Constrained simulated annealing for optimised radiation therapy treatment planning *Comp. Methods Prog. Biomed.* **33** 135–144

Morrill S M, Lane R G, Wong J A and Rosen I I 1991c Dose–volume considerations with linear programming optimisation *Med. Phys.* **18** 1201–1210

Niemierko A 1992 Random search algorithm (RONSC) for optimisation of radiation therapy with both physical and biological endpoints and constraints *Int. J. Rad. Oncol. Biol. Phys.* **23** 89–98

Niemierko A, Urie M and Goitein M 1992 Optimisation of 3D radiation therapy with both physical and biological endpoints and constraints *Int. J. Rad. Oncol. Biol. Phys.* **23** 99–108

Nilsson M and Knöös T 1991 Application of the Fano theorem in inhomogeneous media using a convolution algorithm *Proc. 1st biennial ESTRO meeting on physics in clinical radiotherapy (Budapest, 1991)* p 70

Peterson O S, Foley J C and Mosher R F 1955 Device for control of radiation distribution in moving beam therapy *Radiology* **64** 412–416

Pfalzner P M 1955 Rotation therapy with a cobalt 60 unit: 2: Transit dose measurements as a means of correcting dose for non-waterequivalent absorbing media *Acta Radiologica* **45** 62–68

Pfalzner P M and Inch W R 1956 Rotation therapy with a cobalt 60 unit: 1: Physical aspects of circumaxial rotation *Acta Radiologica* **45** 52–61

Portes F and Chausse M 1921 Procédé pour la mise au point radiologique sur un plan sécant d'un solide, ainsi que pour la concentration sur une zone déterminée d'une action radiothérapeutique maximum, et dispositifs en permettant la ré alisation *French Patent* 541914

Powlis W D, Altschuler M D and Censor Y 1985 Teletherapy treatment planning with physician requirements included in the calculation: 2. Clinical Applications *Optimisation of cancer radiotherapy* ed B R Paliwal, D E Herbert and C G Orton (New York AIP (AAPM)) pp 453–461

Proimos B S 1960 Synchronous field shaping in rotational megavolt therapy *Radiology* **74** 753–757

—— 1961 Synchronous protection and field shaping in cyclotherapy *Radiology* **77** 591–599

—— 1963 New accessories for precise teletherapy with ^{60}Co units *Radiology* **81** 307–316

—— 1969 Shaping the dose distribution through a tumour model *Radiology* **92** 130–135

Proimos B S, Tsialas S P and Coutroubas S C 1966 Gravity oriented filters in arc cobalt therapy *Radiology* **87** 933–937

Radcliffe N and Wilson G 1990 Natural solutions give their best *New Scientist* (14th April 1990) pp 47–50

Rawlinson J A and Cunningham J R 1972 An examination of synchronous shielding in ^{60}Co rotational therapy *Radiology* **102** 667–671

Rosen I I, Lane R G, Morrill S M and Belli J A 1991 Treatment plan optimisation using linear programming *Med. Phys.* **18** 141–152

Smith R M, Galvin J M and Chen X G 1991 The use of a multileaf collimator to optimise dose distributions *Med. Phys.* **18** 613

Smithers D 1953 Collected papers and correspondence of Sir David Smithers held in the library archives of the Institute of Cancer Research, Sutton

Söderström S and Brahme A 1992 Selection of suitable beam orientations in radiation therapy using entropy and Fourier transform measures *Phys. Med. Biol.* **37** 911–924

Starkschall G and Eifel P J 1992 An interactive beam-weight optimisation tool for three-dimensional radiotherapy treatment planning *Med. Phys.* **19** 155–163

Takahashi S, Kitabatake T, Morita K, Okajima S and Iida H 1961 Methoden zur besseren Anpassung der Dosisverteilung an tiefliegende Krankenheitsherde bei der Bewegungsbestrahlung *Strahlentherapie* **115** 478–488

Trump J G, Wright K A, Smedal M I and Salzman F A 1961 Synchronous field shaping and protection in 2-million volt rotational therapy *Radiology* **76** 275

Webb S 1988 The mathematics of image formation and image processing *The Physics of Medical Imaging* ed S Webb (Bristol: Adam Hilger) p 560

—— 1989 Optimisation of conformal radiotherapy dose distributions by simulated annealing *Phys. Med. Biol.* **34** 1349–1369

—— 1990a Inverse tomograph *Nature* **344** 284

—— 1990b A new dose optimising technique using simulated annealing *Proc. 9th Ann. Meeting of ESTRO (Montecatini, 1990)* p 263

—— 1990c *From the watching of shadows: the origins of radiological tomography* (Bristol: Adam Hilger) p 125 et seq

—— 1991a Optimisation by simulated annealing of three-dimensional conformal treatment planning for radiation fields defined by a multileaf collimator *Phys. Med. Biol.* **36** 1201–1226

—— 1991b Optimisation of conformal radiotherapy dose distributions by simulated annealing 2: inclusion of scatter in the 2D technique *Phys. Med. Biol.* **36** 1227–1237

—— 1992a Optimisation by simulated annealing of three-dimensional conformal treatment planning for radiation fields defined by a multileaf collimator: 2. Inclusion of two-dimensional modulation of the X-ray intensity *Phys. Med. Biol.* **37** 1689–1704

—— 1992b Optimising dose with a multileaf collimator for conformal radiotherapy *Proc. 50th Ann. Congress of the British Institute of Radiology (Birmingham, 1992)* (London: BIR) p 15

—— 1992c Optimised three dimensional treatment planning for volumes with concave outlines, using a multileaf collimator *Proc. ART91 (Munich)* (abstract book p 66) *Advanced Radiation Therapy: Tumour Response Monitoring and Treatment Planning* ed A Breit (Berlin: Springer) pp 495–502

Wheatley B M 1955 A method of dose calculation with applications to moving-field therapy *Brit. J. Radiol.* **28** 566–573

Wheatley B M, Steed P R, Savage E W, King J H, Forster E W, Hodt H J, Jones

I R and Smithers D W 1953 The two million volt Van de Graaf generator installation for rotation therapy at the Royal Cancer Hospital *Brit. J. Radiol.* **26** 57–72

Zhu Y and Boyer A 1990 X-ray dose computations in heterogeneous media using 3-dimensional FFT convolution *Phys. Med. Biol.* **35** 351–368

Zhu Y, Boyer A L and Desobry G E 1992 Dose distributions of x-ray fields as shaped with multileaf collimators *Phys. Med. Biol.* **37** 163–173

CHAPTER 3

STEREOTACTIC RADIOSURGERY AND RADIOTHERAPY

3.1. INTRODUCTION

Tumours of the brain are regarded as special candidates for precision conformal therapy. Indeed any interventional procedure, be it a biopsy, surgery, the implantation of radioactive material or external beam therapy must be carried out as accurately as possible. Stereotactic procedures have developed to meet these needs. The term arises from stereo (the Greek word for three-dimensional) and either taxic (the Greek word for system or arrangement), or the Latin verb tactus (meaning to touch) (Galloway and Maciunas 1990). The special need for precision arises because the brain is unique in being almost entirely surrounded by bone, with the associated requirement for the surgeon to preserve this protection and the surgical constraint of preserving the function of tissues surrounding the surgical site. For these reasons using radiation to achieve the same aims as conventional surgery is attractive (section 3.2).

Three-dimensional tomographic imaging is specially essential to these activities. Indeed neurological procedures have been transformed by the availablility of a multiplicity of imaging modalities such as CT, MRI and PET.

It follows that accurate registration of image data from different modalities and between these and radiotherapy and surgical equipment is a sine qua non. Whilst it is sometimes possible to achieve this with landmark-based correlations, computing rigid and/or elastic transformations (see chapter 1), a class of procedures has developed based on the use of the stereotactic frame which 'locks' the head into the same position on imaging and treatment machines (figure 3.1).

3.1.1. Stereotactic positioning devices

It would appear that the first stereotactic frame was designed for investigating animal brains by the British surgeon Robert Henry Clarke and built by James Swift in London in 1905 (Horsley and Clarke 1908). The apparatus was

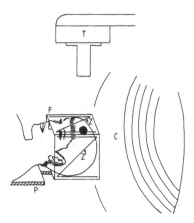

Figure 3.1. *The principle of stereotactic irradiation. The patient wears a frame (F) which may be bolted to the calvarium and is attached to the patient support system (P). The frame is worn both for three-dimensional medical imaging (C could be an x-ray* CT, MRI, SPECT *or* PET *scanner) and, later, at the time of treatment (T could be a treatment linac or a gamma knife). The important function of the Z-bars (Z) is described in the text.*

patented by Clarke in 1914 (Clarke 1914) and can still be seen in the museum of University College Hospital (Fodstad *et al* 1990). Not until 1947 was apparatus of this kind used for stereotactic intracranial surgery in man (Spiegel et al 1947). Leksell (1949) revitalized interest by designing a new stereotactic frame for precision cerebral surgery which formed the basis for his innovative proposal for conducting radiosurgery (see section 3.2) shortly afterwards (Leksell 1951). Modern developments in frame construction stem from this initiative. These are numerous and to avoid tedium only some representative examples are reviewed. In particular we may note that the well known stereotactic frames were developed for conventional surgery, biopsy, the placement of electrodes etc, and whilst some workers have used them for radiosurgery there has been a tendency to develop special frames for this purpose. So, important and numerous though they are, we simply note in passing the names of the four best known of these frames—Leksell apparatus, Riechert–Mundinger apparatus, Todd–Wells apparatus, Brown–Roberts–Wells (BRW) apparatus. These are described in detail by Galloway and Maciunas (1990). Instead we turn to some of the special-purpose constructions. Not all these have been used for stereotactic radiotherapy—some were just for image registration—but it is useful to group them here when considering stereotactic procedures in general.

The principles of stereotactic image registration are well illustrated by Bergström *et al* (1981). Their stereotactic frame comprised an aluminium base-plate with various knobs and holes which during scanning was bolted to a special head-holder on the scanner couch. Four aluminium bars were attached to the base-plate in an adjustable way so they could be positioned as close to the head as

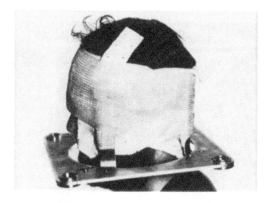

Figure 3.2. *Showing the patient head immobilization with base-plate, aluminium bars and glass-fibre mould, as developed by Bergström. (From Bergström et al (1981).)*

possible. Either small aluminium screws then attached these bars to the calvarium or the head was immobilized by a glass-fibre mesh ('Lightcast'), hardened by ultraviolet light, which wrapped around the aluminium bars (figure 3.2), and by incorporating a dental bite (Kingsley *et al* 1980). Wrapping the head is not a disadvantage for radiosurgery, unlike for stereotactic surgery where the method of fixing with screws is preferred, leaving a wide area of the skull available for access.

A special localization frame made of perspex and aluminium was then attached to the base-plate (figure 3.3). This frame contained thin aluminium strips running at 45° and 90° to the baseplate. Running parallel to these strips were thin plastic tubes into which radioactive fluid could be placed. This arrangement is sometimes referred to as a 'Z' bar or 'N bar' because the positions of the strips and tubes have the shape of these letters.

The patient wore the device during scanning (figure 3.4). The reproducibility of patient positioning was better than 3 mm (Kingsley *et al* 1980). Both the CT scanner and the PET scanner had equivalent head-holders to which the base-plate could be attached. The material of the head holder and the localization frame did not produce image artefacts. The thin aluminium strips showed up in the CT images and the images of the radioactive fluid in the plastic tubes showed up in the PET images. In this way the position of each transaxial slice was coded into each image.

Stereotactic coordinates were established as follows: The vertex of the 'N' bar defined the $y = 0$ plane. This was a fixed distance above the baseplate. The distances between the (point) images of the two strips at 45° to each other in any transaxial slice gave the y coordinate of this transaxial slice. More generally, for an arrangement where the angle between horizontal and diagonal bars is not 45°, if l_1 is the distance between the images of the two strips which meet at the baseplate and l_2 is the distance between the images of the other two strips meeting more

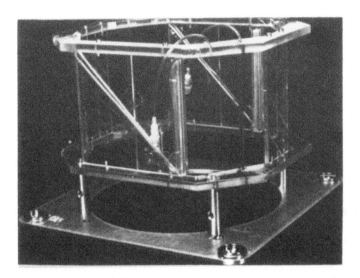

Figure 3.3. *A localization frame made from perspex with embedded aluminium strips. The two strips at 45° are used to find the slice level (y coordinate). The corner of the angle is the zero reference plane. A thin plastic tube is attached to the frame to take a radioactive fluid for when the system is used during PET scanning. It could also take a fluid such as copper sulphate, if the frame were used during magnetic resonance imaging. (From Bergström et al (1981).)*

anteriorly, then

$$y = L \left(\frac{l_1}{l_1 + l_2} \right) \tag{3.1}$$

where L is the length of the longitudinal strips at 90° to the baseplate. For a 45° arrangement this reduces to $y = l_1$. The (x, z) coordinates within the transaxial slice were centred on the midpoint of the localization frame, which again could be established from the images of the thin aluminium strips. Some workers have used three such frames, one either side and the other across the back of the head for increased accuracy.

Greitz *et al* (1980) showed many photographs of this system of fixation in use with six different diagnostic units (two CT scanners, a PET tomograph, a gamma camera and two x-ray systems) and three therapy units (a linac, the Leksell multicobalt device and a stereotactic biopsy unit).

Stereotactic coordinates are directly transferable from one 3D imaging modality to another. The method is as applicable for the abnormal brain as for the normal brain, unlike techniques (see later) which aim to correlate functional images with anatomical images from a stereotactic atlas.

Zhang *et al* (1990) used a commercially available stereotactic system made of non-ferrous, non-electrically conducting material and aluminium. Plexiglas plates were attached to the sides of the frame with a 'Z'-shaped channel pattern which

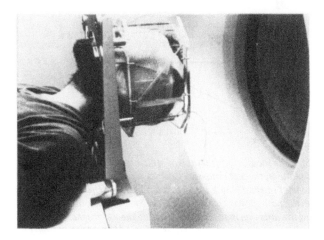

Figure 3.4. *The patient positioned in the CT scanner wearing the immobilization device and the localization frame, as developed by Bergström. The base of the frame is attached to a special head-holder. (From Bergström et al (1981).)*

provided the spatial localization in the same way as the system of Bergström *et al* (1981). The 'Z'-shaped channel was filled with copper sulphate solution for MR imaging, with a ^{68}Ga positron emitter for PET and with aluminium rods for x-ray CT. These imaged to points in transaxial planes enabling the data to be registered. The main feature of the development of Zhang *et al* (1990) was the construction of a sophisticated computer display and analysis environment. This was VAX based, used common image formats for data from each of the modalities, provided for point and region-of-interest transfer from one set of images to another and control via pull down menus. Multiple images were viewable simultaneously. Zhang *et al* (1990) claimed this to be a significant advantage over other systems developed for assisting neurological stereotactic surgery.

One of the disadvantages of a system such as that of Bergström *et al* (1981) is that the patient may be uncomfortable and may not be able to see or hear. Some PET studies are carried out to investigate the response of the brain to visual and auditory stimulae which would be difficult in such a frame. With this in mind Mazziotta et al (1982) designed a head-holder providing a stable enclosure, but not screwed to the skull. The anterior part of the holder was a pair of hinged wings which provided an emergency escape in case of nausea or breathing difficulties, and left the eyes and ears free. The rear of the head-holder was fixed to the scanning couch. Reproducible positioning within the PET scanner was thus achieved. Also correlated studies between x-ray CT and PET tomographic studies were performed.

This system had no localization frame and a different method of registering image data from the two tomographic modalities and with planar x-rays was used: With the patient in the PET scanner the laser which localized the slice was tracked

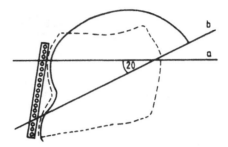

Figure 3.5. *The localization method of Meltzer et al. The line 'a' defines the transaxial plane in the MRI data-set. Line 'b' defines the orientation of some oblique plane of interest which intersects a particular (here the 4th) localizer tube. When the patient was transferred to the PET scanner, the localizing laser was pointed towards this marker, thus ensuring the PET transaxial planes were parallel to the chosen MRI oblique planes. (From Meltzer et al (1990).)*

on to the head-holder and marked with a waxed pencil. A thin aluminium wire was then put in this place to mark the tomographic slice. This wire was clearly imaged on plane x-rays and knowing the magnification of the x-rays and the interplane tomographic spacing, all transaxial PET images were related to the anatomy as depicted on the plane x-rays. By aligning the laser system of the CT scanner with the wax mark, it was ensured that x-ray CT tomograms could be matched to the corresponding PET data.

Miura *et al* (1988) developed a similar but not identical system for image registration. The patient wore a mask with eye-holes which was not fixed to the couch. The orbitomeatal line was marked with a lead string and used to position the patient within the PET or x-ray CT scanner so the transaxial direction aligned with this. Either side of the mask were two aluminium tubes at 45° to each other, able to contain a thin steel wire or a tube of positron-emitting isotope. These imaged to four points in each transaxial modality and CT-based anatomical images were aligned with PET-based functional images by scaling, rotating and translating one data-set relative to the other until the trapezia defined by the four points overlapped, at which stage the data were registered and regions of interest were transferable between them. The geometrical transformation was needed because of the lack of tying the frame to the couch, unlike the technique of Mazziotta *et al* (1982).

Meltzer *et al* (1990) used a head-holder, not locked to the couch, to correlate MRI, CT and PET data. The novel feature of this system was a 'localizing bar' fixed centrally over the nose region containing 20 evenly spaced markers of petroleum jelly. The markers showed as 20 bright dots on the mid-sagittal MRI image. A particlular oblique slice of interest was then defined by measuring the angle between the transaxial MRI plane and the oblique plane and finding which marker this oblique plane intersected. When the patient was transferred to the PET scanner the laser alignment system was directed to intercept this marker, thus assuring that the PET transaxial slices were then parallel to the required oblique plane on the MRI

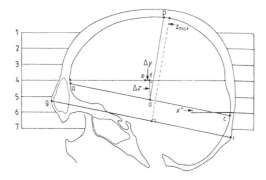

Figure 3.6. *The relationship between tomographic coordinates and stereotactic coordinates, as used in atlases. The method of establishing this is described in the text. 'ac' is the line between the anterior and posterior commissures (AC–PC line), 'gi' is the glabella–inion (GI) line (visible on a head radiograph). The two lines are parallel and separated by 0.21 times the perpendicular distance from the GI line to the vertex of the skull. 'O' is the centre of the tomographic data, 'e' is the centre of the stereotactic coordinate frame. Seven tomographic slices are shown (from the PETT6 scanner). $X°$ is the angle between the GI line and the transaxial direction. The brain size can be measured from the radiograph (length of AC–PC line and distance Z_{max}) and corrected for magnification to provide the scale factor between the real brain and the atlas. (From Fox et al (1985).)*

images (figure 3.5).

Fox *et al* (1985) used a head-holder (comprising a lucite hemicylinder, shape-assuming foam and a plastic face mask) during PET scanning with the PETT6 emission tomograph. The aim of their work was to correlate PET images with images from a stereotactic atlas rather than patient-specific CT or MRI images. A lucite plate, this time containing seven wires at the interslice spacing of the tomograph, was mounted on the head-holder. The wall-mounted laser, defining the first tomographic slice, was aligned with the first wire and a plane x-radiograph was taken showing all seven wires defining the tomographic slices. The origin of the tomographic data was taken to be the middle pixel of the fourth (i.e. middle) slice. In order to interpret the PET data, Fox *et al* (1985) established the relationship between this coordinate system and the stereotactic coordinate system used by neurological atlases (figure 3.6). The origin of the stereotactic coordinate system was the mid-point of the (AC–PC) line between anterior and posterior commissures of the brain in the mid-sagittal plane of the brain. Since this line cannot be seen on a plane x-ray image, it was drawn parallel to the glabella-inion (GI) line connecting the most anterior point of the glabella (g) to the most posterior point of the base of the inion (i) at a distance of 0.21 times the perpendicular distance (Z_{max}) from the GI line to the skull vertex (p). The three assumptions that the GI line can be accurately and reproducibly found from a plane x-ray image, that the GI line is parallel to the AC–PC line and at this precise distance, were extensively validated by Fox *et al* (1985). The angle between the

AC–PC line and the transaxial direction can then be established. It then becomes a relatively simple matter to relate the two sets of coordinates via this angle and the three translations between the origins. Specific measures were taken to ensure no other rotations occurred. All that remained was to scale the recorded tomographic data to the scale of the atlas.

Fox *et al* (1985) demonstrated the usefulness of their technique by sucessfully locating in the stereotactic atlas the anatomic features responsible for observed and measured functional response, in particular the cavernous sinuses (associated with high local cerebral blood volume) and the visually responsive cortex (associated with increased regional cerebral blood flow).

The main limitation of atlas-based correlations with functional images is that the method is really only valid for normal anatomy. Several large atlases of the anatomy of the brain are useful for this purpose, including Talairach and Szikla (1967), Matsui and Hirano (1978), Hanaway *et al* (1980) and Talairach and Tournoux (1988).

Wilson and Mountz (1989) have developed a frame for registering functional and anatomical images which overcomes some limitations of the others. The frame comprises two equilateral triangles of nylon carrying thin tubing in which x-ray attenuating material, positron emitting isotope or nickel chloride fluid can be placed for CT, PET and MR imaging respectively. There were two such triangular frames and they were positioned either side of the head and located firmly using stethoscope earplugs in the auditory meatus and fixation to the lateral canthus. Each frame also had a nylon bar with tubing running from base to vertex (figure 3.7). The contrast materials imaged to six bright dots in each tomographic plane, three either side of the transaxial head image. The angle θ of the image plane to the base of the triangles is a simple function of the ratio A/B, where A and B are the distances between the point images in that plane, i.e.

$$\tan \theta = \left[\frac{(1 - A/B)}{(1 + A/B)} \right] \tan 60°. \tag{3.2}$$

The 'level' L of the image plane (where the plane meets the vertical from the left-hand corner of the triangle) depends again on A and B and on 't' (the half-side of the equilateral triangle) via

$$L = \frac{(t - B \cos \theta) 2A \tan 60°}{(B + A)}. \tag{3.3}$$

So from simple measurements and these formulae, the tomographic plane can be specified. By aligning the images from different modalities such that the six point images coincided, the images were registered. The algorithm for transferring one 3D image set into the frame of the other was driven by the images of the planes in which the dots were furthest apart for accuracy (i.e. the most inferior planes towards the base of the triangles).

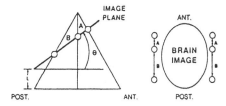

Figure 3.7. *One of the triangular frames in the method of Wilson and Mountz, showing the equilateral triangle with the line from the vertex to the base. Any particular image plane intersects this line and the two sides of the triangle in three points (as shown to the right). By measuring the distances A and B on the right-hand diagram, the angle θ and the 'level' L of this tomographic plane may be determined by equations (3.2) and (3.3). (From Wilson and Mountz (1989).)*

A relocatable stereotactic frame comprising a baseplate to which is secured a dental impression of the patient's upper jaw has been designed by Gill (1987). The baseplate and dental plate are secured to the head with adjustable straps. Together with an individually moulded occipital impression, this provides a non-invasive system of support. The dental impression is taken in polyethylmethacrylate polymer which sets hard in a few minutes with negligible shrinkage. The straps are of woven nylon with buckles of plastic to avoid artefacts when imaging. The head support is hard-setting rubber and the baseplate is made of epoxy. The accuracy of positioning is 1 mm (Graham *et al* 1991b), significantly better than moulded head casts. The frame can be used with a Brown–Roberts–Wells (BRW) stereotactic frame (Heilbrun *et al* 1983) attached for positional localization (Thomson *et al* 1990, Thomas *et al* 1990, Gill *et al* 1991) in both diagnostic imaging units and at the time of therapy.

A stereotactic frame such as that described by Brown (1979) essentially provides a means of relating three-dimensional *image coordinates* to corresponding three-dimensional locations in *frame coordinates*. This is achieved by imaging the points at which the components of one or more frame-constituent 'Z' or 'N'-shaped tubes (whose locations in frame-space are precisely known) cross the transaxial CT slices, giving characteristically elliptical-shaped dots. The measured coordinates of these points in the two spaces determines the matrix transformation which then allows any points in one space to be known in the other space. Under these conditions, the stereotactic frame need not be attached to the imaging device or its couch, provided it is reproducibly attached to the patient's skull.

One of the main uses of this arrangement was to determine, on the CT data, the target point for a rigid intracranial probe and its entry point into the brain and thus specify where to insert the probe in frame coordinates. The computer can then indicate the point of intersection of this probe with each transaxial CT slice and if the probe trajectory were unsatisfactory a new one could be determined. In the context of radiosurgery (see section 3.2) it would equally be possible to investigate the 'trajectory' of a radiation beam in the same way.

3.2.   RADIOSURGERY AND STEREOTACTIC RADIOTHERAPY

Historically the term 'radiosurgery' was introduced by Leksell (1951) to describe the method by which many narrow radiation beams were used to irradiate intracranial structures from many angles as an alternative to surgery. For many years ^{60}Co sources were used and commercial units were produced. Charged-particle irradiations (especially protons) have been used, but facilities are very expensive. Adapted CT scanners and radioactive implants have also been used. More recently, isocentric linear accelerators have taken over this role and corresponding interest in the technique has greatly increased.

The terms stereotactic radiosurgery and stereotactic radiotherapy should be distinguished; the former is applicable to a single-fraction treatment equivalent to a surgical operation, whereas the latter involves therapy by multiple fractions. Radiosurgery is considered for intracranial lesions considered inoperable or carrying excessive risk from conventional surgery. Unlike conventional radiotherapy, the lesion is not 'treated' so much as 'destroyed'. Indeed, as well as applications in cancer medicine, one of the commonest applications of radiosurgery is the treatment of benign arteriovenous malformations (AVM). Stereotactic radiosurgery obliterates tissue by inducing gliosis or fibrosis within it. The therapeutic principle is quite unlike radiation therapy, which aims to preserve healthy cells within the target volume. Radiosurgery is appropriate for single-shot small-volume irradiation where precision is meticulously sought and high dose gradients are required.

Barrow et al (1990) point out however that the vast majority of neurological lesions are best treated by conventional surgery and that the indications and contra-indications for radiosurgery must be considered very carefully before this procedure is used. It is not expected to replace surgery and its role must be kept in perspective. Barrow *et al* (1990) sum up many of the clinical arguments. The physicist might remark that it is unlikely that radiosurgery, important though it is, would ever provide the *unique* justification for purchasing a cyclotron or a linac and this may explain why the special purpose devices (see section 3.2.1) are also so rare.

Whatever radiation-based treatment is selected, precise knowledge of the location of the target is essential. Techniques for registering 3D MRI, CT and PET data with digital subtraction angiograms have been developed by Henri *et al* (1989). Bova and Friedman (1991) have explained why biplane angiography alone might be less than satisfactory. Herbert and Fröder (1990) have developed digital tumour fluoroscopy (DTF), analogous to digital subtraction angiography, whereby an x-ray generator and image intensifier are used to create images in the operating theatre itself. Localization requires complete tomographic datasets in general. All radiosurgery involves the use of some kind of immobilization and localization frame of the type discussed above. When such a stereotactic frame is screwed to the skull it is used in the 'target-centred' mode (as opposed to 'frame-centred mode'—see Galloway and Maciunas 1990)—that is, the target is placed at the centre of the pointing apparatus and all trajectories reach the target regardless of orientation.

3.2.1.   Radiosurgery with ^{60}Co sources; the gamma knife

Apart from the early use of protons for radiosurgery (Larsson *et al* 1958) and Leksell's original apparatus, which had an x-ray set attached to a stereotactic frame (Leksell 1951), historically the technique has relied on ^{60}Co radiation (Leksell 1971, Goodman 1990, Barrow 1990). The first so-called 'gamma knife' (figure 3.8) was constructed by AB Motala Verkstad, Motala and installed at Sweden's Sophiahemmet Hospital in 1968. It was later moved to UCLA School of Medicine (Leksell 1971, To *et al* 1990). The Karolinska Hospital in Stockholm developed the second gamma knife and its arrival in 1975 coincided with the widespread use of computed tomography, which gave impetus through improved accuracy to the technique of stereotactic radiosurgery (Leksell 1987). To date there have only been nine installations in the world (Maitz *et al* 1990, Goodman 1990) (see table 3.1).

The first gamma knife had 179 sources; later models have 201. The Leksell 201 source ^{60}Co gamma knife is shown in figure 3.9 as installed at the facility at the University of Pittsburgh, Pennsylvania (Wu *et al* 1990). The 201 sources are contained in the upper hemispherical shield. All sources are focused to a single point at a source-to-focus distance of 403 mm. The central beam is at 55° to the horizontal plane. The sources lie in an arc of ±48° about this central ray in the long axis of the treatment unit and at ±80° along the transverse axis to the patient couch. The sources are contained in two hemispherical shields made of iron. The slideable entrance door is of steel. The whole unit weighs some 18 000 kg. The total source actiity was 219 TBq in July 1987 delivering 398 *c*Gy min^{-1}. Unlike previous units, which had been installed to their hospital sites with the sources in position, the unit at Pittsburgh was loaded on site for reasons and by methods discussed by Maitz *et al* (1990). Electrically generated hydraulic pressure is used to move the treatment couch and to close the shielding door. In the event of failure several back-up systems are provided.

The final collimation is achieved by means of one of four interchangeable collimator helmets with 201 channels aligning with the sources and removable apertures that produce 4, 8, 14 or 18 mm diameter fields at the focus: 4 mm is probably the smallest beam any machine can produce (Wu et al 1991). On the other hand, the 18 mm size of the largest collimator hole limits the usefulness of the device for irradiating large tumours. Individual plugs can stop-off selected holes if required to shield the eyes or to change the dose distribution. Flickinger *et al* (1990b) show suggestions for which holes to block and the effects which this has on the three-dimensional dose distribution. Since the dose gradients are very high with the gamma knife, precise conformation of the high-dose region to a possibly irregularly shaped target volume is important.

A stereotactic frame is attached to the patient under local anaesthetic and in

(a)

(b)

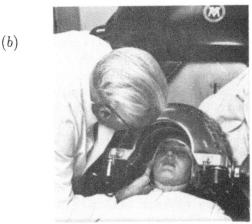

Figure 3.8. *(a) The first Leksell 'gamma knife'. The sources are contained in the domed structure to the right of the picture. The collimating helmet can be seen on the couch next to the two staff. (b) Positioning the patient in the helmet for the Leksell gamma knife. (From Leksell (1971).) (Courtesy of Charles C Thomas, Publisher, Springfield, Illinois, USA.)*

turn the frame is positioned into the collimating helmet so that the tumour to be treated is at the focus. To begin treatment, the couch with the patient so positioned is hydraulically driven into the gamma knife unit and the helmet mates with the primary collimation. After a prescribed time the couch reverses out. Typically the unit can deliver 300–400 cGy min^{-1}, so several minutes treatment time is required for a 'single shot' (Coffey *et al* 1990,1991, Kondziolka *et al* 1990, Kondziolka and Lunsford 1991, Lunsford *et al* 1990a,b). If the lesion is large, several 'shots' (i.e. treatments with the focus at different brain locations and possibly with different helmets) may be necessary.

Table 3.1. *Gamma knife facilities worldwide (data from Maitz (1990) and Goodman (1990)).*

Date	Installation
1968	Sophiahemmet Hospital, Stockholm, Sweden
1975	Karolinska Hospital, Stockholm, Sweden
	(1981 1st Stockholm device donated to UCLA)
1983	Buenos Aires, Argentina
1985	Sheffield, England
1987	University of Pittsburgh, School of Medicine, PA
1989	Charlottesville, Virginia
1989	Chicago, Illinois
1989	Piedmont Hospital, Atlanta, Georgia
1991	Lawrence Livermore National Laboratory

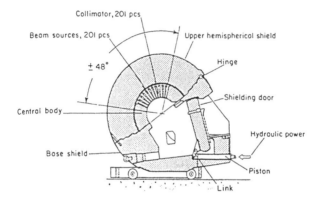

Figure 3.9. *Schematic of the Leksell 'gamma knife' unit used for stereotactic radiosurgery (see text for details). (From Maitz et al (1990).) (Reprinted with permission from Pergamon Press Ltd, Oxford, UK.)*

The gamma knife at Sheffield, UK, constructed by Nucletec SA (Switzerland), is similar (but not identical), comprising 201 sources of total activity 209 TBq distributed over a spherical sector of 60° by 160° with a source-to-focus distance of 395 mm. There are three helmets whose holes are 4, 8 and 14 mm diameter (Walton *et al* 1987). Dose distributions for each helmet were computed by integrating over the sources using the measured dose profile transverse to each beam. They were also measured and found to be in good agreement. With the 14 mm helmet assigned an output factor of unity, giving 295 cGy min^{-1} in Sept 1985, the output factors of the 8 mm and 4 mm helmets were 0.964 and 0.925, respectively. The unit is used as described above for the Pittsburgh unit.

The main disadvantages of the gamma knife are that it cannot be 'turned off' and once the shield is open, the patient receives whole-body dose before the helmet locks into the primary collimator. The cobalt needs replenishing periodically and

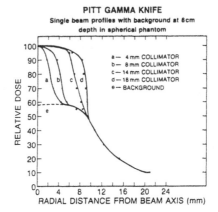

Figure 3.10. *Raw dose profiles at 8 cm depth of a single beam from the Pitt gamma knife, measured with all but one of the 201-collimator-holes collimator plugged. Due to the transmission through the 200 collimator shieldings, the profiles show a considerable background, which appears in the form of broad shoulders on either side of the peak. The four curves correspond to the different helmets (collimators). (From Wu et al (1990).) (Reprinted with permission from Pergamon Press Ltd, Oxford, UK.)*

the unit is very expensive ($3.5 million according to Barrow 1990).

Flickinger and Steiner (1990) and Flickinger *et al* (1991) discuss the dose-response predictions for stereotactic radiosurgery with different collimation helmets.

Dose planning for treatments with the Leksell gamma knife requires knowledge of the dose profile from a single beam. Wu *et al* (1990) measured the dose profile by blocking off 200 of the 201 sources. The raw profiles obtained (figure 3.10) display a substantial shoulder representing almost 60% of the maximum intensity of a single beam. This is due to the background radiation from the plugged sources. The background was measured by plugging all 201 sources and subtracted from the profiles in figure 3.10 to give the 'true' single-beam dose profiles, shown in figure 3.11, for the various collimators. The shape of these gives the off-axis ratio (OAR) needed for planning.

Figure 3.12 shows how the treatment planning program computes the dose to a point P in the patient. The manufacturer specified the dose rate at the reference point R, 1 cm deep in a spherical phantom of radius 8 cm, i.e. 7 cm from the focus of all the sources. Hence the distance SR is 333 mm. The on-axis dose rate at the point Q, $(403 - z)$ mm from the source, can be established from the inverse-square law and the linear exponential attenuation. Finally, the dose-rate at an off-axis point P follows by applying the OAR from the dose profile. This procedure is repeated for all sources and the results summed to give a three-dimensional dose distribution on a 31^3 dose matrix spanning the target volume. The necessary measurements of source-to-skull distance (which affect the exponential term in the calculation) are made by attaching a special plastic helmet to the baseplate of

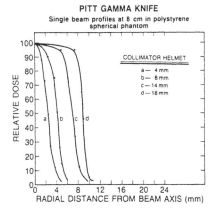

Figure 3.11. *'True' dose profiles for single beams in the Pitt gamma knife for four different collimator sizes. These sizes are defined by the full-width at half-maximum of the 50% isodose line. These curves are derived from the data in figure 3.10. (From Wu et al (1990).) (Reprinted with permission from Pergamon Press Ltd, Oxford, UK.)*

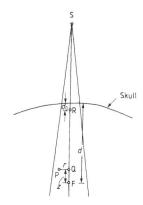

Figure 3.12. *A schematic diagram of a simple dose calculation algorithm developed by the manufacturers of the gamma knife. A dose reference point R was chosen 1 cm from the surface of an 8 cm spherical phantom. The dose at any general point Q in the calculation matrix is then obtained using the dose profiles from figure 3.11 and the inverse-square law and the linear-attenuation exponential formula. (From Wu et al (1990).) (Reprinted with permission from Pergamon Press Ltd, Oxford, UK.)*

the stereotactic frame and using a 'dip-stick' at selected locations on the surface of the helmet to make measurements from which the computer calculates the three-dimensional surface of the skull for treatment planning.

Some gamma knife facilities do not have the 18 mm diameter collimated helmet, restricting the range of size of targets. Even the 18 mm collimator is not large enough for some treatments. The problem can be overcome by treating with two or more isocentres displaced from each other. This generates non-spherical dose

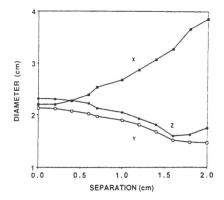

Figure 3.13. *Showing the change in the diameter of the volume defined by the 50% isodose in three orthogonal directions as a function of the separation between isocentres for a radiosurgery gamma-knife unit with an 18 mm diameter collimator helmet. The isocentres are separated along the x-direction. (From Flickinger et al (1990a).) (Reprinted with permission from Pergamon Press Ltd, Oxford, UK.)*

distributions. Flickinger *et al* (1990a) plotted the shape of the distributions for two-isocentre treatments as the separation between the isocentres was increased. For example, if the isocentres were separated along the x-direction the diameter of the volume enclosed by the 50% isodose line expanded in the x-direction and contracted in the orthogonal directions with increasing separation (figure 3.13). Flickinger *et al* (1990a) computed a catalogue of 3D dose distributions with multiple isocentres of varying separation in order to guide the treatment planning process. Each plan was customized from this start.

Some targets could be planned with one- or two-isocentre treatments adequately encompassing the target within the 50% isodose surface. The problem then arose of which was the better treatment. Flickinger *et al* (1990a) generated dose–volume histograms in surrounding normal brain, showing the improvement obtained with two isocentres and used a biological model to show the NTCP was reduced, justifying the increased effort (figure 3.14).

The method of using multiple isocentres to treat larger brain lesions with the gamma knife has many similarities with the corresponding technique for multiple-arc radiotherapy with multiple isocentres (see section 3.2.4).

3.2.2. *Stereotactic multiple-arc radiotherapy with a linear accelerator*

The isocentrically mounted linear accelerator is a natural choice for the x-ray source for radiosurgery. Whilst an x-ray source was Leksell's first choice for stereotactic radiosurgery (Leksell 1951), it was abandoned because of the difficulties in aligning the beam. The move towards using linear accelerators is a consequence of the much improved bearings (Barrow *et al* 1990, Podgorsak *et al* 1989) of modern accelerators. All the experience from the gamma knife is relevant

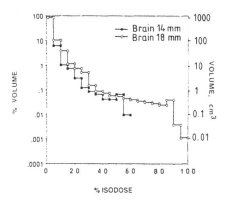

Figure 3.14. *Dose–volume histograms for normal brain tissue comparing treatments of a 21 mm by 14 mm by 14mm tumour with either a 18 mm collimator and a single isocentre or a 14 mm collimator with two isocentres separated by 9 mm. The latter is to be preferred, verified by a biological model giving the NTCP. (From Flickinger et al (1990a).) (Reprinted with permission from Pergamon Press Ltd, Oxford, UK.)*

since there is no reason to believe these different systems will produce different outcomes in similar groups of patients. Further advantages include the lower cost of a linac and its availability for 'conventional' radiotherapy when not used for radiosurgery. Betti and Derechinsky (1982) and Heifetz *et al* (1984) were among the first to study the potential, including making the suggestion that multiple arcs could be made by rotating the couch to different positions. Many authors have since discussed linac-based arc therapy including: Betti and Derechinsky (1982), Betti et al (1989), Hartmann *et al* (1985), Colombo *et al* (1985a,b, 1987, 1990), Patil (1985, 1989), van Buren *et al* (1986), Chierego *et al* (1988), Hariz et al (1988, 1990), Lutz *et al* (1988), Saunders *et al* (1988), Friedman and Bova (1989), Pastyr *et al* (1989), Casentini *et al* (1990), Croft (1990), Delannes *et al* (1990), Engenhart *et al* (1990), Loeffler *et al* (1990a,b), Schwade *et al* (1990), Thomson *et al* (1990), Graham *et al* (1991a), Gill *et al* (1991), Schell *et al* (1991), and Serago *et al* (1991).

By arranging for the centre of the tumour to be at the isocentre, treatment becomes possible using the beam from a number of non-coplanar arcs by combining gantry rotation with a finite number of couch positions with twist. Fractionated radiotherapy also becomes possible, potentially increasing the therapeutic ratio (Graham *et al* 1991a). The success of fractionated radiotherapy depends on the reproducibility of patient position, both between diagnostic imaging sessions and between the treatment fractions. This generally demands use of a reliable head-holder and stereotactic frame, although Heifetz *et al* (1989) suggest that using external scalp landmarks may be acceptable for large-volume irradiations. Thomson (1990) and Thomson et al (1990) used the Gill frame for localization and five arcs of 140° of rotation. A Clinac 6/100 6 MV treatment unit was equipped with a special applicator to minimize penumbra. Planning software was written in house (Thomson 1990). They proposed the replacement of the term 'radiosurgery' by 'stereotactic multiple-arc radiotherapy (SMART)'.

Gill *et al* (1991) report that, in their use of a prototype head-frame constructed with a dental impression, occipital support and straps and with a BRW frame for external beam radiotherapy, the positioning uncertainty was less than 1 mm. This allowed fractionated radiotherapy to be confidently practised. Dose planning was performed with an IGE TARGET system. Because the dental impression acted only on the upper jaw, the patient was comfortable and free to talk. The securing straps were of woven nylon, which produced no artefacts in diagnostic imaging. The straps were secured with a plastic buckle, which gave a quick-release mechanism in case of emergency.

Hartmann *et al* (1985) and Pastyr *et al* (1989) developed the 'Heidelberg system', based on the use of a Riechert–Mündinger head-frame and 11 rotational non-coplanar arcs.   Friedman and Bova (1989) developed a linac-based radiosurgery facility in which an extra mechanical system was provided which functioned independently of the proprietry rotation bearings for couch and gantry. This enabled an accuracy of 0.2 ± 0.1 mm to be achieved, comparable with the accuracy of a gamma knife. An in-house-developed planning system also allowed almost real-time visualization of dose contours relative to anatomy.

Barrow *et al* (1990)and McGinley *et al* (1990) described the 'Emory x-ray knife' at the Emory Clinic, Atlanta, Georgia, USA (figure 3.15). This utilized a BRW immobilization scheme, with the patient seated upright on a rotatable turntable. The centre of the lesion being treated was arranged to coincide with the centre of rotation of the turntable and also with the axis of rotation of a 6 MV linear accelerator fitted with a circular collimator to define a small beam of diameter 5 to 30 mm. The gantry could take up angles from $35°$ to $80°$ to the vertical. The gantry did not move during irradiation, thus reducing the risks of collisions in the event of electrical failure. During irradiation, the patient turntable rotated slowly at 3–4 RPM. The dose rate of the accelerator was sufficiently low to allow at least four complete rotations of the turntable at each gantry angle. By using several gantry angles with rotation at each angle, near-spherical high-dose volumes were created. Experiment showed (McGinley *et al* 1990) that the 90% isodose was approximately 4 mm smaller than the collimator diameter and the 80% isodose was 2 mm smaller than the collimator diameter. The shape and size of the high-dose region was independent of position within the brain. Non-spherical dose distributions could be obtained by differentially weighting the number of monitor units for each gantry angle.

Stereotactic radiosurgery with a linear accelerator requires a special-purpose collimator to reduce penumbra and so sharpen the dose gradient at the edge of the target volume. Circular collimators were described by Hartmann et al (1985) of size 5 mm to 20 mm diameter for use in Heidelberg. Croft (1990) shows circular collimators of 12.5 mm to 30 mm diameter. Figure 3.16 shows one such collimator designed by Winston and Lutz (1988) for the Boston system, which generally uses four arcs. The circular assembly extends to within 23 cm of the isocentre and will accept a range of eight different cerrobend inserts with different conically drilled

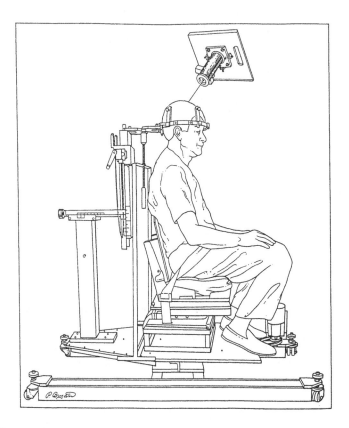

Figure 3.15. *The Emory x-ray knife. The patient is seated on a rotating turntable. The linear accelerator gantry rotates about an axis coinciding with that of the turntable. The intersection point is the centre of the brain lesion. The gantry is stationary during irradiation. Notice the special collimator. (From Barrow et al (1990).)*

openings, whose diameters, projected to the isocentre, range from 12.5 to 30 mm in steps of 2.5 mm (Lutz *et al* 1988). Figure 3.17 shows the static beam profiles for three diameters of collimator measured experimentally. When several arcs of radiation are combined, the three-dimensional dose distributions show that the volume of non-target tissue receiving a high amount of radiation increases rapidly as the diameter of the collimator increases, but nevertheless is confined close to the target volume (for detailed calculations see Winston and Lutz 1988).

The potential of linear-accelerator-based radiosurgery depends on accurate verification. The delivery of single shots of very high dose to small volumes of brain tissue will be intolerant of error. For this reason radiographic verification must be performed to ascertain that the collimator is accurately positioned relative to the target, at all orientations of the gantry. Each development should be

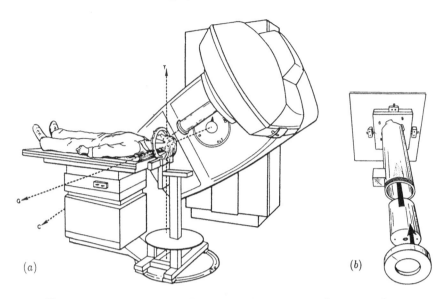

Figure 3.16. *(a) Linear-accelerator-based stereotactic radiosurgery, show-ing the coincidence of the axes of rotation of the gantry (G), the turntable (T) and the collimator (C) at the isocentre of the machine. The radiation col-limator will extend to close to the isocentre. (b) The small-field collimator supplements the internal rectangular collimator. The aperture of the collima-tor can be adjusted by inserting different cerrobend castings. Each collimator is labelled by the diameter of the field at the isocentre. (From Winston and Lutz (1988).)*

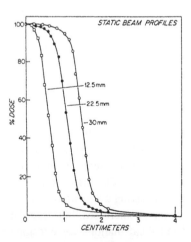

Figure 3.17. *Static beam profiles for three circular collimators used for stereotactic radiosurgery. Notice the very rapid decline at the edge of the field and the nearly identical shape of these edges. (From Winston and Lutz (1988).)*

critically assessed with this in mind. Winston and Lutz (1988), Lutz *et al* (1988) and Saunders *et al* (1988), reporting the Boston system, have provided a particularly rigorous technique incorporating verification and error assessment. Tsai *et al* (1991) have discussed extensively the Boston quality assurance programme, which underpins the safety of the stereotactic irradiation technique. Drzymala (1991) also addresses issues of quality assurance. Arndt (1989), however, raises doubts concerning the accuracy of linac-based radiosurgery compared with the accuracy of the gamma knife, whose errors are smaller than 0.3 mm.

Hartmann *et al* (1985) used a dose computation algorithm based on integration of superpositions of beams whose radial and depth-dose profiles had been measured experimentally. Dose distributions were computed in orthogonal planes through the target volume. Consideration of integral dose to the normal brain led Hartmann *et al* (1985) to conclude that stereotactic radiosurgery should be confined to tumours of 3 cm diameter or less. This is now a commonly accepted limitation for single-shot irradiations. Heifetz *et al* (1989) suggested that larger volumes could be treated by fractionated radiosurgery. Multiple-arc techniques have also been applied to stereotactic radiotherapy of head and neck lesions (Hartmann *et al* 1987).

For those implementations of the method in which the head is held in a frame fixed to the floor, it is necessary to deactivate the motors which drive the couch to protect the patient during therapy (Croft 1990, Lutz *et al* 1988). The total time required for treatment is a function of the output of the linear accelerator in monitor units (MU) per degree, and can typically be some 0.5–0.75 hours (Croft 1990). However, the time taken to attach the frame, to take 3D images, perform planning and set up the treatment can easily be a whole day. Stereotactic radiosurgery is very time consuming.

3.2.2.1. Dynamic stereotactic radiosurgery.   If a single transaxial arc is used, the dose gradient in the plane of the arc will not be great because, for a full 2π rotation (see e.g. chapter 2), each beam has a parallel-opposed counterpart. The dose gradient in the orthogonal direction, normal to the plane of rotation, will be as good as for the static beam. Multiple arcs help to overcome this problem. An alternative solution was provided by Pike et al (1987) and Podgorsak et al (1987, 1988, 1989), called 'dynamic stereotactic radiosurgery'. The patient was positioned isocentrically so that the axis of the gantry and the axis of couch rotation passed through the centre of the spherical target volume. The gantry was rotated from 30° to 330° whilst the couch *simultaneously* rotated from −75° to +75° with the beam switched on. The entrance trajectory thus describes a complex curve on the patient's head in which the beam always enters from the upper hemisphere and parallel-opposed fields do not occur. The situation is summarized in figure 3.18. Podgorsak *et al* (1988) showed that this leads to dose gradients which improve upon those from single-arc rotations for some planes of measurement, but are worse for others. When a comparison is made with the dose profile for a single

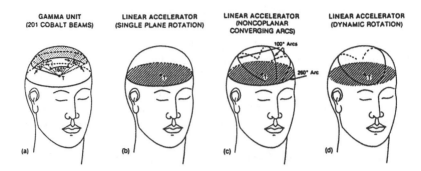

Figure 3.18. *A comparison of points of beam entry into the patient's skull for various competing radiosurgical techniques. (a) The Leksell gamma knife, (b) the single transaxial plane rotation with a linear accelerator (simple rotation therapy with a very small area beam), (c) a multiple non-coplanar converging-arc technique with a linear accelerator, (d) dynamic stereotactic radiosurgery with a linear accelerator. (From Podgorsak et al (1989).) (Reprinted with permission from Pergamon Press Ltd, Oxford, UK.)*

arc in the plane of its rotation, dynamic radiosurgery is always superior, whatever the radius of the beam. Dynamic radiosurgery cannot, however, improve the ratio of integral (brain) dose to the target dose.

One of the major disadvantages of using the complicated entrance trajectory is that treatment planning becomes very difficult and Podgorsak *et al* (1988) derive the target dose even for this trajectory from a single-arc calculation. This does not get the correct dose distribution outside the target volume so this is estimated from profiles of dose gradients.

Podgorsak *et al* (1988) use an in-house-constructed stereotactic frame with 3D image-based dose calculations. Corrections are made for the photon attenuation in the frame. Circular collimators shape the field. The clinical experience at the McGill University facility has been summarized by Souhami *et al* (1990a,b).

A detailed comparison of the dose fall-off outside the target volume for the gamma knife, the single-plane rotation, the Heidelberg multiple-converging-arc technique, the Boston multiple-converging-arc-technique and the McGill University dynamic stereosurgery method has been provided by Podgorsak (1989). Although measurements of dose gradient are not true volumetric assessments (Graham et al 1991a), this information (see figure 3.19) leads to the conclusion that the dynamic stereosurgery technique can rival the gamma knife, which is seen as a gold standard. Since none of the methods for linac-based radiosurgery provide a uniform distribution of entry points over the surface of the skull, all dose distributions are slightly spherically asymmetric. Since the dose gradient thus depends on the plane of measurement through the target volume, figure 3.19 displays the best and worst gradients obtained with each method. All other gradients lie within the extremes.

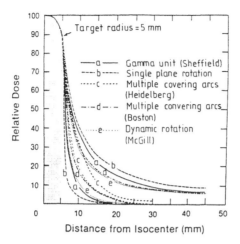

Figure 3.19. *A comparison of dose gradient for different techniques for stereotactic radiosurgery with a 1 cm diameter spherical target covered by the 90% isodose surface. Since all techniques provide an anisotropic distribution of entry portals on the surface of the patient's skull, the dose distributions are asymmetric and dose profiles depend on the chosen measurement plane. In this figure the steepest and the shallowest profiles are shown for each method: (a) for the gamma knife obtained from Walton et al (1987), and (b), (c), (d), and (e) for a 10 MV linear accelerator; (b) single plane rotation, (c) 11 multiple converging arcs (Heidelberg system), (d) four multiple converging arcs (Boston system), and (e) the dynamic stereotactic radiosurgery technique (McGill). (From Podgorsak et al (1989).) (Reprinted with permission from Pergamon Press Ltd, Oxford, UK.)*

3.2.2.2. The dose calculation algorithm in linac-based stereotactic radio-surgery. Kubsad *et al* (1990) used the EGS4 Monte Carlo code to compute the beam profile for a linear accelerator fitted with a circular collimator. Profiles were computed by a full-scale simulation of particle transport through all the machine components (primary collimation, flattening filter, secondary collimation, moving jaws, and stereotactic collimator). The same profiles were also independently computed by the convolution method (equation (2.50)) and shown to be in good agreement with the full calculation (and with experimental measurements). The dose, computed this way, is

$$D_E(r) = \int \int \int T_E(s)\, h(E, r - s)\, \mathrm{d}^3 s \tag{3.4}$$

where T_E is the terma (the product of the mass attenuation coefficient $\mu/\rho\,(E, r)$ and the energy fluence $E\Gamma_E(r)$) and h is the convolution kernel.

The dose distribution in multiple-arc linac-based stereotactic radiosurgery is constructed as the superposition of the doses from each arc. In turn, each arc is represented as a sum of fixed 'stationary' beams. Because these beams are small,

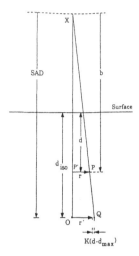

Figure 3.20.   *Conceptual diagram showing the dose calculation for a single beam as part of a stereotactic multiple-arc therapy. The dose is calculated at point P. The x-ray source is at X and the isocentre is at O. All other quantities are defined in the text. (From Luxton et al (1991).)*

there is no need to correct for the shape of the patient contour when computing off-axis points; this contour can be considered locally flat. Luxton *et al* (1991) provide the formalism needed for each of these elemental beams for each arc:

- Let the beam monitor be calibrated so the dose per monitor unit (MU) is 1 *c*Gy at a depth d_{max} in water for a 10×10 cm field at a distance of 1 m $+ d_{max}$ from the source.
- With the circular collimator in place, let $OF(S)$ be the equivalent quantity for a field of diameter S projected to the isocentre. This is called the output factor.
- Let $TMR(S, d)$ be the tissue-maximum ratio at depth d, field diameter S.

Then the single-beam dose per monitor unit to a point P a depth d from the surface and distance r from the central axis of the beam is given by

$$D(d, r, S, d_{iso}) = TMR(S, d)\, OF(S)\, R(r')\, (SAD + d_{max})^2\, /b^2 \qquad (3.5)$$

where d_{iso} is the depth of the isocentre O, SAD is the source-to-isocentre distance and $b = (SAD - d_{iso} + d)$ is the distance from the x-ray source to the perpendicular onto the beam axis from the point at which the dose is calculated. r' is related to r via

$$r' = rSAD/b - K(d - d_{max}) \qquad (3.6)$$

(see figure 3.20) and K is a small constant which accounts for scatter, the first term being the correction for beam divergence. $R(r')$ is the off-axis ratio at distance r' from the central ray at a depth of d_{max} in the plane of the isocentre and this may be found by measurement of the profile.

Luxton *et al* (1991) measured beam profiles at the isocentre for a number of depths in scattering material. As the depth increased the profiles broadened by a small but measurable amount, allowing calculation of the parameter K by fitting the data. Experimental measurements of the TMR showed that this could be represented in the region $d \geqslant d_{max}$ by an analytic function

$$TMR(d) = \exp\left(-\mu\left(d - d_{max}\right)\right) \tag{3.7}$$

where the coefficient μ depended on beam energy and collimator diameter. They also gave a formula for calculating the TMR for $d \leqslant d_{max}$, but this is not an issue in stereotactic radiosurgery. Luxton *et al* (1991) verified the accuracy of the calculation algorithm by performing TLD and film-dose measurements for irradiations of test phantoms. Gehring *et al* (1991) discuss a planning system in which the weights of the arcs may be adjusted.

3.2.2.3. Dynamic collimation for stereotactic radiosurgery with multiple arcs.

The limitations of single-isocentre multiple-arc radiotherapy/surgery have already been discussed. In order to generate non-spherical dose distributions, several workers have developed multiple-arc therapy with multiple isocentres. However, this can lead to difficulties in obtaining a uniform distribution of dose within the target.

Leavitt *et al* (1991) have presented a very elegant method of obtaining a uniform dose to an irregularly shaped and possibly quite large target in the brain with multiple-arc therapy by dynamic field shaping. Figure 3.21 shows how they added four independent block collimators to the usual circular collimator. These were 7 cm high, 4.1 cm wide and 2.1 cm thick. Each one could be independently translated and rotated about the mid-line under microprocessor control. Each pair lay in a plane and anti-collision software ensured that no two collimators in a plane ever collided, although two collimators not in the same plane could overlap in the beam's-eye-view. Sufficient range of movement was engineered so that, in the limit,the field could be either circular (all four vanes backed off beyond the diameter of the circle), have four straight sides (all four vanes within the circle),or one to three straight with three to one circular sides.

These eight degrees of movement (four translations; four rotations) enabled the field to be dynamically shaped to the beam's-eye-view of the target at different positions within any one arc and for each arc, thus protecting normal non-target brain tissue. The concept is analogous to the idea behind the (large-field) multileaf collimator (chapter 5).

McGinley *et al* (1992) have made a similar,but not identical,development. The group at the Emory Clinic have added the possibility of incorporating blocks into the cylindrical collimator. These blocks are set by hand,rather than under computer control.

Figure 3.21. *Perspective view of the 'dynamic collimator' for stereotactic radiosurgery with the field shaped to the beam's-eye-view of the target at each arc position. The usual circular collimator is supplemented with four independently positioned block collimators to define an eight-sided field (see text). (From Leavitt et al (1991).) (Reprinted with permission from Pergamon Press Ltd, Oxford, UK.)*

3.2.3. Stereotactic multiple-arc radiotherapy with a CT scanner

Iwamoto *et al* (1990) propose to use the CT scanner itself to perform multiple-arc radiotherapy. By adding a secondary collimator to the CT scanner to define a pencil beam, the machine's scanning movements naturally provided for rotation-arc therapy. By tilting the gantry by up to $\pm 20°$ from the vertical plane, non-coplanar arcs were possible. The advantage of this is that the need for a stereotactic frame, or precise correlation between the geometries of imaging and therapy, disappears. Images may be taken *during* therapy, confirming that the tumour is being irradiated correctly. It turned out that a CT image could be made with the radiation which leaked into the full fan through the collimator, showing the target volume as a black circle with a white halo and some detail of the rest of the slice. Iwamoto *et al* (1990) compared the dose profile delivered this way with that from two gamma knives in Stockholm and UCLA and noted little difference when nine arcs were used. A typical dose of 2 Gy can be given in 20 rotations of 4 s duration with the tube operating at 170 mA without overheating problems. Perhaps this implies that the machine is more a rival for SMART than for the gamma knife, since to deliver say 10 Gy for radiosurgery would require nearly an hour with time for tube cooling. Account has to be made of the different RBE (1.5) of 140 kVp radiation.

3.2.4. Comparison of different radiation types and geometries

Which radiation type (charged particles or photons) and which geometry are best for stereotactic radiosurgery? This is clearly an important question, given that the technologies are expensive and that each has developed in its own separate way at different treatment centres. Phillips *et al* (1990) studied this question. They compared charged-particle irradiations from protons, carbon, helium and neon ions with photon irradiations in the 'gamma-knife' geometry and with photon irradiation from multiple arcs on a linear accelerator.

Dose distributions were developed for irradiating spherical tumours of diameters 1, 2, 3, 4, 5 cm and for a range of tumours from clinical examples. For the charged-particle irradiations the regular treatment geometries for these were adopted. An optimal treatment modality is that which delivers a uniform dose throughout the target volume whilst minimizing the dose to surrounding tissue, and so the problem was analysed by computing the dose–volume histograms in the target volume, in the 'normal brain' (being all the brain minus the tumour) and for selected annular volumes surrounding the tumour. Additionally, the figure of merit—the 'localization factor'—was computed, being the fraction of the radiation energy deposited in the patient which was delivered to the target volume. The higher the localization factor, the better the treatment.

For very small tumours (of diameter 2 cm or less) virtually no differences were observed between photon, proton and carbon irradiations. For medium size (diameters 2–3 cm) and large size tumours (diameter $\geqslant$ 3 cm) photon irradiations were increasingly inferior to charged-particle irradiations. For a tumour of volume 56 cm^3 the localization factor for photons (0.20) was just over half that (0.38) for protons, and for such large tumours the dose–volume histograms showed considerably more normal-tissue (e.g. brainstem) dose with photons than with protons. Only small differences were noted between the different charged-particle modalities for stereotactic radiosurgery, these differences being greater for the tumours with largest volumes. The biological effects of the major difference between photons and protons has yet to be worked out, and Phillips *et al* (1990) conclude that the efficacy of one type of radiation or irradiation scheme over another is not yet clinically established.

Graham *et al* (1991a,b) used the Boston 3D planning system, JCPLAN, to assess the relative merits of using 1, 2, 3, 4, 5, and 9 arc stereotactic irradiations compared with static coplanar three-field and static non-coplanar six-field techniques. The dose distributions were characterized and ranked in terms of the integral dose–volume histogram in the normal brain lying in an annulus of thickness 2 cm around spherical target volumes of size 10, 20, 40, and 55 mm diameter. It had already been established that the DVH for these annulae were identical to the whole-brain DVH, except at low doses, and the use of the annulus speeded the computations. These target volumes matched the diameters of available collimators. The different irradiation geometries were also ranked in terms of the volume receiving 50% of the isocentre dose. (This value was chosen since the brain may tolerate 10 Gy in a single fraction when the target dose—for a cerebral AVM—may be 20 Gy.) Graham *et al* (1991a) concluded that (i) the distributions with multiple arcs were superior to the three-static-field arrangements for all target sizes, (ii) there was no increased sparing of normal brain tissue above the 50% isodose with more than three arcs of rotation, and (iii) for the larger tumours, two arcs were marginally better than three and the six-field static method also gave a very similar distribution of dose to multiple-arc irradiations.

Serago *et al* (1992) also studied the effect of multiple arcs on the dose–volume histogram for the normal brain surrounding spherical targets of diameter 1, 2 and 4 cm. They carefully quantitated the volume of normal brain inside each isodose contour as a function of number of arcs for each target size and concluded that multiple arcs beyond four provided no significant improvement. There was a marginal reduction in the volume of tissue enclosed by the 10% isodose contour for 2 and 4 cm diameter targets, but this is clinically insignificant. Serago *et al* (1992) concluded there may be a small safety advantage using multiple arcs, in that the dose per arc is less and hence failure of the gantry to rotate would be less serious. However, this event is unlikely to occur if proper safety precautions are taken. Schell *et al* (1991) characterized dose distributions using the dose–volume histogram, showing that, although differences were observed between 4, 5 and 10-arc therapy, these were insignificant clinically.

Chierego *et al* (1988) decided, on the basis of measuring, with film, the dose distribution on selected planes (not volumes), that the best energy for use in stereotactic radiotherapy was between 4–6 MV and that between 9 and 17 arcs were generally needed. This conflict with the conclusions of Graham et al (1991a) and Serago *et al* (1992) may arise because conclusions were made from planar rather than volumetric dose measurements. Chierego *et al* (1988) measured the absorbed dose to critical organs when a lesion sited in a deep medial cerebral plane was irradiated and found the doses were negligible (all $\leqslant$ 0.2% of the isocentre dose.)

Pike *et al* (1990) showed that the dose distributions obtained with multiple-arc therapy were similar to those from the gamma knife. It was also demonstrated that the distributions were virtually independent of photon energy in the range 4–25 MV, concurring with results by Mazal *et al* (1989).

In principle, the spherical sector which may be covered by a multiple-arc technique can be larger than for the gamma knife. For example, Colombo *et al* (1985a) created spherical sectors of 110° × 160° compared with the 60° × 160° of the gamma knife. How important this might be must be quantified for each situation. Another potential advantage of using a linear accelerator is the ability to use fields of larger area than provided by the collimators of the gamma knife. Colombo *et al* (1990) introduced a modification to multiple-arc radiosurgery enabling non-spherical dose distributions to be obtained. The radiation beam was directed to the isocentre of the gantry and couch and, for a number of gantry angles β, the couch was dynamically rotated through an angle α symmetrically placed about the long axis of the patient (figure 3.22). By also arranging for couch *translation*, cigar-shaped isodose distributions were obtained, more suitable when the target volume was non-spherical. Casentini *et al* (1990) used the same apparatus for arcing in the angle β at a set of fixed angles α.

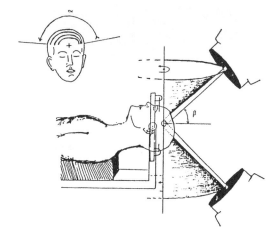

Figure 3.22. *Stereotactic conical irradiation with a linear accelerator arranged so that multiple-arc therapy would lead to a non-spherical dose distribution. The gantry was set at a number of fixed angles β and the patient couch was moved through an angle α. By also arranging for couch translation, elongated dose distributions were obtained. (From Colombo et al (1990), permission of S Karger AG, Basel, Switzerland.)*

3.3.   STEREOTACTIC INTERSTITIAL IMPLANT THERAPY

For many tumour sites the high-dose volume can be conformed to the target by interstitial radiotherapy; a brain tumour is one such candidate. This greatly reduces the dose to non-target tissue, since external rays do not pass through this tissue en route to the target. Stereotactic procedures are again needed for accuracy. Welsh and Chapman (1990) described procedures exemplifying the method. The physical basis is as follows.

A stereotactic frame is attached to the head of the patient non-invasively. The frame is also attached to the couch of a CT scanner and all procedures take place in the CT scanner suite. To the frame is attached a template comprising a matrix of holes spaced at 1.5 mm intervals. These holes are candidate drilling sites for catheterizing the brain. CT images are taken with the frame and template in place; the template appears suspended above the skull and each aperture appears as an air-filled column on the template. From these images, a suitable set of apertures can be selected for approaching the target. The skull is drilled and the catheters are introduced to a depth of 1 cm beyond the contrast-enhanced border of the tumour. When all catheters are in place, a repeat set of CT scans is made to establish that the catheters are all parallel, correctly placed and that there has been no intraprocedural haemmorhage. Finally, the catheters are secured externally. The whole procedure takes place under general anaesthetic and takes about 2 hours. A widely used radioisotope is ^{125}I, in the form of seeds. This isotope has a half-life of 60 days and emits x-rays of 28 keV (among other less abundant radiations).

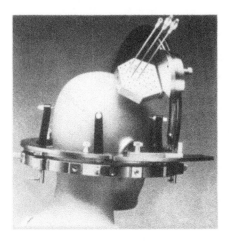

Figure 3.23. *The BRW stereotactic system in place, with the template guiding the placement of several catheters. These could be catheters holding radioactive seeds (^{125}I, ^{192}Ir) or ferromagnetic seeds for inductive hyperthermia. (From Lulu et al (1990).) (Reprinted with permission from Pergamon Press Ltd, Oxford, UK.)*

The catheters with the seeds are removed after the required dose has been given, and the patient is CT scanned once again. Lulu *et al* (1990) and Marchosky *et al* (1989) have described a similar CT-based stereotactic procedure for guiding the placement of catheters and the associated treatment planning. Figure 3.23 shows the BRW stereotactic system in place with a template and several catheters.

Hawliczek *et al* (1991) have presented a technique for implanting seeds permanently in the brain. Planning methods have been evolved to display the isodoses from candidate seed positions superposed on anatomical images (Weaver et al 1990) and then to calculate through coordinate transformations the entry trajectories in the frame coordinates for catheters. Ling *et al* (1985) and Schell *et al* (1987) have provided two-dimensional dose matrices for ^{125}I dosimetry calculations.

3.4. SUMMARY

Stereotactic radiotherapy and radiosurgery were proposed over four decades ago. Although the earliest treatments were made with x-rays, the specialist gamma knife was for many years the only reliable means for treatment. However, the achievement of stereotactic conformal radiotherapy and radiosurgery using a linear accelerator with special fixation devices has recently given the subject an enormous boost, and today there is widespread interest in the technique. In turn this has led to characterizing the effectiveness of the method in terms of well understood outcomes such as the dose–volume histogram of dose to target and surrounding tissue and plots of the dose profiles in selected planes. These have led many workers to draw conclusions on the optimum number of arcs as a function of target size. Special techniques are being developed to generate non-spherical treatment volumes, where required. Image-guided stereotactic implants are also performed. In many circumstances these methods can be as satisfactory as charged-particle irradiation.

REFERENCES

Arndt J 1989 Use of the linear accelerator for neurosurgery *Neurosurgery* **24** (4) 641

Barrow D L, Bakay R A E, Crocker I, McGinley P and Tindall G T 1990 Stereotactic radiosurgery *J. Med. Assoc. of Georgia* **79** 667–676

Bergström M, Boëthius J, Eriksson L, Greitz T, Ribbe T and Widén L 1981 Head fixation device for reproducible position alignment in transmission CT and positron emission tomography *J. Comp. Assist. Tomog.* **5** (1) 136–141

Betti O O and Derechinsky Y E 1982 Irradiations stéreotaxiques multifaisceaux *Neurochirurgie* **28** 55–56

Betti O O, Munari C and Rossler R 1989 Stereotactic radiosurgery with the linear accelerator: treatment of arteriovenous malformations *Neurosurgery* **24** 311–321

Bova F J and Friedman W A 1991 Stereotactic angiography: an inadequate database for radiosurgery? *Int. J Rad. Oncol. Biol. Phys.* **20** 891–895

Brown R A 1979 A stereotactic head frame for use with CT body scanners *Investigative Neurology* **14** 300–304

van Buren J M, Landy H J, Houdek P V and Ginsberg M S 1986 CT-directed stereotactic fractionated rotational radiotherapy by linear accelerator *Neurosurgery* **19** (1) 149

Casentini L, Colombo F, Pozza F and Benedetti A 1990 Combined radiosurgery and external radiotherapy of intracranial germinomas *Surgical Neurology* **34** (2) 79–86

Chierego G, Marchetti C, Avanzo R C, Pozza F and Colombo F 1988 Dosimetric considerations on multiple arc stereotaxic radiotherapy *Radiother. Oncol.* **12** 141–152

Clarke R H 1914 Stereotactic instrument *UK Patent no* 11711

Coffey R J, Flickinger J C, Bisonette D J and Lunsford L D 1991 Radiosurgery for solitary brain metastases using the cobalt-60 gamma unit: methods and results in 24 patients *Int. J Rad. Oncol. Biol. Phys.* **20** 1287–1295

Coffey R J, Lunsford L D, Bisonette D and Flickinger J C 1990 Stereotactic gamma radiosurgery for intracranial vascular malformations and tumours: report of the initial North American Experience in 331 patients *Stereotactic and functional neurosurgery* ed C Ohye, P L Gildenberg and P O Franklin (Basel: Karger) pp 535–540

Colombo F, Benedetti A, Casentini L, Zanusso M and Pozza F 1987 Linear accelerator radiosurgery of arteriovenous malformations *Appl. Neurophysiol.* **50** 257–261

Colombo F, Benedetti A, Pozza F, Avanzo R C, Marchetti C, Chierego G and Zanardo A 1985a External stereotactic irradiation by linear accelerator *Neurosurgery* **16** 154–160

Colombo F, Benedetti A, Pozza F, Marchetti C, Chierego G and Zanardo A 1990 Linear accelerator radiosurgery of three-dimensional irregular targets *Stereotactic and functional neurosurgery* ed C Ohye, P L Gildenberg and P O Franklin (Basel: Karger) pp 541–546

Colombo F, Benedetti A, Pozza F, Zanardo A, Avanzo R C, Chierego G and Marchetti C 1985b Stereotactic radiosurgery utilising a linear accelerator *Applied Neurophysiol.* **48** 133–145

Croft M J 1990 Stereotactic radiosurgery of arteriovenous malformations *Radiologic Technol.* **61** 375–379

Delannes M, Daly N, Bonnet J, Sabatier J and Trémoulet M 1990 Le stéréo-adapteur de Laitinen *Neurochirurgie* **36** 167–175

Drzymala R E 1991 Quality assurance for linac-based stereotactic radiosurgery *Quality assurance in radiotherapy physics* ed G Starkschall and J Horton (Madison, WI: Medical Physics Publishing) pp 121–129

Engenhart R, Kimmig B N, Hover K H, Wowra B, Sturm V, Kaick G and Wannenmacher M 1990 Stereotactic single high dose radiation therapy of benign intracranial meningiomas *Int. J. Rad. Oncol. Biol. Phys.* **19** 1021–1026

Flickinger J C, Lunsford L D, Wu A and Kalend A 1991 Predicted dose- volume isoeffect curves for stereotactic radiosurgery with the ^{60}Co gamma unit *Acta Oncol.* **30** 363–367

Flickinger J C, Lunsford L D, Wu A, Maitz A H and Kalend A M 1990a Treatment planning for gamma knife radiosurgery with multiple isocentres *Int. J. Rad. Oncol. Biol. Phys.* **18** 1495–1501

Flickinger J C, Maitz A, Kalend A, Lunsford L D and Wu A 1990b Treatment volume shaping with selective beam blocking using the Leksell Gamma Unit *Int. J. Rad. Oncol. Biol. Phys.* **19** 783–789

Flickinger J C and Steiner L 1990 Radiosurgery and the double logistic product formula *Radiother. Oncol.* **17** 229–237

Fodstad H, Hariz M and Ljunggren B 1990 History of Clarke's stereotactic instrument *Stereotactic and functional neurosurgery* ed C Ohye, P L Gildenberg and P O Franklin (Basel: Karger) pp 242

Fox P T, Perlmutter J S and Raichle M E 1985 A stereotactic method of anatomical localization for positron emission tomography *J. Comp. Assist. Tomog.* **9** (1) 141–153

Friedman W A and Bova F J 1989 The University of Florida radiosurgery system *Surg. Neurol.* **32** 334–342

Galloway R L and Maciunas R J 1990 Stereotactic neurosurgery *Critical Reviews in Biomedical Engineering* **18** (3) 181–205

Gehring M A, Mackie T R, Mehta M P, Kubsad S S, Sanders C A and Reckwerdt

P S 1991 A 3D treatment planning system for stereotactic radiosurgery *Proc. 1st biennial ESTRO meeting on physics in clinical radiotherapy (Budapest, 1991)* p 29

Gill S S 1987 *UK Patent application no* 8728150

Gill S S, Thomas D G T, Warrington A P and Brada M 1991 Relocatable frame for stereotactic external beam radiotherapy *Int. J. Rad. Onc Biol. Phys.* **20** 599–603

Goodman M L 1990 Gamma knife radiosurgery: current status and review *Southern Medical J.* **83** (5) 551–554

Graham J D, Nahum A E and Brada M 1991a A comparison of techniques for stereotactic radiotherapy by linear accelerator based on 3-dimensional dose distributions *Radiother. Oncol.* **22** 29–35

Graham J D, Warrington A P, Gill S S and Brada M 1991b A non-invasive relocatable stereotactic frame for fractionated radiotherapy and multiple imaging *Radiother. Oncol.* **21** 60–62

Greitz T, Bergström M, Boëthius J, Kingsley D and Ribbe T 1980 Head fixation system for integration of radiodiagnostic and therapeutic procedures *Neuroradiology* **19** 1–6

Hartmann G H, Schlegel W, Sturm V, Kober B, Pastyr O and Lorenz W J 1985 Cerebral radiation surgery using moving field irradiation at a linear accelerator facility *Int. J. Rad. Oncol. Biol. Phys.* **11** 1185–1192

Hartmann G H, Treuer H, Boesecke R, Bauer B, Sturm V and Lorenz W J 1987 'Problems of dose calculation applying stereotactic convergent beam irradiations at head neck region *Computer Assisted Radiology* ed H U Lemke, M L Rhodes, C C Jaffee and R Felix (Berlin: Springer) pp 328–334

Hanaway J, Scott W R and Strother C M 1980 *Atlas of the human brain and the orbit for computed tomography* (St Louis, Missouri: Warren H Green Inc)

Hariz M I, Henriksson R, Löfroth P O, Laitinen L V and Säterborg N E 1990 A non-invasive method for fractionated stereotactic irradiation of brain tumours with linear accelerator *Radiother. Oncol.* **17** 57–72

Hariz M I, Laitinen L V Löfroth P O and Säterborg N E 1988 A non-invasive adapter in the stereotactic irradiation of brain tumours with linear accelerator *Acta Neurochirur.* **91** (3/4) 162

Hawliczek R, Neubauer J, Schmidt W F O, Grunert P and Coia L R 1991 A new device for interstitial ^{125}Iodine seed implantation *Int. J. Rad. Oncol. Biol. Phys.* **20** 621–625

Heifetz M D, Wexler M and Thompson R 1984 Single-beam radiotherapy knife *J. Neurosurgery* **60** 814–818

Heifetz M D, Whiting J, Bernstein H, Wexler M, Rosemark P and Thomson R W 1989 Stereotactic radiosurgery for fractionated radiation: a proposal applicable to linear accelerator and proton beam programs *Stereotact. Func. Neurosurgery* **53** 167–177

Heilbrun M P, Roberts T S, Apuzzo M L J, Wells Jr T H and Sabshin J K 1983 Preliminary experience with the Brown–Roberts–Wells (BRW) computerized tomography stereotaxic guidance system *J. Neurosurgery* **59** 217–222

Henri C, Collins L, Peters T M, Evans A C and Marrett S 1989 Three-dimensional interactive display of medical images for stereotactic neurosurgery planning *Proc. SPIE* **1092** (Medical Imaging 3: Image Processing) 67–74

Herbert M H and Fröder M 1990 Digital tumour fluoroscopy (DTF)—a new direct imaging system in the therapy planning for brain tumours *Int. J. Rad. Oncol. Biol. Phys.* **18** 221–231

Horsley V and Clarke R H 1908 The structure and functions of the cerebellum examined by a new method *Brain* **31** 45–124

Iwamoto K S, Norman A, Kagan A R, Wollin M, Olch A, Bellotti J, Ingram M and Skillen R G 1990 The CT scanner as a therapy machine *Radiother. Oncol.* **19** 337–343

Kingsley D P E, Bergström M and Berggren B M 1980 A critical evaluation of two methods of head fixation *Neuroradiology* **19** 7–12

Kondziolka D and Lunsford L D 1991 Stereotactic radiosurgery for squamous cell carcinoma of the nasopharynx *Laryngoscope* **101** 519–522

Kondziolka D, Lunsford L D, Coffey K J, Bisonette D J and Flickinger J C 1990 Stereotactic radiosurgery of angiographically occult vascular malformations: indications and preliminary experience *Neurosurgery* **27** (6) 892–900

Kubsad S S, Mackie T R, Gehring M A, Misisco D J, Paliwal B R, Mehta M P and Kinsella T J 1990 MonteCarlo and convolution dosimetry for stereotactic radiosurgery *Int. J. Rad. Oncol. Biol. Phys.* **19** 1027–1035

Larsson B, Leksell L, Rexed B, Sourander P, Mair W and Andersson B 1958 The high-energy proton beam as a neurosurgical tool *Nature* **182** 1222–1223

Leavitt D D, Gibbs F A, Heilbrun M P, Moeller J H and Takach G A 1991 Dynamic field shaping to optimise stereotactic radiosurgery *Int. J. Rad. Oncol. Biol. Phys.* **21** 1247–1255

Leksell D G 1987 Stereotactic radiosurgery *Neurological Research* **9** 60–68

Leksell L 1949 A stereotaxic apparatus for intracerebral surgery *Acta Chirurg. Scand.* **99** 229–233

—— 1951 The stereotaxic method and radiosurgery of the brain *Acta Chirurg. Scand.* **102** 316–319

—— 1971 *Stereotaxis and radiosurgery* (Springfield, IL: Charles C Thomas)

Ling C, Schell M, Yorke E, Palos B and Kubiatowicz D 1985 Two-dimensional dose distribution of ^{125}I seeds *Med. Phys.* **12** 652–655

Loeffler J S, Rossitch E Jr, Siddon R, Moore M R, Rockoff M A, Alexander III E 1990a Role of stereotactic radiosurgery with a linear accelerator in treatment of intracranial arteriovenous malformations and tumours in children *Pediatrics* **85** 774–782

Loeffler J S, Siddon R L, Wen P Y, Nedzi L A and Alexander III E 1990b Stereotactic radiosurgery of the brain using a standard linear accelerator: a study of early and late effects *Radiother. Oncol.* **17** 311–321

Lulu B A, Lutz W, Stea B and Cetas T C 1990 Treatment planning of template-guided stereotaxic brain implants *Int. J. Rad. Oncol. Biol. Phys.* **18** 951–955

Lunsford L D, Coffey R J, Cojocaru T and Leksell D 1990a Image-guided stereotactic surgery: a 10 year evolutionary experience *Stereotactic and functional neurosurgery* ed C Ohye, P L Gildenberg and P O Franklin (Basel: Karger) pp 375–387

Lunsford L D, Flickinger J and Coffey R J 1990b Stereotactic gamma knife radiosurgery *Arch. Neurol.* **47** 169–175

Lutz W, Winston K R and Maleki N 1988 A system for stereotactic radiosurgery using a linear accelerator *Int. J. Rad. Oncol. Biol. Phys.* **14** 373–381

Luxton G, Jozsef G and Astrahan M A 1991 Algorithm for dosimetry of multiarc linear-accelerator stereotactic radiosurgery *Med. Phys.* **18** 1211–1221

Maitz A H, Lunsford L D, Wu A, Lindner G and Flickinger J C 1990 Shielding requirements on-site loading and acceptance testing of the Leksell Gamma Knife *Int. J. Rad. Oncol. Biol. Phys.* **18** 469–476

Marchosky J A, Moran C, Fearnot N E, Welsh D M and Zumwalt C B 1989 A stereotactic system for volumetric implantation of the brain *Proc. 8th Ann. Meeting North American Hyperthermia Group* **1** 47

Matsui T and Hirano A 1978 *An atlas of the human brain for computerized tomography* (Tokyo: Igaku-Shoin)

Mazal D A, Rosenwald J C, Gaboriaud G and Porcher J B 1989 Implication of basic physical phenomena on the dose distribution in stereotactic x-ray external irradiation *Proc. 17th Int. Congress of Radiology* p 217

Mazziotta J C, Phelps M E, Meadors A K, Ricci A, Winter J and Bentson J R 1982 Anatomical localisation schemes for use in positron computed tomography using a specially designed headholder *J. Comp. Assist. Tomog.* **6** (4) 848–853

Meltzer C C, Bryan R N, Holcomb H H, Kimball A W, Mayberg H S, Sadzot B, Leal J P, Wagner H N and Frost J J 1990 Anatomical localisation for PET using MR imaging *J. Comp. Assist. Tomog.* **14** (3) 418–426

McGinley P H, Butker E K, Crocker I R and Aiken R 1992 An adjustable collimator for stereotactic radiosurgery *Phys. Med. Biol.* **37** 413–419

McGinley P H, Butker E K, Crocker I R and Landry J C 1990 A patient rotator for stereotactic radiosurgery *Phys. Med. Biol.* **35** 649–657

Miura S, Kanno I, Iida H, Marakami M, Takahashi K, Sasaki H, Inugami A, Shishido F, Ogawa T and Uemura K 1988 Anatomical adjustments in brain positron emission tomography using CT images *J. Comp. Assist. Tomog.* **12** (2) 363–367

Pastyr O, Hartmann G H, Schlegel W, Schabbert S, Treuer H, Lorenz W J and Sturm V 1989 Stereotactically guided convergent beam irradiation with a linear accelerator: localisation technique *Acta Neurologica* **99** 61–64

Patil A A 1985 Isocentric placement of the target in the linear accelerator using CT stereotaxis *Acta Neurochirurgica* **78** 168–169

Patil A A 1989 Radiosurgery with the linear accelerator *Neurosurgery* **25** 143

Phillips M H, Frankel K A, Lyman J T, Fabrikant J I and Levy R P 1990 Comparison of different radiation types and irradiation geometries in stereotactic radiosurgery *Int. J. Rad. Oncol. Biol. Phys.* **18** 211–220

Pike B, Peters T M, Podgorsak E, Pla C, Olivier A and de Lotbinière A 1987

Stereotactic external beam calculations for radiosurgical treatment of brain lesions *Appl. Neurophysiol.* **50** 269–273

Pike G B, Podgorsak E B, Peters T M, Pla C and Olivier A 1990 Three-dimensional isodose distributions in stereotactic radiosurgery *Stereotactic and functional neurosurgery* ed C Ohye, P L Gildenberg and P O Franklin (Basel: Karger) pp 519–524

Podgorsak E B, Olivier A, Pla M, Hazel J, de Lotbinière A and Pike B 1987 Physical aspects of dynamic stereotactic radiosurgery *Appl. Neurophys.* **50** 263–268

Podgorsak E B, Olivier A, Pla M, Levebvre P Y and Hazel J 1988 Dynamic stereotactic radiosurgery *Int. J. Rad. Oncol. Biol. Phys.* **14** 115–126

Podgorsak E B, Pike B, Olivier A, Pla M and Souhami L 1989 Radiosurgery with high energy photon beams: A comparison among techniques *Int. J. Rad. Oncol. Biol. Phys.* **16** 857–865

Saunders W M, Winston K R, Siddon R L, Svensson G H, Kijewski P K, Rice R K, Hansen J L and Barth N H 1988 Radiosurgery for arteriovenous malformations of the brain using a standard linear accelerator: Rationale and technique *Int. J. Rad. Oncol. Biol. Phys.* **15** 441–447

Schell M C, Ling C, Gromadzki Z and Working K 1987 Dose distributions of model 6702 I-125 seeds in water *Int. J. Rad. Oncol. Biol. Phys.* **13** 795–799

Schell M C, Smith V, Larson D A, Wu A and Flickinger J C 1991 Evaluation of radiosurgery techniques with cumulative dose volume histograms in linac-based stereotactic external beam irradiation *Int. J. Rad. Oncol. Biol. Phys.* **20** 1325–1330

Schwade J G, Houdek P V, Landy H J, Bujnoski J L, Lewin A A, Abitol A A, Serago C F and Pisciotta V J 1990 Small-field stereotactic external-beam radiation therapy of intracranial lesions: fractionated treatment with a fixed halo device *Radiology* **176** 563–565

Serago C F, Houdek P V, Bauer-Kirpes B, Lewin A A, Abitol A A, Gonzales-Arias S, Marcial-Vega V A and Schwade J G 1992 Stereotactic radiosurgery: dose-volume analysis of linear accelerator techniques *Med. Phys.* **19** (1) 181–185

Serago C F, Lewin A A, Houdek P V, Gonzales-Arias S, Hartmann G H, Abitol A A and Schwade J G 1991 Stereotactic target point verification of an x-ray and CT localiser *Int. J. Rad. Oncol. Biol. Phys.* **20** 517–523

Souhami L, Olivier A, Podgorsak E B, Hazel J, Pla M and Tampieri D 1990a Dynamic stereotaxic radiosurgery in arteriovenous malformation-preliminary treatment results *Cancer* **66** (1) 15–20

Souhami L, Olivier A, Podgorsak E B, Pla M and Pike G B 1990b Radiosurgery of cerebral arteriovenous-malformations with the dynamic stereotaxic irradiation *Int. J. Rad. Oncol. Biol. Phys.* **19** 775–782

Spiegel E A, Wycis H T, Marks M and Lee A J 1947 Stereotaxic apparatus for operations on human brain *Science* **106** 349–350

Talairach J and Szikla G 1967 *Atlas d'anatomie stéréotaxique du télencéphale: études anatomio-radiologiques* (Paris; Masson et Cie)

Talairach J and Tournoux P 1988 *Co-planar stereotaxic atlas of the human brain* (Stuttgart: Georg Thieme)

Thomas D G T, Gill S S, Wilson C B, Darling J L and Parkins C S 1990 Use of relocatable stereotactic frame to integrate positron emission tomography and computed tomography images: application in human malignant brain tumours *Stereotactic and functional neurosurgery* ed C Ohye, P L Gildenberg and P O Franklin (Basel: Karger) pp 388–392

Thomson E S 1990 Stereotactic brain radiotherapy *PhD thesis* University of London

Thomson E S, Gill S S and Doughty D 1990 Stereotactic multiple arc radiotherapy *Brit. J. Radiol.* **63** 745–751

To S Y C, Lufkin R B, Rand R, Robinson J D and Hanafee W 1990 Volume growth rate of acoustic neuromas on MRI post-stereotactic radiosurgery *Comp. Med. Imag. Graph.* **14** 53–59

Tsai J S, Buck B A, Svensson G K, Alexander III E, Cheng C-W, Mannarino E G and Loeffler J S 1991 Quality assurance in stereotactic radiosurgery using a standard linear accelerator *Int. J. Rad. Oncol. Biol. Phys.* **21** 737–748

Walton L, Bomford C K and Ramsden D 1987 The Sheffield stereotactic radiosurgery unit: physical characteristics and principles of operation *Brit. J. Radiol.* **60** 897–906

Weaver K, Smith V, Lewis J D, Lulu B, Barnett C M, Leibel S A, Gutin P, Larson D and Phillips T 1990 A CT-based computerized treatment planning system for ^{125}I stereotactic brain implants *Int. J. Rad. Oncol. Biol. Phys.* **18** 445–454

Welsh D and Chapman D 1990 A stereotactic technique for volumetric interstitial implantation in the brain *J. Neuroscience Nursing* **22** (4) 245–249

Wilson M W and Mountz J M 1989 A reference system for neuroanatomical localisation on functional reconstructed cerebral images *J. Comp. Assist. Tomog.* **13** (1) 174–178

Winston K R and Lutz W 1988 Linear accelerator as a neurosurgical tool for stereotactic radiosurgery *Neurosurgery* **22** (3) 454–464

Wu A, Lindner G, Maitz A H, Kalend A M, Lunsford L D, Flickinger J C and Bloomer W D 1990 Physics of gamma knife approach on convergent beams in stereotactic radiosurgery *Int. J. Rad. Oncol. Biol. Phys.* **18** 941–949

Wu A, Maitz A H and Kalend A M 1991 The role of the gamma knife in radiosurgery-physicist's perspective *Proc. 1st biennial ESTRO meeting on physics in clinical radiotherapy (Budapest, 1991)* p 30

Zhang J, Levesque M F, Wilson C L, Harper R M, Engel J, Lufkin R and Behnke E J 1990 Multimodality imaging of brain structures for stereotactic surgery *Radiology* **175** 435–441

CHAPTER 4

THE PHYSICS OF PROTON RADIOTHERAPY

4.1. INTRODUCTION: ELEMENTARY PHYSICS OF PROTON BEAMS

All radiotherapy aims to achieve a high therapeutic ratio, defined as the ratio of the probability of tumour control to the probability of normal tissue complication. Ways to do this with x-rays rely on arranging that the dose to organs-at-risk is significantly lower than the dose to the target volume by suitable multiport, rotational, dynamic or other techniques. These include those involving shaping the geometric field to the beam's-eye-view of the target, whose ultimate aim is dose conformation to the target volume. However, all radiotherapy with x-rays is frustrated by the physics of the interaction of radiation with matter. This determines that when a single beam irradiates a target volume, the dose on the proximal side of the target volume is always higher than that in the target volume and the dose on the distal side, although lower, is certainly not zero. For this reason, even combinations of beams do not always give acceptable dose distributions.

Radiotherapy with charged particles is fundamentally different and electrons, protons, heavy negative ions and helium ions have all been used. When a proton beam interacts with tissue, most of its energy is lost by collision with atomic orbital electrons. Since the proton is some 1835 times more massive than the electron, it does not significantly deviate from a straight-line path as it interacts, slows down and eventually stops. The proton deposits energy with an energy loss inversely proportional to the square of its speed. The typical distribution of dose with depth (figure 4.1) exhibits a strong Bragg peak towards the very end of the range, beyond which the dose very rapidly falls to zero and on the proximal side of which, the dose is only some 20% of the peak dose. Figure 4.2 emphasizes inherent differences between photon and proton irradiation. The pattern of dose deposition was first observed by Bragg and Kleeman (1904). In principle this is an ideal arrangement for radiotherapy if a target can be arranged to coincide with the Bragg peak. To give a quantitative feel for this, the range of 160 MeV protons in tissue is about 16 cm and the width of the Bragg peak at the 80% level is only

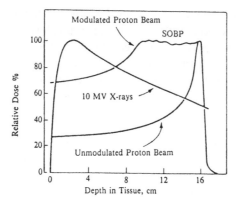

Figure 4.1. *Depth-dose curve for an unmodulated and a modulated 160 MeV proton beam, compared with that for a 10 MV x-ray beam. The x-ray beam falls exponentially after the initial build-up, whereas the maximum in the proton curve rises slowly to the Bragg peak where the protons stop. If the proton beam is modulated, the Bragg peak spreads out but the dose on the proximal side of the Bragg plateau is then larger than for monoenergetic protons. The advantage of the proton beam is the complete absence of dose beyond the Bragg peak and the avoidance of surface dose. (From Verhey and Munzenrider (1982).) (Reproduced, with permission, from the Annual Review of Biophysics and Bioengineering; Vol 11 ©1982 by Annual Reviews Inc.)*

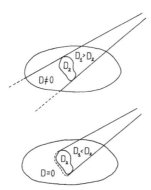

Figure 4.2. *Illustrates the fundamental advantage of proton therapy. In the upper figure a target is irradiated to dose D_2 by photons. The dose proximal to the target is higher than the target dose and the distal dose is non-zero. In the lower figure the same target is irradiated to dose D_2 by protons. The dose proximal to the target is now lower than the target dose and the distal dose is absolutely zero when the beam is shaped to the distal surface of the target volume. Combinations of photon beams can improve on the upper situation of course, but the single proton beam possesses this fundamental advantage.*

7 mm (Verhey and Munzenrider 1982).

About 1–2% of the energy loss arises from collisions with the atomic nuclei during which a much larger fraction of the proton's energy is transferred to the medium. This causes a high linear-energy-transfer (LET) and in turn a higher radiobiological effectiveness (RBE) than the electron-collision component which is identical to that of x-rays (RBE=1). A commonly adopted value for the RBE of a proton beam is about 1.1, determined from clinical studies with fractionation. This value is however not certain and Raju (1980) provides a large table of the biological results to that date. Values ranging between 0.6 and 1.4 have been found. A commonly used unit is the 'Cobalt Gray Equivalent (CGE)', being the dose in proton gray multiplied by 1.1, which gives the dose in cobalt gray (Cole 1990). Because the RBE for protons is similar to that of x-rays, 'conventional' x-ray clinical experience can be translated into proton-therapy management. So protons present a challenging opportunity to improve conformal radiotherapy because of essentially the better dose distributions which can be obtained (Kantor *et al* 1985) and the slightly enhanced RBE. Because dose distributions can be more tightly conformed to the target volume, dose escalation can be practised. At the Harvard Cyclotron Laboratory radiation doses of 70 CGE are regularly prescribed. Koehler and Preston (1972) show many examples of comparative dose distributions with x-rays and protons showing the advantages of the latter. The advantages of proton irradiation over photons for stereotactic radiosurgery are discussed in chapter 3 (section 3.2.4).

The stopping power of particles depends on the square of the ratio of their charge to their velocity, so, once the stopping power of protons is established, the stopping power of other particles follows from this law. For example, at the same velocity, the values for helium ions (doubly charged alpha particles) are four times larger but the values are the same for protons and deuterons, which have the same charge.

The range of protons in tissue depends on their energy (see Janni 1982, Bichsel 1972). A convenient formula to use is

$$R = 1.11 \times 10^{-1} \, (E/9.29)^{1.8} \tag{4.1}$$

where R is the proton range in tissue in cm and E is the proton energy in MeV (Wilson 1946). Some specific values are in table 4.1 and figure 4.3 shows range in water as a function of energy. The range of two different elementary particles of the same velocity but different charge and mass is proportional to the mass and inversely proportional to the square of the charge (Wilson 1946, Raju 1980). Hence figure 4.3 can be used for deuterons if the energy axis and the range axis are scaled by a factor of two. For example, a 200 MeV deuteron has a 16 cm range. Figure 4.3 can be used for alpha particles if the energy axis is scaled by a factor of four and the range axis stays the same. For example, a 400 MeV alpha particle has a range of 8 cm.

Low-energy protons can be used for sites such as the treatment of occular melanoma. High-energy protons are needed for deep sites. The Bragg peak of monoenergetic protons is too narrow to be of use unmodified. Practical proton beams for radiotherapy have range modifiers in the beam-line which stretch out the Bragg peak. These are usually rotating propellers or paddles of variable thickness,

Table 4.1. *The variation of proton range in muscle with proton energy (from Bonnett (1991)).*

Energy (MeV)	Range in ICRU Muscle (cm)
50	2.2
100	7.7
150	15.8
200	26.0
250	38.0

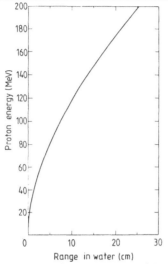

Figure 4.3. *The variation with proton energy of the proton range in water (from Raju (1980)).*

made of plastic, which develop a spectral beam (see section 4.3.1). The so-called 'Spread Out Bragg Peak' (SOBP) can be arranged to straddle a width appropriate to the target volume by constructing the appropriate beam modifier. Unfortunately, spreading the Bragg peak also causes the dose on the proximal side of the target volume to be significantly increased (see figure 4.1). In order to achieve a high ratio between the dose to the target volume and the surrounding tissues, multiple-proton fields can be utilized.

The large mass of the proton relative to the electron leads to a smaller range straggling ($\simeq 1\%$) than for the latter, a penumbra which is comparable with that of x-radiation and much tighter than for electrons. Typically the 20% and 80% isodose contours on the lateral edge are only 6 mm apart at 160 MeV. Penumbra depends on the proton collimator and increases with depth. Coutrakon *et al* (1991a) have measured the penumbra (90%–10%) for the Loma Linda proton facility as 6 mm at a depth of 3 cm in water and 15 mm at a depth of 25 cm. Electron radiotherapy does not enjoy these advantages. Heavy-ion and negative-pion therapy have a different biological effect and are not considered here.

4.1.1. Protons for clinical problems

Clinically, proton-beam radiotherapy is worth considering when the organs at risk are extremely proximal to the target volume, circumstances when x-radiotherapy is problematic because it is difficult to arrange to irradiate the target volume without also irradiating the organs at risk. With x-radiation, even if the primary beams do not irradiate the organs at risk, there is dose from scattered radiation. Examples of such circumstances are choroidal melanomas and sarcomas abutting critical central nervous structures.

For these problems, proton therapy has been considered justified without the need to perform randomized trials comparing it with x-radiotherapy. Proton therapy can treat occular melanoma without the need for enucleation with disfigurement. Proton therapy is a strong candidate for treating chordoma, soft-tissue sarcoma of the retroperitoneum, paranasal tissues and head and neck, glioblastoma multiforme and meningioma (Cole 1990). Experience at the Massachusetts General Hospital and Harvard Cyclotron Laboratory at these clinical sites has led Suit *et al* (1990) to write 'Our assessment.....with proton-beam therapy over the period of 1974 to the present (1990) is that for the anatomic sites tested there have been increases in the frequency of tumour control, lower frequencies of treatment-related morbidity, and no increases in marginal failures.' Since proton therapy is very expensive (and has its critics) such remarks are valuable and the paper by Suit *et al* (1990) documents clinical experience in detail.

For many other clinical problems it is relatively easy to show that better dose distributions can be obtained by proton therapy compared with x-radiotherapy (figure 4.4) but the clinical benefit in terms of improved therapeutic ratio is less obvious. In these cases it is necessary to arrange clinical randomized trials. Since the number of proton facilities worldwide is quite small, the necessary clinical data take time to accumulate (Antoine 1987). It is nearly 40 years since the first treatment with protons and the worldwide number of patients treated stood at (only) 10 358 in Jan 1992. Indeed Goitein and Munzenrider (1984) argue that such trials would have to be multicentric and international and various efforts are underway to formulate a framework for such collaboration (Antoine 1991). In particular, the PTCOG (Proton Therapy Cooperative Group) (Gottschalk 1987) meets regularly. In a sense there is a chicken-and-egg situation where more proton facilities are needed just to do the trials to determine whether proton irradiation is worthwhile.

Raju (1980) gives three reasons why he feels that proton therapy has not really 'taken off' clinically:

1. proton therapy was first mooted in the 1940's just as megavoltage photon therapy was being considered for the first time,

2. high-energy nuclear physics was so well funded in the 1950's and 1960's that accelerators were in full-time use for nuclear physics and beam time could not be spared for radiotherapy,

3. poor experience with neutron therapy clouded the issues.

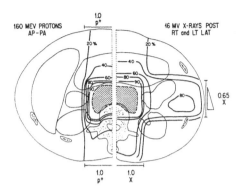

Figure 4.4. *Comparative dose distributions for parallel-opposed proton portals (left) and for a three-field 16 MV x-ray plan with lateral weighting and wedging (right). The target volume is the hatched region. Notice how the isodose contours relate to the location of the spinal chord. The proton plan produces a distribution which irradiates normal structures much less than for the x-ray plan. (From Verhey and Munzenrider (1982).) (Reproduced, with permission, from the Annual Review of Biophysics and Bioengineering; Vol 11 ©1982 by Annual Reviews Inc.)*

Planning for proton therapy has many features similar to planning external-beam x-radiotherapy. It must be image-driven with excellent delineation of target volumes and organs at risk. Generally, CT images provide the necessary data. Interestingly, because some facilities use stationary beam-lines in the horizontal position and the patient must be correspondingly seated or standing for treatment, two special-purpose CT scanners have been constructed which rotate in the horizontal plane. These have been installed at the Harvard Cyclotron Centre/Massachusetts General Hospital (HCC–MGH) and at the Lawrence Berkeley Laboratory (LBL) (Goitein *et al* 1982, Chen 1983). That the early proton facilities have static beam-lines is reminiscent of a similar situation in the early days of photon linear accelerators when the accelerator guide was large and vacuum systems were a further limitation (Sharma *et al* 1991).

Some differences in proton therapy planning include the need to select optimum entrance ports and to tailor beam-defining collimators from bronze blanks or Cerrobend alloy. The patient is bolussed with material designed to create a compensator which accounts for the different paths of the beam through the patient, since tissue inhomogeneities, having electron densities different from water, will modify the proton range. It has to be arranged that the combined path length of bolus and different tissues between the beam port and the proximal edge of the target volume is the minimum proton range in the spectral beam. Compensators may be fabricated from lucite (Munzenrider *et al* 1985) (see section 4.4.2). Similar considerations apply for radiotherapy with other heavy ions (Chen *et al* 1979).

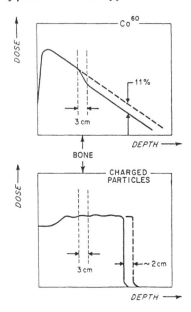

Figure 4.5.   *The effect of a 3 cm slab of bone on the depth-dose distribution of a ^{60}Co beam compared with a proton beam. In the former the intensity of the radiation is reduced; in the latter, the intensity is substantially unchanged but the range is reduced. (From Goitein (1982).)*

Tissue inhomogeneities present a different problem for proton therapy than for x-ray therapy. In the latter it is the *intensity* of x-rays which is modulated by the presence of an inhomogeneity; in the former the intensity is virtually unaltered but the *range* changes. It is vital to account for this, far more so than for x-ray therapy where combining beams can lead to tissue inhomogeneity effects being no more than a few percent. For proton therapy the presence of a range-shortening tissue inhomogeneity would reduce the target dose from the value in the stretched Bragg peak to essentially zero (figure 4.5), which would be extremely serious (Goitein 1982). Goitein (1982) identified three phenomena caused by tissue inhomogeneities: range modification, edge scattering (if the inhomogeneity does not fill the beam; see section 4.4.3) and dose modification from thin slivers. It turns out that if slivers are small, multiple scattering wipes out their effect, which is fortunate because otherwise impossibly accurate CT data would be required for treatment planning. The range problem was recognized before the advent of commercial CT and the group at LBL considered reconstructing from proton projections to get the necessary CT images (Goitein 1972).

Since proton therapy is indicated when the highest conformation is needed, precise patient positioning is required and patients are generally immobilized comfortably in face masks (for head therapy) or shells (for body therapy). Goitein *et al* (1982) report the amazing accuracy of 0.1 mm of the Lawrence Berkeley immobilization devices. HCC–MGH assessed their mean error to be some 0.6

to 0.9 mm. Historically, stereotactic devices have been used (see chapter 3) for proton neurosurgery (Larsson *et al* 1958).

Proton therapy is one of the techniques which, if more widely available, will increase precision in radiation therapy, along with advances in three-dimensional planning, on-line portal imaging, multileaf collimation and dynamic photon therapy. Suit and Verhey (1988) have placed these new developments in context with older developments (such as supervoltage radiation, portal radiographs, simulators, isocentric gantries, secondary collimation) which over the last 50 years have transformed radiotherapy. They write, prophetically, 'It is to be hoped that, by the year 2000, most of the treatment methods of 1987 will be judged as obsolete'.

4.2. PROTON-THERAPY FACILITIES

As early as just after the second world war, Wilson (1946) wrote a historic paper fortelling the use of protons in radiotherapy. It is instructive to reproduce a calculation (in modern units) which he made to evaluate whether the proton beam current would be large enough for medical purposes. In the last centimetre of range a proton loses 30.1 MeV of energy in tissue (water) (put another way, the range of a 30.1 MeV proton is 1 cm). A flux of F protons per cm^2 deposits $F \times (30.1 \times 10^6)(1.6 \times 10^{-19})$ Joules (J) in 1 cm^3. Assuming 1 cm of tissue has a mass of 1 g, this energy deposition is $F \times (4.8 \times 10^{-9})$ J kg^{-1}. So to create a dose of 10 Gy (10 J kg^{-1}) requires $10^{10}/4.8$ protons. Since each proton carries a charge 1.6×10^{-19} Coulombs, a 1 s exposure requires a proton beam current of 3.33×10^{-10} A. Wilson noted that the machines under construction could easily achieve this and persevered with advocating proton radiotherapy.

Worldwide, the number of clinical proton therapy centres is not large (14 centres in eight countries) (Chen and Goitein 1983, Slater *et al* 1988, Bonnett 1991, Sisterson 1990, 1991, 1992). The centres (see table 4.2) are at the Harvard Cyclotron Centre–Massachusetts General Hospital (HCC–MGH) (where by far the largest number of patients in the world have been treated), at the Lawrence Berkeley Laboratory–University of California at San Fransisco (LBL–UCSF), at the Loma Linda Medical Centre, at the Gustaf Verner Institute in Uppsala, and at three centres in the (former) USSR, one in Switzerland, one in the UK, two in France, one in Belgium and two centres in Japan.

The first proposal for the medical use of protons was made by Robert Wilson, the founder of Fermilab, just after the second world war (Wilson 1946). He proposed utilizing the accelerators which were being developed for high-energy physics research. The first patients were treated in 1955 at the Lawrence Berkeley Laboratory (Tobias *et al* 1956, 1964), soon followed in Uppsala, Sweden in 1957 (Larsson *et al* 1963, 1974). The HCC–MGH began treating patients in 1961, although large-field irradiation did not begin until 1974 (Suit *et al* 1975). Proton therapy began in Dubna, (former) USSR in 1964, Moscow, (former) USSR in

Table 4.2. *Proton particle facilities for medical radiation therapy (condensed from Slater et al (1988), Sisterson (1991, 1992) and from Graffman (1985)).*

Installation	Location and Country	Accelerator type	First date of therapy use	Energy
Lawrence Berkeley Laboratory, University of California	Berkeley, CA, USA	Synchrocyclotron	1955	910 MeV
Gustav Werner Institute	Uppsala, Sweden	Synchrocyclotron	1957/1988	185 MeV, 100 MeV
Harvard Cyclotron Laboratory	Cambridge, MA, USA	Synchrocyclotron	1961	160 MeV
Joint Institute for Nuclear Research	Dubna, (former) USSR	Synchrocyclotron	1964/1987	70–200 MeV
Institute of Theoretical and Experimental Physics	Moscow, (former) USSR	Synchrotron	1969	70–200 MeV
Konstantinov Institute of Nuclear Physics	Gatchina, Leningrad, (former) USSR	Synchrocyclotron	1975	1 GeV
National Institute of Radiological Sciences	Chiba, Japan	Sector-focused cyclotron	1979	70 MeV
Particle Radiation Medical Centre	Tsukuba, Japan	Synchrotron	1983	250 MeV
Swiss Institute for Nuclear Physics (Paul Scherrer Institute)	Villigen, Switzerland	Synchrotron	1984	70 MeV
Clatterbridge Hospital	Clatterbridge, UK	Cyclotron	1989	62 MeV
Loma Linda University, California	Loma Linda, USA	Synchrotron	1991	70–250 MeV
Université Catholique de Louvain	Louvain la Neuve, Belgium	Isochronous Cyclotron	1991	90 MeV
Centre Antoine Lacassagne	Nice, France	Cyclotron	1991	65 MeV
Centre de Protonthérapie	Orsay, France	Synchrocyclotron	1991	200 MeV

1969 and at Gatchina, (former) USSR in 1975 (Ambrosimov *et al* 1985). In Japan proton therapy began at Chiba in 1979 and in Tsukuba in 1983 (Tsunemoto *et al* 1985, Kitigawa 1988, Tsunemoto 1988). Swiss experience commenced in 1984 (Blattman 1986). A facility opened at the Clatterbridge Hospital, UK in 1989 (Bonnett 1991), at Loma Linda University in 1991 (Slater *et al* 1990), at Nice in 1991 (Sisterson 1992), at Orsay in 1991 (Sisterson 1992), and at Louvain la Neuve in 1991 (Sisterson 1991). The worldwide clinical experience is summarized in table 4.3. The main source of these data are the twice-yearly reports issued by Sisterson at the HCC–MGH called the 'Particles' newsletter. Some starting dates should not be interpreted too strictly as some of the above papers suggest slightly different dates.

Proton radiotherapy began at Uppsala in 1957 using a 185 MeV synchrocyclotron. Protons of this energy have a range in tissue of 23 cm. Only a very small number of patients were treated, generally those with advanced cancer or where the tumour was very close to sensitive structures. The programme stopped in 1968. The pencil beam was swept in a reticular pattern by two variable crossed magnetic fields giving a transverse homogeneity of ± 5% over a 20 by 20 cm field (Graffman et al 1985). The Bragg peak was spread out by first a variable water absorber, replaced by ridge filters which introduced different thicknesses of absorbing material into the beam. Beams of cross section 0.1 to 20 cm in diameter were achieved with dose rates (depending on field size) between 1 and 100 Gy min^{-1}. The Uppsala team worked closely with all three proton accelerator sites in the (former) USSR and shared experience. In 1985 plans were well advanced for a new cyclotron facility in Uppsala with a variable energy accelerator and an isocentric gantry and the first patients were treated in 1988 (see table 4.3). Clinical experience has been reported by Montelius *et al* (1991).

HCC–MGH has a 160 MeV synchrocyclotron with two horizontal beamlines, one of 5 cm diameter with a doserate of 8–10 Gy min^{-1} and the other of 30 cm diameter with a dose rate of 0.8–1.5 Gy min^{-1}. The machine was constructed between 1946–1949 (Blosser 1989) for physics research and started medical therapy in 1961. The patient is arranged on a couch with precisely controllable movements with a precision of ±1 mm. The patient is either sitting, standing or decubitus. The treatment room is equipped with three orthogonally arranged diagnostic x-ray sets for positioning. Figure 4.6 shows the geometry of the broad-field proton beam. A double-scattering system with an annular occluding plate can give a uniform dose across a 30 cm field width (figure 4.7) (see also section 4.3.2). The range modulator wheel (see section 4.3.1) is constructed of lucite, different wheels being used to spread out the peak from 0.5 cm to 16.3 cm at the 90% isodose level (Sisterson and Johnson 1985, 1986 and 1987, Sisterson 1990).

The proton facility at the Institute for Theoretical and Experimental Physics (ITEP) in Moscow is equipped with a 10 GeV proton synchrotron from which protons can be extracted with energies between 70 and 200 MeV for medical purposes. Only one in sixteen pulses are extracted by a beam-kicker, so medical work can go on without interrupting high-energy physics activity. The accelerator

Table 4.3.   *Clinical experience to date with proton therapy facilities (table condensed from Slater et al (1988), Bonnett (1991) and Sisterson (1990, 1991, 1992)). This table is updated every six months in 'Particles', newsletter of PTCOG (January and June). This table is as of January 1992.*

Installation	Location and Country	Patients treated	Year started
Lawrence Berkeley Laboratory University of California	Berkeley, CA, USA	30 by 1957	1955
Gustav Werner Institute (2nd installation)	Uppsala, Sweden	73 by 1976 23 by 1991	1957 1988
Harvard Cyclotron Laboratory	Cambridge, MA, USA	> 3000 by 1986 4841 by 1990 5419 by 1991	1961
Joint Institute for Nuclear Research (2nd installation)	Dubna, (former) USSR	84 by 1980 13 by 1991	1964 1987
Institute of Theoretical and Experimental Physics	Moscow, (former) USSR	1250 by 1986 1993 by 1990 2200 by 1991	1969
Konstantinov Institute of Nuclear Physics	Gatchina, Leningrad (former) USSR	381 by 1985 719 by 1991	1975
National Institute of Radiological Sciences	Chiba, Japan	29 by 1984 74 by 1991	1979
Particle Radiation Medical Centre	Tsukuba, Japan	158 by 1990 242 by 1991	1983
Swiss Institute for Nuclear Physics	Villigen, Switzerland	121 by 1986 719 by 1990 1150 by 1991	1984
Clatterbridge Hospital	Clatterbridge, UK	83 by 1990 189 by 1991	1989
Loma Linda University, California	Loma Linda, USA	76 by 1991	1991
Université Catholique de Louvain	Louvain la Neuve, Belgium	8 by 1991	1991
Centre Antoine Lacassagne	Nice, France	45 by 1992	1991
Centre de Protonthérapy	Orsay, France	13 by 1992	1991
Total proton beams worldwide:		10 358 by end 1991	

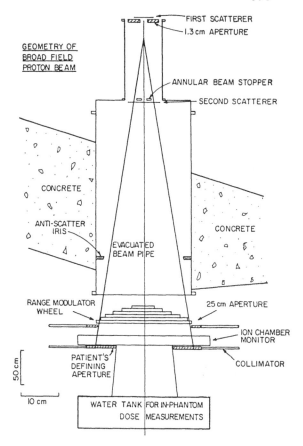

Figure 4.6. *Schematic of the spread-out proton beam, designed by the Harvard Cyclotron Laboratory, showing the location of the scattering foils in the passive scattering system, lucite range-modulator wheel and the patient's beam-defining aperture. The individual patient compensator (bolus) would be placed after this aperture. The beam is monitored by an ion chamber. The scale of this diagram is distorted for convenient display (see key bottom left). (From Sisterson and Johnson (1986).)*

feeds three treatment rooms equiped with stereotactic localization apparatus for neurological procedures. Lasers and x-ray based positioning equipment is also in use. The beam energies are degraded and large fields produced by scattering in thin foils (see section 4.3.2). Dose planning would appear to be by hand with a 'dose field catalogue' created by dosimetric instruments, claimed to be accurate to 7–9% (Chuvilo *et al* 1984, Khoroshkov *et al* 1969).

The proton facility at the National Institute of Radiological Sciences, Chiba, Japan has a cyclotron-generated 70 MeV beam only suitable for superficial radiotherapy, penetrating 3.8 cm in water. The Bragg peak is only 1.3 mm wide and is stretched by a rotating plastic degrader to 2.5 cm. The proton beam is

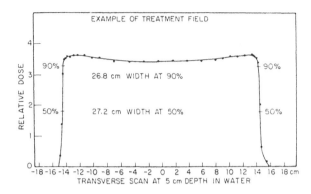

Figure 4.7. *A transverse scan in a water phantom for the 160 MeV proton beam at the Harvard Cyclotron Laboratory showing the uniformity of the beam and the sharp lateral fall-off of the dose. (From Sisterson and Johnson (1986).)*

defocused electromagnetically and collimated mechanically into a 1 cm^2 field. This is swept horizontally and vertically with two perpendicular swing-magnets to create a uniform field up to 20 by 20 cm^2.

Proton therapy is practised in the UK at the Clatterbridge Hospital (Bonnett 1991) where, being a 'low'-energy facility it is used exclusively for treating occular melanoma. There are plans to introduce an 'add-on' linear accelerator to boost the energy to 180 MeV, due to be completed in 1994 (Sisterson 1992).

A proton therapy facility has just been completed at Loma Linda University (Antoine 1987, Awschalom *et al* 1987, Miller *et al* 1987, Slater *et al* 1988, 1990, 1992) having a 70–250 MeV synchrotron directed to one stationary beam-line and three rotation gantries. Each gantry can rotate the beam through a full revolution about a patient on a couch as in x-radiation therapy. The synchrotron is the world's smallest, built by Fermi National Accelerator Laboratory; it is a ring of magnets some 20 feet in diameter through which protons circulate in a vacuum tube. The energy of the protons increases as the magnetic field is increased reaching 250 MeV in $\frac{1}{2}$ s. The beam is steered by giant magnets to gantries weighing 95 tons and standing three stories high. These gantries are 35 feet in diameter supporting the bending magnets and focusing magnets. All the patient sees is a small tunnel through the facility. The details of the accelerator construction and beam transport systems have been provided by Slater *et al* (1988). The horizontal beam-line can also be operated at 155 MeV and 200 MeV and there are plans to produce a continuously variable energy beam in steps of 5 MeV (Sisterson 1992). This is a major investment in technology; as well as the proton facility, all ancillary imaging equipment (CT, PET and SPECT) and no less than eight treatment planning workstations are installed (Potts *et al* 1988). The Loma Linda facility achieves 2.5×10^{10} protons per spill, a repetition rate of 30 spills min^{-1} corresponding to 50 cGy min^{-1} for a 20 cm diameter area and 9 cm range modulation (Coutrakon *et al* 1991a). The first occular treatment was on October 23rd 1990 (Slater *et al*

1992). Treating brain tumours began in March 1991 and treatments began with the rotating gantry beam-line on June 26th 1991. 76 patients had been treated by the end of December 1991 (Sisterson 1992).

The history of the Loma Linda facility has been published by Slater *et al* (1992) including an account of how the Proton Therapy Cooperative Group (PTCOG) formed. Loma Linda is unique in that the building and accelerator were planned together for the efficient use of both. The structure of the building even accommodated for the relatively high seismic activity in Southern California.

The latest proton therapy installations include Louvain la Neuve in Belgium (Sisterson 1991). It has a proton energy of 90 MeV with a maximum water penetration of 5.5 cm. The first treatment was in January 1991 and eight patients had been treated by September 1991 at a dose rate of 1–2 Gy min^{-1}. The proton facility in Nice has a 65 MeV cyclotron-generated beam. Treatments started on June 17th 1991. The second French facility at Orsay has a synchrocyclotron and started treating patients on September 17th 1991 (Sisterson 1992).

Most proton accelerators used for radiotherapy are cyclotrons or synchrotrons (Beeckman *et al* 1991, Gottschalk 1987). Variations for cyclotrons include the synchrocyclotron and the isochronous cyclotron with magnets at room temperature or superconducting (Blosser 1989). The synchrocyclotron is a particularly reliable machine; Sisterson *et al* (1991) have quantified the relatively small 'downtime' at the HCC machine. A room-temperature cyclotron is too massive to rotate and to provide a rotating beam, some beam-transport mechanics are needed, as in the design by Beeckmann *et al* (1991). Recognizing that the ability to rotate the proton beam about a stationary patient is highly desirable, Blosser (1989) has designed a superconducting isochronous cyclotron which could be rotated about the patient as shown in figure 4.8. The magnets in superconducting cyclotrons are some 17 times less heavy than their room-temperature counterparts for the same bending power, allowing the cyclotron to itself rotate about the patient. This removes the need for the elaborate beam transport needed for the rotating gantry for room-temperature cyclotrons or for the proton synchrotron as at Loma Linda.

Recently Lennox (1991) has proposed that a single proton linear accelerator could be used with the appropriate beam-switching techniques to achieve the following four purposes at the same clinical site: (i) proton therapy, (ii) fast-neutron therapy (iii) boron-capture neutron therapy (using epithermal—1 keV—neutrons from 70 MeV protons striking a lead target surrounded by an iron moderator) and (iv) isotope production for PET imaging.

Features of the other existing proton facilities are discussed elsewhere in this chapter (section 4.3). There are many proposals for future proton therapy installations. These have been summarized by Bonnett (1991) and Sisterson (1992) and include the British proposal by AEA Technology (Oxford UK) for a linac-based facility delivering protons of energies 100, 150, 200 and 250 MeV. Sisterson (1992) has recently reported pre-clinical measurements at the Indiana University Cyclotron Facility proton beam-line (185–200 MeV).

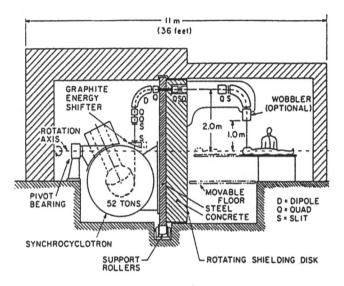

Figure 4.8. *Showing a likely layout for a one-treatment-room proton therapy facility. The cyclotron and beam-transport system rotate as a unit so that intricate out-of-plane bending of the beam is not required. The figure shows an optional wobbler magnet, although moving the patient relative to the beam which has been focused into a perpendicular vertical line is more attractive. The isochronous cyclotron has a superconducting magnet which reduces weight compared with its room temperature counterpart and allows the rotation. (From Blosser (1989).)*

4.3.   RANGE MODULATION AND PRODUCTION OF LARGE-AREA BEAMS

The spreading out of the Bragg peak or range modulation can be achieved in a number of ways (Bonnett 1991, Graffman *et al* 1985, Goldin and Monastyrsky 1978, Koehler *et al* 1975). The Bragg peak is spread out by introducing extra absorbing material before the beam enters the patient. If different thicknesses of such absorber are present for different fractions of the irradiation time, the narrow monoenergetic peak can be spread into a useful plateau. This allows the high-dose deposition region to straddle the target volume containing the tumour. Ways to do this are described in section 4.3.1.

As well as spreading the peak, the final range itself must be shaped to the distal surface of the target volume. Bonnett (1991) sums up ways to do this:

1. Fixed range modulation (see section 4.4.2): a compensator is constructed to shape the distal surface of the target volume to the maximum proton range.

2. Variable range modulation and spot scanning: these techniques (Chu *et al* 1989) to spare normal tissue proximal to the target volume are under development.

The production of a large-area flat beam can be achieved by (Chu *et al* 1989):

1. Scattering from foils (see section 4.3.2): this was the first way to be used clinically (Koehler *et al* 1977) and is still the main method in use (see figure 4.6). The main disadvantage is the energy loss in scattering and the reduction of beam output intensity. The penumbra is also broadened.

2. Pixel or spot scanning (see section 4.3.3): the target area is divided into pixels, scanned separately. The scanning method does not degrade beam energy nor output. The sharp penumbra is retained and large field sizes are attainable.

4.3.1. Spreading out the Bragg peak

The Bragg peak can be spread out to a useful plateau by the use of a rotating stepped absorber. The idea was first suggested by Wilson (1946) and Koehler *et al* (1975) provide details of the implementation at the HCC–MGH proton facility. The idea is elegantly simple. It relies on time modulation, presenting different thicknesses of absorber to the beam for different fractions of the irradiation time (figure 4.9). A range modulator wheel (figure 4.10) is constructed from a pyramid of blades each in the shape of a fan, of different thicknesses and different fan angles. The thinnest part of the stack has the widest fan angle. The modulator rotates at 90 RPM in the beam-line (figure 4.11).

The thickest part of the modulator pulls back the Bragg peak furthest. A computer program determines the amplitudes of elemental Bragg curves with

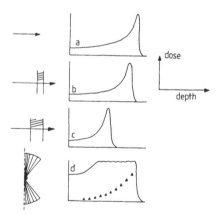

Figure 4.9. *Shows the principle of range modulation to produce the spread-out Bragg peak (SOBP). Curve (a) shows the depth dose of an unmodulated proton beam. Curves (b) and (c) show the effect of introducing absorber upstream of the patient. As the thickness of the absorber increases, the Bragg peak becomes more proximal to the surface. Curve (d) is an SOBP created by the range-modulator wheel (shown end on schematically). The small triangles indicate the fraction of the time for which each thickness of absorber should be in the beam. The triangles are at the depths of the elemental proton ranges. This in turn determines the differential fan angles for the rotating wheel.*

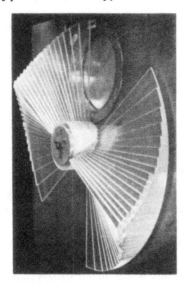

Figure 4.10. *The 14.5 cm-range, rotating range-modulator wheel of the HCC–MGH proton facility. The wheel is constructed from a pyramid of plexiglass fans of different opening angle. The wheel is 82 cm in diameter and the circular beam-defining aperture visible through the wheel is 27 cm in diameter. (From Koehler et al (1975).)*

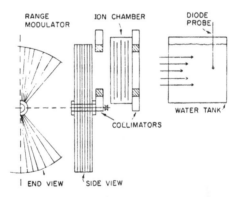

Figure 4.11. *Layout of the proton-beam treatment area at the HCC–MGH proton facility showing (left) the range-modulator wheel mounted on to the collimator. An arrangement is shown on the right for measuring the depth-dose curve. (From Koehler et al (1975).)*

different ranges which, when added, would give the required plateau. These amplitudes are then converted to a set of differential fan angles, normalized so the integral fan angle (the sum of the differentials) is 180°. The differential fan angle is that part of 180° for which a particular thickness of modulator is in the beam. From this set of differential fan angles the integral opening angle or fan angle of that blade corresponding to the ith Bragg sub-peak is then the sum of the integral fan angle for the $(i + 1)$th Bragg sub-peak and the differential fan angle for the ith Bragg sub-peak (the sub-peaks are labelled so that '0' labels the unmodulated peak and the label increases in the direction of decreasing range or increasing thickness of modulator).

The modulator is normalized to 180° rather than 360° so that each blade appears twice, diagonally opposed. This ensures dynamic equilibrium as the modulator rotates. As can be seen, part of the time the beam passes unmodulated. The HCC–MGH wheels (of which there were three for different plateaux) were constructed from plexiglass, a low-Z, little-multiple-scattering material. The wheels were 82 cm in diameter.

Measurements on the Loma Linda proton facility showed the range of the unmodulated proton beam was 32.7 cm which, taking account of material upstream of the measuring water tank, corresponded to a beam energy of 235 MeV. When 5 mm of lead were upstream, the range shifted to 29.5 cm. In both cases the peak-to-entrance dose was approximately 3:1 and the distal edge fall off from 90% to 10% was 7 mm. A propeller, rotating at 310 RPM, to spread out the Bragg peak was designed by the Harvard Cyclotron Medical Facility for Loma Linda, comprising nine sections of lucite, each 8 mm thick reducing the range by 1 cm of water. This created an SOBP of width 9 cm with uniformity ±5%. The distal dose fall-off was then 8 mm with a distal edge at a depth of 27 cm in water, relecting the range reduction introduced by the scatterers (see section 4.3.2). The peak-to-entrance dose ratio for the SOBP fell to 1.3:1 (Coutrakon *et al* 1991a). The Indiana University proton beam-line uses a similar mechanism to produce an SOBP with a 1.3% uniformity (Sisterson 1992).

The 'propeller' method of creating an SOBP is applicable to a beam which has been spread out by a double-scattering-plus-annulus arrangement. It *could* be used with a small-area beam but generally is not. When a small-area proton beam is used to create a large-area field by spot scanning, another method of spreading out the Bragg peak is needed. In some installations this is achieved by rapidly shooting the appropriate thickness of absorber into the beam under pneumatic control.

Another very elegant method which has been developed by the company Ion Beam Applications (IBA: Belgium) is to make use of a rotating hollow cylinder of diameter 20 cm with variable wall thickness. The beam is directed at right angles to the long axis of the cylinder and is much narrower than its diameter. The beam intersects the axis of rotation of the cylinder. The walls of the cylinder increase from some minimum thickness to a maximum thickness diametrically opposite, returning immediately to the minimum thickness and thereafter increasing again to the maximum, 180° further round. This cylinder rotates at some 1500–3000 RPM, presenting a variable thickness of absorber which is precisely known as a function of the angular orientation of the cylinder. This orientation is monitored by an optical encoder which can give signals to the beam control so that the beam is only switched on for specific chosen parts of each rotation. If the beam is on continuously then the dose in the patient will exhibit the maximum width of SOBP. If the beam is switched off for a selected portion of the rotation between when the cylinder presents some thickness of wall greater than the minimum up to when the maximum thickness is presented (corresponding to removing the contribution from selected lower energy protons), the width of the SOBP can be reduced. In this way the width of the SOBP can be controlled. The distal extremum of the SOBP is separately varied by a stationary absorber upstream towards the isochronous cyclotron (Jongen 1992a,b, Jongen *et al* 1992).

4.3.2.   *Theory of flattening proton dose distributions for large-field radiotherapy*

When a thin foil is placed in a narrow beam of protons the resulting projected distribution on the treatment, or on some measurement, plane a distance Z_1 from the foil is given by the radially symmetric function of radius r_1:

$$f\,(r_1) = \left(1/\pi\,R_1^2\right) \exp\left(-r_1^2/R_1^2\right) \tag{4.2}$$

where R_1 is the root-mean-square radius of multiple scattering, related to the root-mean-square scattering angle $\langle\theta_1^2\rangle$ by

$$R_1 = Z_1\langle\theta_1^2\rangle^{1/2}. \tag{4.3}$$

The multiplicative constant in equation (4.2) normalizes the distribution such that

$$\int f\,(r_1)\,\mathrm{d}r_1 = 1. \tag{4.4}$$

This is easy to prove if r_1^2 is written as x^2+y^2, the integral is split into two integrals over x and y, variables are changed to $t = x/R_1$ and $t = y/R_1$ and we recall that

$$\int_0^\infty \exp\left(-t^2\right)\,\mathrm{d}t = \sqrt{\pi}. \tag{4.5}$$

The proton distribution given by equation (4.2) is far too centrally peaked to be useful for wide-field irradiation and is too non-uniform. Figure 4.12 shows the Gaussian distribution. Koehler *et al* (1977) pioneered techniques for flattening the beam as follows. A second scatterer was placed downstream from the first along with a beam-stopper on the central axis. The purpose of this beam-stopper is to expose the second scattering foil to just part of the first-scattered beam and this creates the flat distribution, as will be seen in a moment. Photons which interact on the second scatterer at positions which project to r_1 on the measurement plane give rise to a distribution $f\,(r_2)$, where r_2 is the vector from the position r_1 (see figure 4.13). The final distribution as a function of radius r at the measurement plane is then

$$f\,(r) = f\,(r_1)\,f\,(r_2). \tag{4.6}$$

Vectorially

$$r = r_1 + r_2 \tag{4.7}$$

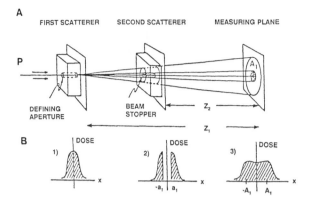

Figure 4.12.   *The geometry of double scattering with a beam-stopper in front of the second scatterer. (A) shows the defining aperture and the two scattering foils. The beam-stopper projects to a radius A_1 at the measurement plane. (B) shows the beam profiles (1) after the first scattering foil, (2) after the beam-stopper, (3) at the measurement plane (where the distribution has been flattened; see figure 4.7). (From Koehler et al (1977).)*

i.e.

$$r_2^2 = r^2 + r_1^2 - 2rr_1 \cos\theta \qquad (4.8)$$

and so combining equations (4.2), (4.6) and (4.7)

$$f(r, r_1) = \frac{2}{\pi^2 R_1^2 R_2^2} \exp\left(-\left(\frac{r_1^2}{R_1^2} + \frac{r^2 + r_1^2}{R_2^2}\right)\right) \int_0^\pi \exp\left(\frac{2rr_1 \cos\theta}{R_2^2}\right) d\theta. \qquad (4.9)$$

This is the probability of scattering into a radius r at the measurement plane Z_1 via a projected radius r_1 from the first foil, where r_1 are all the positions on the circumference of a circle with this radius. The integral over θ performs this task and the 2 on the numerator is because the integral should be over 2π but is symmetric so can be reduced to half the range. The parameter R_2 is given by (analogous to equation (4.3))

$$R_2 = Z_2 \langle \theta_2^2 \rangle^{1/2} \qquad (4.10)$$

where $\langle \theta_2^2 \rangle^{1/2}$ is the RMS scattering angle for the second foil. The two scattering angles depend on the material and thickness of the foil. Once these are known, R_1 and R_2 follow from the geometry.

Now make the substitutions

$$x = r_1 / R_2 \qquad (4.11)$$

$$k_1 = \left(1 + R_2^2 / R_1^2\right) \qquad (4.12)$$

and

$$k_2 = 2r / R_2 \qquad (4.13)$$

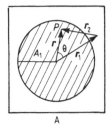 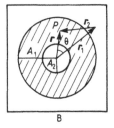

A   B

Figure 4.13. *The projection of scattered rays from the two scatterers on to the measurement plane at Z_1 from the first scatterer. The left diagram (A) shows a solid beam-stopper. Values of R_1 inside A_1 are forbidden but protons scatter from values $r_1 > A_1$ back inside A_1 to fill up the space. The right diagram (B) shows an annular beam-stopper. Here the disallowed values of r_1 are between A_1 and A_2. Small values of r_2 from r_1 fill in the shadowed region (see figure 4.12). (From Koehler et al (1977).)*

to obtain

$$f(r, x) = \frac{2}{\pi R_1^2 R_2^2} \exp\left(-\left(k_1 x^2 + \frac{k_2^2}{4}\right)\right)(1/\pi) \int_0^\pi \exp(k_2 x \cos\theta)\, d\theta.$$

(4.14)

But by definition of the Bessel function of an imaginary argument

$$I_0(Z) = (1/\pi) \int_0^\pi \exp(Z \cos\theta)\, d\theta$$

(4.15)

and so

$$f(r, x) = \frac{2}{\pi R_1^2 R_2^2} \exp\left(-\left(k_1 x^2 + \frac{k_2^2}{4}\right)\right) I_0(k_2 x).$$

(4.16)

To obtain the distribution over r alone we need to integrate over the whole *area* in which r_1 (i.e. in which x) can lie. So we need to integrate with respect to $x\, dx$ to perform the integration over the slim annular area bounded by circles of radius x and $x + dx$. Thus

$$f(r) = \frac{2}{\pi R_1^2 R_2^2} \exp\left(\frac{-k_2^2}{4}\right) \int \exp(-k_1 x^2) I_0(k_2 x)\, x\, dx.$$

(4.17)

This is the complete expression for the two-dimensional function $f(r)$. It is given in terms of just two parameters R_1 and R_2, which, as we have seen, depend on the scattering foil.

The key to understanding the importance of Koehler *et al*'s (1977) development is recognizing that the remaining free choices are the limits of the integral. These limits are in terms of x or (from equation (4.11)) r_1, the projected position of the first scatter at the measurement plane. If a radially symmetric beam-stopper is placed centrally in the beam just in front of the second scattering foil (figure 4.12)

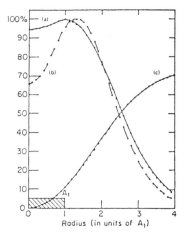

Figure 4.14. *A calculated scattered distribution of protons with a central beam-stopper of projected radius A_1 when 'good' parameters $R_1 = 1.7A_1$ and $R_2 = 1.3A_1$ were chosen (curve (a)). When 'poor' parameters were chosen, $R_1 = 1.7A_1$ and $R_2 = 1.0A_1$, the distribution is not flattened (curve (b)). Curve (c) shows the total dose within a given radius. The beam-stop is shown hatched bottom left. (From Koehler et al (1977).)*

this disallows a range of values for r_1 (and so for x). Two types of beam-stopper were proposed by Koehler *et al* (1977):

1. A single central stopper of radius A_1 (projected at the measurement plane) blocking the central portion of the beam (i.e. $r_1 > A_1$ only). The stopper was a brass block, tapered to the divergence of the beam and of thickness 3.7 cm which completely stops protons of 160 MeV (figures 4.12 and 4.13).

2. An annular stopper, open to the central portion but stopping the beam between r_1 values of A_1 (projected outer radius) and A_2 (projected inner radius). This was used for very large fields (figure 4.13).

Figure 4.14 shows a calculated scattered distribution of protons with a central beam-stopper of projected radius A_1 when 'good' parameters $R_1 = 1.7A_1$ and $R_2 = 1.3A_1$ were chosen. Notice the beam is above 95% out to about 1.5 A_1 (curve (a)). When 'poor' parameters were chosen, $R_1 = 1.7A_1$ and $R_2 = 1.0A_1$, the distribution is not flattened (curve (b)). Curve (c) shows the integrated dose within a given radius.

Figure 4.15 shows a calculated scattered distribution of protons with an annular beam stopper of projected radii A_1 and A_2 when 'good' parameters $R_1 = 1.7A_1$ and $R_2 = 0.8A_1$ were chosen and $A_2 = 0.38A_1$. When 'poor' parameters were chosen, $R_1 = 1.7A_1$ and $R_2 = 0.7A_1$ and $A_2 = 0.5A_1$, the distribution is not flattened (curve (b)). Curve (c) shows the integrated dose within a given radius.

The first blocking technique with a single stopper was used at HCC–MGH from May 1973 for intracranial targets. The second technique with the annular blocker was used from December 1973 for large-field irradiations. Koehler *et al* (1977)

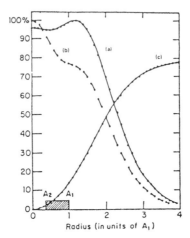

Figure 4.15. *Shows a calculated scattered distribution of protons with an annular beam-stopper of projected radii A_1 and A_2 when 'good' parameters $R_1 = 1.7A_1$ and $R_2 = 0.8A_1$ were chosen and $A_2 = 0.38A_1$ (curve (a)). When 'poor' parameters were chosen, $R_1 = 1.7A_1$ and $R_2 = 0.7A_1$ and $A_2 = 0.5A_1$, the distribution is not flattened (curve (b)). Curve (c) shows the total dose within a given radius. The beam-stop is shown hatched bottom left. (From Koehler et al (1977).)*

christened these systems 'proton nozzles' and they have proved very effective. The double-scattering-plus-annulus method was also used at the newest proton facility in Belgium (Sisterson 1991) giving field sizes up to 10 by 10 cm^2, at Tsukuba, Japan (Tsunemoto *et al* 1985) and at the Indiana University proton beam-line (Sisterson 1992).

A double-scattering technique similar to this has been installed at the Loma Linda proton facility. The first foil is 6.75 mm of lead and the second foil is 0.5 mm of lead on a $\frac{1}{4}$ in lucite plate. The second scatterer was immediately followed by *two* concentric brass occluding rings of diameters 0.77 to 1.45 cm and 2.4 to 3.05 cm and total thickness 7cm. Measurements with both a small silicon diode and with a 400-element ion chamber showed the proton flux was uniform to within 2% over a 20 cm diameter area, but that this uniformity depended on good alignment of the occluding rings (Coutrakon *et al* 1991a).

At the National Institute of Radiological Sciences (NIRS), Chiba, Japan, a quite different method of achieving large uniform fields is used. The elemental proton beam is some 1 cm^2 and this is swept horizontally and vertically with two perpendicular swing magnets to create a uniform field up to 20 by 20 cm^2 (Kanai *et al* 1991). We now turn attention to the alternative method.

4.3.3. Spot scanning

The method of spot scanning has been used at the NIRS 70 MeV proton facility at Chiba, Japan (Kanai *et al* 1980, 1991, Tsunemoto *et al* 1985, Kawachi *et al* 1983).

The beam from the cyclotron is shaped to a 10 mm by 10 mm square field by a series of apertures and magnets. Range modulation is achieved by a rotating lucite disk of variable thickness. The elementary beam is moved in space by horizontal and vertical scanning magnets capable of creating a field of size 18 cm by 18 cm. When the elementary fields were all of the same duration, the uniformity of the large field so created was 2.5%. The position of the elementary field was switchable by 1 cm in 1 ms. Hence to achieve 2% accuracy the exposure time at any particular location should not fall below 50 ms. In practice the beam delivered some 1 Gy s^{-1} over a 1 cm by 1 cm field, so it took about 3 minutes to scan a 10 cm by 10 cm field delivering 2 Gy.

The advantage of spot scanning over the use of scattering foils is that highly non-uniform irradiation patterns are achievable. Figure 4.16 shows such a complex irradiation field and how well this can be achieved. The scanning pattern is prestored into a computer along with the time which the beam should spend at each location. A multileaf collimator comprising 40 1 cm^2 brass rods provides back-up collimation. Tsunemoto *et al* (1985) report on a fully three-dimensional spot-scanning technique. Each CT slice is treated separately and within the slice, the beam is scanned incrementally. At each position within the slice a different energy degrader was placed in the beam to tailor the Bragg peak to the local distal edge of the target.

It is important to realize that this method of spatially varying the proton beam intensity is quite different from how an x-ray field can be modulated with compensators. For protons, compensators shift the *range* of the protons, not their intensity. To modulate the proton intensity, the duration time has to be modulated.

The raster-scanning method of creating large-area fields can also used for the proton facility at Loma Linda (Awschalom *et al* 1987, Miller *et al* 1987, Coutrakon *et al* 1991b). The elementary beam is 2 cm in diameter and can be swept to fields as large as 40 by 40 cm at a maximum sweep speed of 2 cm ms^{-1}.

Phillips *et al* (1992) report the spot-scanning method at the Paul Scherrer Institute (PSI), Villigen, Switzerland. The small-area beam can be steered in the (xy) plane at right angles to its direction by sweeper magnets. To move the Bragg peak in the patient along the direction (z) of propagation of the beam, range-shifter plates of different thicknesses can be inserted. The order of scanning is x, z, y. This method of covering the three-dimensional volume of a target would allow the intensity of the beam to be varied as a function of the *three-dimensional* location of the Bragg peak by methods such as that described by Webb (1989) for photons (see chapter 2).

Respiratory motion complicates the technique and was extensively evaluated by Phillips *et al* (1992). Respiration was simulated by perturbing the base beamspot position by a sinusoidal motion whose amplitude, frequency and phase could be varied. For single fractions, the dose inhomogeneity increased as the amplitude of respiration increased as expected. By keeping the amplitude and frequency constant and varying the phase, the effects of fractionation were studied and shown to 'smooth over' otherwise irregular distributions of dose. The

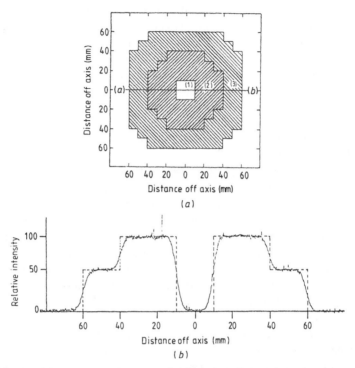

Figure 4.16. *(a) A two-dimensional modulation of proton intensity across a large field created by a spot-scanning method. The dose to area 2 is twice the dose to area 3. The dose to area 1 is zero. (b) A profile across the line a–b of the measured distribution of dose. (From Kanai et al (1980).)*

percentage standard deviation decreased roughly as the square root of the number of fractions. Respiration frequency (between 6 and 60 breaths per minute) had little effect on dose homogeneity. There was a strong correlation between the dose inhomogeneity and the relationship between the movement direction and the direction of scanning. The fastest scan direction (x) should be chosen to correspond to the axis of the largest respiration motion. The slowest scan direction (y) should be along the direction of smallest motion.

4.3.4. Monitoring the beam intensity

Any beam-intensity monitor must satisfy the following design constraints:

1. it must present minimal mass to the beam in order not to degrade the energy (and hence the proton range), nor greatly degrade the beam size (which would affect the transport to the treatment room),

2. it must be radiation hard,

3. it should ideally measure radiation intensity to 1% accuracy in about 1 ms if the beam is swept magnetically.

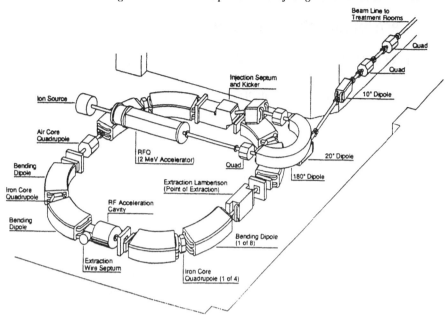

Figure 4.17. *The variable energy proton synchrotron for the Loma Linda Cancer therapy facility. The air-core quadrupole in the accelerator regulates the beam extraction using the beam intensity monitoring signal from the treatment room. The intensity monitor is described in section 4.3.4. (From Coutrakon et al (1991b).)*

A beam-intensity monitor satisfying these requirements has been developed for the Loma Linda facility (figure 4.17) (Coutrakon *et al* 1991b). It comprises a 5 cm thick gas cell coupled to a photomultiplier tube. The gas cell was filled with xenon at 1.2 atmospheres pressure with 12.7 μm thick titanium windows. This presents 0.14 g cm^{-2} total mass to the beam, degrading the energy by 70 keV corresponding to a loss of 1.4 mm range in water. The beam increases in size by 6mm at the patient site after transport through 3 m, an extra angular spread of 1.88 milliradians.

Each passing proton produces about 2000 photons (70 000 eV of energy are deposited per proton; about 20% of the energy deposited in a noble gas is dissipated radiatively, in this case as 7 eV photons). Of these, about 0.1% (i.e. two photons per proton) are detected by the photomultiplier tube whose spectral response overlaps the xenon scintillation spectrum by some 50%. The proton beam can operate at upwards of 10^{10} protons per 1 s 'spill'. Thus in the 1 ms that the beam is at some particular geometric location in its sweep, some 10^7 photoelectrons are produced, which is adequate to measure the intensity to better than 1% accuracy.

4.4. PROTON TREATMENT PLANNING

Goitein *et al* (1982) have reviewed all aspects of planning proton therapy. Of particular note are:

1.   the need to consider the full 3D nature of the problem,
2.   the need to account for tissue inhomogeneities (more so than with photon planning),
3.   design of compensators,
4.   special imaging procedures for planning proton therapy,
5.   exquisite patient immobilization,
6.   the need to assess and plan for unknowns,
7.   the need to compare plans with 'rival' photon plans.

4.4.1.   *Tomographic imaging for proton treatment planning*

As for planning x-ray therapy, CT tomographic data are an essential pre-requisite to proton treatment planning (Chen 1983).   As shown above, the physics determining how proton beams deposit their energy in tissue is different from x-rays and the use made of CT images is therefore quite different. The major concern is that tissue inhomogeneities adjust the proton range (not intensity) and could if improperly handled lead to serious underdosage. The inability to accurate map inhomogeneities prior to commercially available CT was a major hindrance to its acceptability and development.   Goitein (1977) has studied the physical specification of the CT scanner. Consider in turn, density resolution and spatial resolution.

At the end of its range the proton dose falls from 80% to 20% in about 4 mm suggesting that an accuracy of ±2 mm is necessary in specifying the proton range. If this range is say 20 cm, this implies a 1% error would be acceptable. If the pixel size is say 2 mm (i.e. there are 100 pixels along the range) the *random* error per pixel in CT number which leads to this overall error is 10%. This is considerably larger than the 1% density resolution which CT scanners can achieve. However, *systematic* errors must be kept to 1% and this is just at the limit of present CT technology.

Now consider spatial resolution. The effect of a thin sliver inhomogeneity whose thickness is less than the proton range critically depends on its width (very thin slivers hardly cast any shadow because multiple scattering fills in the dose distally—see figure 4.18). However, if the spatial resolution of the CT images were poor (say 2 or 3 mm instead of the usual $\simeq$ 1 mm) such slivers would appear to be larger, and importantly, wider than they really are. If such low-resolution images were used to plan proton treatment, the dose-perturbing effect of small structures would be misjudged. The wider the inhomogeneity appears, the more the range appears foreshortened (see figure 4.18 again). Goitein (1977) thus argues that a spatial resolution of 1–2 mm is needed, fortuitously matched by the actual performance of state-of-the-art CT scanners. Because the range-modifying effects

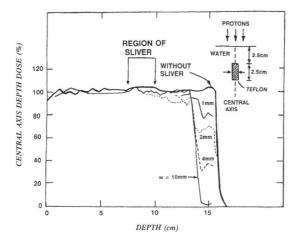

Figure 4.18. *The calculated central-axis dose distribution for an initially parallel energy-modulated proton beam incident on a sliver of teflon of length 2.5 cm and different widths. The graphs show how, for thin slivers, multiple scatter can fill in the shadow region, whereas this effect decreases as the width (w) of the sliver increases. This has important implications for the required resolution of CT images used for planning proton therapy. (From Goitein (1977).) (Reprinted with permission from Pergamon Press Ltd, Oxford, UK.)*

of inhomogeneities are not confined to two-dimensions, it is however important that the *axial* spatial resolution in CT is also 1–2 mm. This is generally not the case for diagnostic scanning so special arrangements must be made if CT data are to be used for planning proton therapy.

High-Z material in the patient can present problems when using CT data for planning proton therapy. The stopping power of protons is determined by the electron density of tissues. For most soft tissues, it is the electron density which maps to the CT number because the dominant mechanism is Compton interaction. Hence for such tissues CT data can be used to predict the range directly. However, the CT number is determined by the *total* linear attenuation coefficient of x-rays (μ in cm^{-1}) at the effective energy of about 80 keV. The fraction of the total attenuation due to Compton interaction falls off with increasing atomic number Z (see figure 4.19). Also the physical density ρ in g cm^{-3} increases with increasing Z. Hence the fraction of the total attenuation coefficient μ/ρ in cm^2 g^{-1} due to the Compton effect falls with increasing Z. Goitein (1977) plotted the dimensionless product of the proton range in g cm^{-2} and this fraction, which is the electron component of linear x-ray attenuation (figure 4.19), and showed how it is therefore not a horizontally straight line, as one would like, but varies with Z. However, he argues that although this appears to be a problem, in practice the volume of high-Z material in the patient is not large and can possibly be ignored to a first approximation. Alternatively, dual-energy CT scanning can be used to address the problem.

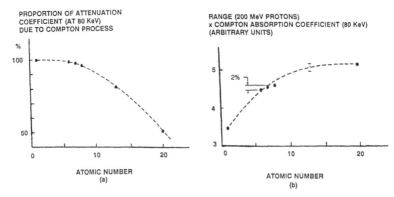

Figure 4.19. *(a) The proportion of x-ray attenuation coefficient at 80 keV due to the Compton effect. (b) The product of proton range and Compton not total x-ray linear attenuation coefficient as a function of the atomic number Z. Note that this graph is not horizontal because both the component of attenuation due to Compton scatter decreases with increasing Z (see (a) above) and also because of the Z-dependence of physical density. (From Goitein (1977).) (Reprinted with permission from Pergamon Press Ltd, Oxford, UK.)*

In conclusion, commercial CT scanners are entirely appropriate for proton therapy planning, provided the slice width is reduced to 1–2 mm.

4.4.2. Theory of range modulation by compensator or bolus

Compensating for tissue inhomogeneities is a vital component of proton treatment planning. The penetration of the beam must be modified to take them into account. Where the integrated stopping power is large the beam must be made correspondingly penetrating and vice versa. Computed tomography provides the necessary data (section 4.4.1). Range modulation may be accomplished by varying the maximum proton energy while scanning a narrow beam across the field. Alternatively, and more commonly, bolus applied externally to the patient achieves the range modulation. In principle the method is no more complicated than applying, at each location in a field, the appropriate thickness of bolus to force the distal dose edge (say the 90% dose contour) to follow the distal edge of the target volume. The bolus is thick where the beam should be less penetrating and vice versa. One consequence of this bolus is that the *proximal* 90% isodose contour will now become a ragged surface in a heterogeneous medium. Figure 4.20 shows in two-dimensions how a compensator can account for a number of factors to match the range to a prescription.

Figure 4.21 shows how to compute the compensator thickness. An element of the beam passes a pre-absorber PA, a defining aperture A, through the compensator C into the treatment volume V. For element $a_{x,y}$ of the compensator the thickness is c_{xy}. If the CT density of the patient is ρ_{xyz} then the required range in the patient to reach a depth z_{max} is

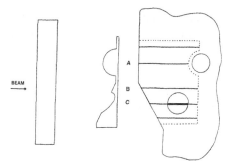

Figure 4.20. *Shows how a range-compensator can adjust the proton penetration so the range matches a prescription. The beam enters at the left and passes through a fixed path-length absorber. The 'patient' is to the right with the compensator in between. Beneath region A there is an organ at risk (shown circular) so the compensator has a circular increase in depth to pull back the range. In region B the patient external contour starts to fall away so the compensator must increase in thickness. Beneath region C there is a circular bone inhomogeneity which causes range shortening. The compensator is correspondingly thinned. In practice all these effects may occur together and the overall shape of the compensator would be as shown. (From Wagner (1982).)*

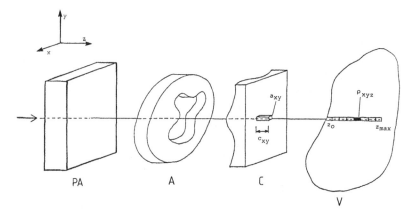

Figure 4.21. *Shows how to compute the compensator thickness. An element of the beam passes a pre-absorber PA, a defining aperture A, through the compensator C, into the treatment volume V. For element $a_{x,y}$ of the compensator the thickness is c_{xy}. The CT density of the patient is ρ_{xyz}. z_0 is the entrance depth of the patient and z_{max} is the required proton range (or where the distal 90% contour should be). (From Wagner (1982).)*

$$RR_{xy} = \int_{z_0}^{z_{max}} \rho_{xyz} \, dz. \tag{4.18}$$

The absorber thickness should then be

$$AT_{xy} = \left(P - RR_{xy}\right)/1.15 \tag{4.19}$$

where P is the maximum proton range in water (e.g. 16 cm at 160 MeV) and the factor 1.15 converts to a thickness in lucite not water. Finally the compensator thickness is

$$c_{xy} = AT_{xy} - PA \text{(mm)} \qquad (4.20)$$

where PA is the thickness of the preabsorber. PA is set to 1 mm less than the minimum value of AT_{xy}. This ensures the minimum compensator thickness is 1 mm rather than 0 mm leaving some material to support the thicker compensator regions (Wagner 1982).

This simple compensation ignores two factors which should not be ignored; multiple scattering and patient movement. To properly account for the first, Monte Carlo studies in three dimensions must be made and this is unrealistic if required for each patient. Such calculations have been done for studying the general properties of introducing absorbers into the beam-line (Kanai *et al* 1991). The second cannot be precisely accounted for as patient motion is almost always unpredictable. Goitein (1978a) thus argues that the aim of the treatment plan should be to predict the *worst situation* that can occur and ensure that even this plan would be acceptable. Consider the two factors in turn:

To account for multiple scattering the maximum and the minimum path lengths which a proton could follow are computed. These would give, respectively, the minimum and maximum proton ranges. In three dimensions two *surfaces* would result between which the actual proton range *must* lie. These path lengths can be estimated by replacing in turn each CT voxel by first the smallest (to get minimum path length) and second the largest (to get maximum path length) of itself and its neighbours up to distances of about ±2 voxels. This procedure is not as good as a full calculation but gives the range of uncertainty. Clinically, the bolus should be adjusted so the most proximal of the two candidate 90% dose surfaces coincides with the distal surface of the target volume. In his paper Goitein (1978a) writes of lines rather than surfaces because only 2D CT data were available to him, but recognized the need to make these computations in 3D, as expressed here.

Movement is similarly accounted for. The worst-case movement is estimated and the elements of patient which *could* have intercepted the proton beam are used to predict minimum and maximum ranges within which the actual range must lie. For example, in the case of a high-density structure the beam must be made additionally penetrating not only where the structure *is* but also where it *might be*. Goitein (1978a) thus argues that single proton plans are not acceptable. Plans should be made with worst-case calculations for movement and multiple-scattering ensuring the required plan lies between the two extreme possible plans (Chen 1983). At HCC–MGH the dose-planning computer is linked to an automatic milling machine to make the compensators (Wagner 1982).

Urie *et al* (1984) present an alternative way to account for tissue movement. The method is now the accepted technique at HCC–MGH, replacing the earlier one above (Goitein 1978a, Urie *et al* 1986a). A so-called 'simple bolus' is made by the method described above using simple path lengths through the CT data.

No account need be taken of adjacent pixels to attempt to account for multiple scattering or movement of tissues. This is done differently. Instead, once the simple bolus has been computed, the thickness at any particular position in the bolus is replaced by the minimum thickness at $\pm d$ from that position, where d is 3–5 mm. The resulting bolus is called the 'expanded bolus'. The design philosophy is that movement of the tissue by $\pm d$ would now 'not matter' in the sense that the beam would still be sufficiently penetrating to generate the required high-dose volume. The worst that could now occur is that the dose might extend a little beyond the target volume. There is no chance that any of the target volume would be *underdosed*.

Urie *et al* (1984) tested this extensively with phantoms worked up in the same way as patients. They found that misregistration of a simple bolus with the tissue led to very bad errors (more than 1 cm misjudgement of proton range). However, when the expanded bolus was also deliberately misregistered, these errors were reduced to at worst 2.5 mm in a phantom situation which was the worst expected clinically. Of course if the expanded bolus and the phantom were *not* misregistered, then the dosimetry was not quite as good as had the correctly registered simple bolus been in use. This is the small price to pay for confidence in accounting for tissue movement. It turns out that the finite size of the milling tool (Wagner 1982) used for bolus construction actually itself introduces a measure of 'expansion'.

Daftari *et al* (1991) have taken a different approach to assessing the importance of random movement during charged-particle therapy. Treatment plans with simulated patient movement were made; the dose–volume histograms were constructed for organs at risk and converted to a measure of NTCP via the Lyman model (see chapter 1). They concluded that random movements with maximum size 2 mm, typical of clinical experience at the Lawrence Berkeley helium-ion treatment facility, led to no appreciable increase in NTCP.

The compensating bolus should preferably be in contact with the patient but in practice this may not be possible and an air gap will intervene (Urie *et al* 1985, Sisterson *et al* 1987). Sisterson *et al* (1989) have analysed the effects of this air gap on the location of the distal dose contours using Monte Carlo methods. Model pairs of compensator plus phantom were created, as shown in figure 4.22. For the simple bolus these pairs lock together with no air gap to give a uniformly thick block. Several model pairs were considered in which the compensator was either 'ridge first', with an increased thickness to shorten the range, or 'valley first', with a decreased thickness to lengthen the range. The valley and ridge could take various apex angles and the air gap was variable. Protons were modelled interacting with this arrangement and also with the 'expanded bolus' arrangement shown in the lower part of the figure.

The behaviour was such that the amplitude of the perturbation on the proton penetration increased with increasing air gap, with increasing pre-absorber thickness (i.e. the uniform base of the bolus) and depended on the apex angle. However, provided the air gap was less than 12 cm, the perturbation was never

'Simple' boluses

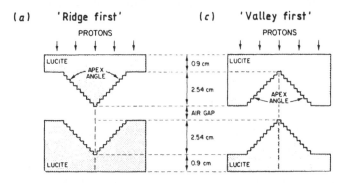

'Expanded' boluses

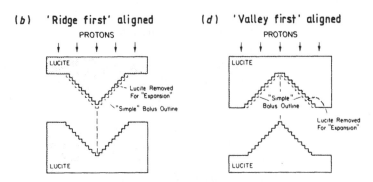

Figure 4.22. *The compensating bolus- ('patient') phantom arrangement used to investigate the importance (or otherwise) of air gaps between the bolus and the patient. The apex angle and air gap are defined in the figure. (a) 'Ridge-first' geometry for the simple bolus; the ridge half simulates the bolus and the valley half the patient; (b) 'ridge first' for the expanded bolus; material is removed from the bolus to 'open it up'; (c) 'valley first' geometry for the simple bolus; the valley half simulates the bolus and the ridge half the patient; (d) 'valley first' for the expanded bolus; material is removed from the bolus to 'open it up'. The direction of protons is shown by the arrows. (From Sisterson et al (1989).)*

more than 4 mm. Since the margin added at the planning stage to account for uncertainties in penetration was typically $\simeq$ 5% of the maximum depth of penetration (say 5% of 12 cm), the perturbation turns out to be less than this uncertainty. It would therefore appear that air gaps are not a problem.

At the Japanese proton facility at Chiba, bolus construction also relies on CT data but is much simpler and ignores tissue inhomogeneities, since the beam can only penetrate 3.8 cm and is used mainly for irradiating the head and neck. At Chiba a dental impression material, alginate, with a range correction factor of 1.06 relative to water is poured into little frames attached to the skin surface just prior to treatment to create a flat proximal surface. Thin acrylic plates are used to fine tune the compensation to account for changes in tumour volume during the course of therapy (Akanuma *et al* 1982).

The sharp penumbra is one of the compelling arguments for proton therapy. However, introduction of scattering and beam-defining components into the beam degrades the penumbra. The degrading components are the pre-absorber, the beam-defining aperture, the compensating bolus and the effects of these depend on the size of the drift (air) spaces. Characterizing all these effects is complex, but Urie *et al* (1986b) have applied Monte Carlo methods to the problem. As one might expect, penumbra increases with:

1. a decrease in proton residual range (increase in thickness of pre-absorber),
2. size of air gap between the beam aperture and the patient,
3. depth into the patient,
4. thickness of bolus.

Detailed curves were given by Urie *et al* (1986b). Particularly useful are a set of 'isopenumbra' curves showing the combinations of circumstances which give the same penumbra.

4.4.3. *Theoretical treatment of thin inhomogeneities*

In section 4.4.1 the influence of thin slivers of tissue inhomogeneity on proton dose distributions was discussed. Depending on the width of the sliver the dose distribution in the shadow of the inhomogeneity may get filled in by multiple scatter. Conversely the main effect of a large slab of tissue inhomogeneity is to alter the proton range. The effect requires compensation by external bolus.

Another concern is the dose-perturbing effect of *thin* inhomogeneities which do not extend across the full cross section of the proton beam. These cause local perturbations to the dose near, and in the shadow of, the lateral margin of the inhomogeneity. Goitein (1978b) has provided a very elegant mathematical description of these effects, which now follows. Firstly, and qualitatively, the effect of an edge discontinuity can be predicted without the use of mathematics. Consider figure 4.23. A proton beam irradiates an edge of a thin inhomogeneity. Particles in group I miss the edge and give rise to an unscattered step-like fluence downstream (shown dotted). Particles in group II however intercept the inhomogeneity, scatter, and give rise to a fluence falling from left to right in the diagram. The resulting total fluence has the very characteristic shape shown as the solid line. The fluence is exactly halved just below the edge and rises to 1.5 times the mean level just to the right of this. This is because the fluence near Q, close to the edge, must be exactly half that at points such as Q' because Q lacks scatter from half the plane. On the other side of the edge-shadow at a close point, such as P, the scatter is the same and adds to the fluence from the group I particles to give an enhancement to 1.5. The total fluence profile is a universal curve if the horizontal axis is considered to be the angular deviation from the forward direction, since the scattering depends only on angular deviation

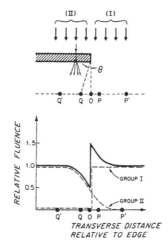

Figure 4.23. *(Upper) shows schematically a parallel beam of protons incident on a thin inhomogeneity only partly filling the field. (Lower) the fluence close to the shadow of the edge. There is a universal curve (solid) for the form of the fluence, being the sum of two components (dotted) from the two groups of protons, shown as I and II. At the edge there is a sudden discontinuity of size unity (see text for explanation). (From Goitein (1978b).)*

of the measurement point from the forward direction (shown as θ) and so the lateral extent of the perturbation increases with increasing distance beneath the inhomogeneity. The above argument assumes the inhomogeneity is sufficiently thin that particles do not loose much energy traversing it.

Goitein (1978b) derives the general formulae which give the edge perturbation and can be applied to any geometry. Consider figure 4.24 in which a parallel beam of protons irradiates a thin slab with $a_i n_0$ protons per unit length. The factor a_i (1 for a uniform beam) can be non-unity to describe a non-uniform beam. The index i labels the position in the scatterer; consider those photons incident on the ith element. Multiple scattering gives rise to an angular distribution

$$f(\theta)\,d\theta = \left[1/\sigma_i\,(2\pi)^{1/2}\right]\exp\left(-\theta^2/2\sigma_i^2\right)d\theta \qquad (4.21)$$

where σ_i is the characteristic projected scattering angle which depends on the atomic number of the material, the charge, velocity and momentum of the proton and on the thickness and radiation length of the scatterer (Rossi (1956) formula). The fluence of particles crossing a test region P of length dl at distance t from a small element of length dx in the ith scatterer is

$$(a_i n_0 dx)\left[1/\sigma_i\,(2\pi)^{1/2}\right]\exp\left(-\theta^2/2\sigma_i^2\right)(dl/t) \qquad (4.22)$$

since

$$d\theta = dl/t \qquad (4.23)$$

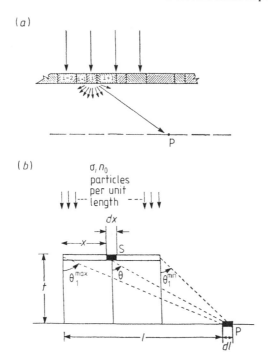

Figure 4.24. *(Upper) showing a parallel beam of protons incident on a thin scatterer, considered to be divided into elements labelled by i. The scattered protons are shown figuratively with some reaching point P. (Lower) a single scattering element i and an element dx within this, scattering to point P. The total fluence scattered to P from this element is the integration over the length of the element. The total fluence scattered to P from the whole scatterer is this quantity integrated over all the elements i. (From Goitein (1978b).)*

is the elemental angle subtended by dl at dx. In the absence of scatterer the fluence at P for a uniform beam ($a_i = 1$) would be simply $n_0 dl$. The ratio F_i of the fluence with scatterer to the fluence without scatterer is the result of taking the ratio of the quantity in equation (4.22) to this value and integrating over the range of x. This ratio is

$$F_i = (1/n_0 dl) \int_x (a_i n_0 dx) \left[1/\sigma_i (2\pi)^{1/2}\right] \exp\left(-\theta^2/2\sigma_i^2\right) (dl/t). \quad (4.24)$$

If we make a small-angle approximation that $\theta = (l - x)/t$ so that $d\theta = -dx/t$, this reduces to

$$F_i = a_i / \left[1/\sigma_i (2\pi)^{1/2}\right] \int_{\theta_i^{min}}^{\theta_i^{max}} \exp\left(-\theta^2/2\sigma_i^2\right) d\theta. \quad (4.25)$$

The fluence at P from all elements labelled by i is then the sum

$$F = \sum_i F_i. \quad (4.26)$$

This can be translated to dose D by a multiplicative constant, the dose-to-fluence ratio at P in the absence of inhomogeneity

$$D = \sum_i d_i F_i. \tag{4.27}$$

Equations (4.25) and (4.26) involve only the specification of the homogeneity of the beam (through a_i) and a knowledge of the characteristic scattering angle (through σ_i). Everything else is a geometrical integration.

The above analysis holds provided the following (very reasonable set of) approximations hold:

1. the material is thin and does not significantly affect proton range,
2. the inhomogeneity of finite thickness is 'condensed' to be considered as an infinitely thin plane for mathematical argument with the same scattering properties,
3. the discontinuity is infinite in the third dimension (realistic because perturbing effects are local to the edge),
4. the projected scattering angle can be represented by a Gaussian distribution,
5. the scattering is sufficiently forward peaked that one can make the small-angle approximation (see above equation (4.25)) that $\theta \simeq \tan(\theta)$
6. the particles are monoenergetic.

Fan-beam geometry can be incorporated easily into this model. Equation (4.25) requires specification of the two limits of integration; the largest and smallest scattering angles. These can be specified equally well in fan-beam geometry (figure 4.25). Goitein (1978) also shows that the effect of tissue inhomogeneities both upstream and downstream from the inhomogeneity in question can be similarly accounted for since they alter the distribution of the angles of incident protons.

Returning to equation (4.25), if the thin inhomogeneity extends for half the proton field, θ_i^{max} is set to $\pi/2$. For a uniformly irradiating proton beam, the calculations are easy to do by hand because equation (4.25) reduces to looking up values of the error function $\mathrm{erf}(x)$. Under other circumstances a computer is needed to evaluate equation (4.25). The formula was applied to tissue–air and tissue–bone interfaces. Whereas the former introduce 50% perturbations close to the edge, it turns out the latter only introduce a maximum of 9% variations. Thus tissue–air boundaries may be much more important in proton therapy than tissue–bone boundaries as far as edge-scattering effects are concerned. Even these values are reduced if the beam is divergent and there is scattering tissue above and below the inhomogeneity (as of course is always the case in clinical reality). This avoids what might otherwise be a disturbing and serious clinical problem.

Goitein *et al* (1978) pursued this further, calculating and measuring the perturbations from a variety of different geometrical interfaces for proton (and incidentally electron) beams. For all situations considered there was excellent agreement between theory and experiment.

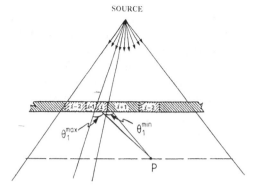

Figure 4.25. *Thin-inhomogeneity scattering of a divergent beam showing the definition of the range of angles in the integration of scatter from a small elemental length of scatterer, labelled by i to the point P. (From Goitein (1978).)*

It is apparent that the formalism presented in this section cannot become the basis of a treatment-planning tool. It would be just too complicated to properly treat the effects of all the body inhomogeneities. Indeed, as Goitein *et al* (1978) showed, tissue inhomogeneities close to each other produce fluence interference effects. From a computational viewpoint there are too many interactive decisions needed in working out equation (4.25) for complex geometries. Instead the formalism and the results therefrom should be used to *estimate* the effects of inhomogeneities in perturbing the distribution of dose in specific clinical situations.

4.4.4. The dose-perturbing effect of thick inhomogeneities

Whilst an analytic technique has been developed for calculating the dose-perturbing effects of thin inhomogeneities, the perturbation from thick inhomogeneities cannot be represented analytically. 'Thick' means 'of comparable dimension to the proton range in the material'. This is because some of the assumptions which went into the 'thin' model are violated by the 'thick' model. For example the 'condensed' material approximation cannot be made and the monoenergetic scattering-angle approximation is invalid. The small angle 'paraxial' approximation still holds.

The answer lies with Monte Carlo techniques. Goitein and Sisterson (1978) have numerically modelled the scattering in two-dimensions from slivers of bone (or teflon substitute) in tissue, when the slivers were comparable to or longer than the proton range. The scattering medium is pixellated and then the Monte Carlo is a two-step process. Firstly, the protons are transported small distances in straight lines with corresponding energy losses and range changes. Secondly, the proton's direction is altered by randomly selecting from the well known Gaussian distribution with standard deviation given by the Rossi formula (see section 4.4.3).

Some 0.1 M histories were followed and flux and dose were plotted as a histogram. A typical result from this work has already been discussed (figure 4.18). They modelled a realistic beam, range-modulated to give a uniform dose over a depth of 16 cm, characteristic of what could be achieved with the HCC–MGH proton facility. The following results, for slivers spanning the central axis of the beam as in figure 4.18, are abstracted from the large volume of Monte Carlo data created:

1.   wide slivers reduce the penetration from the value in homogeneous water,

2.   as the slivers become narrower, central-axis dose builds up beyond the theoretical (wide sliver) range because of 'in-scatter' from the surrounding water,

3.   dose is enhanced laterally to thin slivers in the surrounding water,

4.   if additional angular smearing is added to an otherwise parallel beam, the dose-perturbing effects are considerably reduced,

5.   rocking the inhomogeneity (equivalent to a positioning error) causes similar reductions in dose perturbation,

6.   irradiation of a sliver by a parallel beam *not* parallel to the long edge of the sliver, dramatically reduces the dose perturbation.

Unfortunately, multiple and complex tissue inhomogeneities can have a totally destructive effect on the Bragg peak and Urie *et al* (1986a) have carried out experiments with protons (as well as other heavy ions) to show this. There are circumstances in which multiple scattering cannot be disregarded and cannot be simply accounted for by the methods of bolus construction discussed in section 4.4.2. Urie *et al* (1986a) argue that nothing short of full Monte Carlo computations will really suffice. One must recall that the construction of compensators can only account for the changes in 'areal density', i.e. the product of density and path length, together with methods of attempting simple accounting for movement and multiple scatter. The methods are not assumption-free; there are classes of inhomogeneities which can 'fool' the algorithm for compensator bolus construction. This problem remains the 'Achilles' heel' of proton therapy.

4.4.5.   *The differential pencil-beam model for proton dose calculation*

In recognition of the difficulties of properly accounting for multiple scattering, Petti (1992) has developed a differential pencil-beam model for protons. The philosophy is the same as underlies the differential pencil-beam formalism for photons, namely the important physics is built into the elemental dose distribution and then complex field calculations are made by integrating the elemental distribution in the appropriate geometry (cf section 2.5.2.1). The result is much more accurate than simple ray-tracing, but is far less time consuming than full Monte Carlo calculations. Petti (1992) validated the following formalism by showing that, for five sample geometries, the predicted dose distributions were very close to the results of full Monte Carlo calculation and differed from the results of simple ray-tracing in many important respects. The differences became more pronounced as the complexity of the problems increased. The formalism accounts for tissue inhomogeneities in the path of the protons.

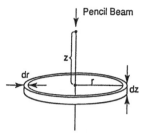

Figure 4.26. *The geometry for defining the differential pencil-beam dose distribution for protons. The beam is of infinitessimal area and directed down the z axis. The DPB dose distribution, F (r, z), is the dose deposited in the annulus shown of radius r to r + dr and depth z to z + dz, normalized to the dose D_0 in the Bragg peak or SOBP. (From Petti (1992).)*

Define the differential pencil-beam dose distribution, $F(r, z)$, as the dose deposited in a homogeneous water phantom in an annular ring between radii r and $r + dr$ and depth z and $z + dz$, normalized to the maximum dose D_0 in the Bragg peak (or SOBP) of the beam (figure 4.26). The pencil beam has infinitessimal area but contributes dose to a finite range of radii because of multiple scattering. $F(r, z)$ may be computed by Monte Carlo methods. The dose deposited at any point (r, z) per unit beam area is $D_0 F(r, z)/2\pi r$. Suppose the proton intensity (protons per unit area) at any position in the beam is $\Phi(r, \theta)$, where (r, θ) is a general point, with origin at the point of dose calculation, in the xy plane normal to the incident direction of the protons. The dose at any cartesian point (x, y, z) is then

$$D(x, y, z) = D_0 \int_0^{r_{max}} \int_0^{2\pi} \frac{F(r, z)\, \Phi(r, \theta)}{2\pi r} r\, dr\, d\theta. \qquad (4.28)$$

In equation (4.28) r is the radius from the centre of the contributing pencil beam at (x', y') to the calculation point (x, y) at depth z (i.e. $(x - x') = r\cos(\theta)$ and $(y - y') = r\sin(\theta)$). Equation (4.28) simply expresses a summation of the dose from the set of elemental pencil beams appropriately weighted for intensity. The upper limit on the radial integral represents the maximum radial distance for which multiply scattered protons can contribute to the dose at point (x, y, z). In practice this is fairly small for protons leading to short computational times. If the proton intensity is uniform, equation (4.28) reduces to

$$D(x, y, z) = D_0 \int_0^{r_{max}} F(r, z)\, dr. \qquad (4.29)$$

Inhomogeneities are accounted for by replacing z by the water-equivalent pathlength z_w to obtain

$$D(x, y, z) = D_0 \int_0^{r_{max}} \int_0^{2\pi} \frac{F(r, z_w)\, \Phi(r, \theta)}{2\pi}\, dr\, d\theta \qquad (4.30)$$

where

$$z_w = \int_0^z \rho\left(r, \theta, z'\right) dz' \qquad (4.31)$$

and ρ is the electron density of tissues. The integration in equation (4.31) is performed along the ray-path of the protons and, if desired, can account for beam divergence. It does *not* account for the position of the tissue inhomogeneity down the ray-path.

Petti (1992) also developed a full Monte Carlo code modelling proton multiple scatter, using it to predict benchmark dose distributions for five model geometries. She then compared the results with predictions from the pencil-beam model with very good agreement. The ray-tracing algorithm which ignores multiple scatter was adequate for homogeneous phantoms but became increasingly more inaccurate by comparison with the gold standard as the complexity increased.

Unlike photon-beam calculations involving pencil beams, the DPB formulation for protons is not enormously slower than simple ray-tracing. Petti (1992) quotes a reduction in speed of about a factor 12. If tissue inhomogeneities have a complicated geometry, the effort is most worthwhile because ray-tracing can give quite erroneous results.

4.5. SUMMARY

Radiotherapy with protons is nearly four decades old but, possibly because its birth coincided with the development of megavoltage photon therapy and in view of the enormous expense and consequent rarity of proton facilities, has not become widely available. However, it has been persued vigorously in a number of centres and now is poised to capture greater interest as more purpose-built centres begin to come on stream.

The inherent physical properties of protons interacting with tissue make them ideal candidates for conformal therapy. The Bragg peak can be spread out by range-modulating wheels and the physical area of the beam can be enlarged from the elementary beam by double-foil-plus-annuli scattering or by spot scanning.

Proton treatment planning is very complex and needs to account for tissue inhomogeneities and the possibility of patient movement. Because of the rapid decline of dose distal to the SOBP, patient movement is potentially more dangerous for proton than photon irradiation because severe underdosage could arise unless the measures described are put into practice. Patient fixation is important.

Many aspects of proton dose distributions can only be studied in model situations because of the complex interaction physics. Monte Carlo studies have been performed to assess the importance of disturbing features such as inhomogeneities and movement.

REFERENCES

Akanuma A, Majima H, Furukawa S, Okamoto R, Kutsutani Y, Nakamura, Tsunemoto H, Morita S, Arai T, Kurisu A, Hiraoka T, Kawachi K and Kanai T 1982 Compensation techniques in NIRS proton beam radiotherapy *Int. J. Rad. Oncol. Biol. Phys.* **8** 1629–1635

Ambrosimov N K, Vorobev A A, Zherbin E A and Konnov B A 1985 Proton therapy at the cyclotron at Gatchina, USSR *Proc. Acad. Sci. USSR* **5** 84–91

Antoine F S 1987 Proton beam therapy—cost versus benefit *J. Nat. Cancer Inst.* **81** 559–562

Antoine F S 1991 International collaboration may enhance proton therapy research *J. Nat. Cancer Inst.* **83** 9–10

Awschalom M W, Cole F T, Lidahl P V, Mills F E and Teng L C 1987 Design of a proton therapy accelerator *Med. Phys.* **14** 468

Beeckman W, Jongen Y, Laisne A and Lannoye G 1991 Preliminary design of a reduced cost proton therapy facility using a compact, high field isosynchronous cyclotron *Nucl. Instrum. Methods in Phys. Res.* **B56/57** 1201–1204

Bichsel H 1972 Passage of charged particles through matter *American Inst of Physics Handbook* section 8 (New York: McGraw-Hill) pp 142–189

Blattman H (ed) 1986 Proton therapy of choroidal melanomas (OPTIS) *Swiss Institute for Nuclear Research Medical Newsletter* **8** 32

Blosser H G 1989 Compact superconducting synchrocyclotron systems for proton therapy *Nucl. Instrum. Methods in Phys. Res.* **B40/41** 1326–1330

Bonnett D E 1991 The role of protons in radiotherapy *Proc. 1990 Int. Conf. on Applications of Nuclear Techniques* ed G Vourvopoulous and T Paradellus (Singapore: World Scientific) pp 109–130

Bragg W H and Kleeman R 1904 On the ionisation curves of radium *Phil. Mag.* **8** 726–738

Chen G T Y 1983 Computed tomography in high LET radiotherapy treatment planning *Computed tomography in radiation therapy* ed C C Ling, C C Rogers and R J Morton (New York: Raven) pp 221–227

Chen G T Y, Singh R P, Castro J R, Lyman J T and Quivey J M 1979 Treatment planning for heavy ion radiotherapy *Int. J. Rad. Oncol. Biol. Phys.* **5** 1809–1819

Chen G T Y and Goitein M 1983 Treatment planning for heavy charged particle beams *Advances in Radiation Therapy Treatment Planning* ed A E Wright and A L Boyer (New York: American Institute of Physics) pp 514–541

Chu W T, Renner T R and Ludewigt B A 1989 Dynamic beam delivery for three-dimensional conformal therapy *Proc. Eulima Workshop on the potential value of light ion beam therapy (Nice, 1988)* ed Chauvel P and Wambersie A

Chuvilo I V, Goldin L L, Khoroshkov V S, Blokhin S E, Breyev V M, Vorontsov I A, Ermolayev V V, Kleinbock L, Lomakin M I, Lomanov M F, Medved' Ya V, Miliokhin N A, Narinsky V M, Pavlonsky L M, Shimchuck G G, Ruderman A I, Monzul G D, Shuvalov E L, Kiseliova V N, Marova E I, Kirpatovskaya L E, Minakova E I, Krymsky V A, Brovkina A K, Zarubey G D, Reshetnikova I

M and Kaplina A V 1984 ITEP synchrotron proton beam in radiotherapy *Int. J. Rad. Oncol. Biol. Phys.* **10** 185–195

Cole D J 1990 The case for proton therapy in the UK *private communication to AEA* September 1990

Coutrakon G, Bauman M, Lesyna D, Miller D, Nusbaum J, Slater J, Johanning J, DeLuca P M Jr, Siebers J and Ludewigt B 1991a A prototype beam delivery system for the proton medical accelerator at Loma Linda *Med. Phys.* **18** 1093–1099

Coutrakon G, Miller D, Kross B J, Anderson D F, DeLuca P Jr and Siebers J 1991b A beam intensity monitor for the Loma Linda cancer therapy proton accelerator *Med. Phys.* **18** (4) 817–820

Daftari I, Petti P L, Collier J M, Castro J R and Pitluck S 1991 The effect of patient motion on dose uncertainty in charged particle irradiation for lesions encircling the brain stem or spinal chord *Med. Phys.* **18** 1105–1115

Goitein M 1972 Three dimensional density reconstruction from a series of two dimensional projections *Nucl. Instrum. Methods* **101** 509–518

—— 1977 The measurement of tissue heterodensity to guide charged particle radiotherapy *Int. J. Rad. Oncol. Biol. Phys.* **3** 27–33

—— 1978a Compensation for inhomogeneities in charged particle radiotherapy using computed tomography *Int. J. Rad. Oncol. Biol. Phys.* **4** 499–508

—— 1978b A technique for calculating the influence of thin inhomogeneities on charged particle beams *Med. Phys.* **5** (4) 258–264

—— 1982 Applications of CT in radiotherapy treatment planning *Progress in medical radiation physics* ed C G Orton (New York: Plenum) pp 195–293

Goitein M, Abrams M, Gentry R, Urie M, Verhey L and Wagner M 1982 Planning treatment with heavy charged particles *Int. J. Rad. Oncol. Biol. Phys.* **8** 2065–2070

Goitein M, Chen G T Y, Schneider R J and Sisterson J M 1978 Measurements and calculations of the influence of thin inhomogeneities on charged particle beams *Med. Phys.* **5** (4) 265–273

Goitein M and Munzenrider J 1984 Proton therapy: good news and big challenges *Int. J. Rad. Oncol. Biol. Phys.* **10** 319–320

Goitein M and Sisterson J M 1978 The influence of thick inhomogeneities on charged particle beams *Rad. Res.* **74** 217–230

Goldin L L and Monastyrsky 1978 Analytical calculation of ridge filter configuration *Medic. Radiol.* **23** (12) 65–68 (in Russian)

Gottschalk B 1987 Design of a hospital-based accelerator for proton radiation therapy *Nucl. Instrum. Methods in Phys. Res.* B24/25 1092–1095

Graffman S, Brahme A and Larsson B 1985 Proton radiotherapy with the Uppsala cyclotron. Experience and plans *Strahlentherapie* **161** 764–770

Janni J F 1982 Proton range-energy tables, 1 keV—10 GeV *Atomic Data and Nuclear Tables* **27** 147–339

Jongen Y 1992a private communication

—— 1992b An improved isocentric gantry with reduced diameter for proton therapy by scanned beam *IBA paper* Chemins du Cyclotron,2 b-1348 Louvain-la-Neuve, Belgium

Jongen Y, Beeckman W and Laisné A 1992 Development of a low-cost compact cyclotron system for proton therapy *IBA paper* Chemins du Cyclotron,2 b-1348 Louvain-la-Neuve, Belgium

Kanai T, Kawachi K and Hiraoka T 1991 An irradiation facility and a beam simulation program for proton radiation therapy *Nucl. Instrum. Methods in Phys. Res.* A302, 158–164

Kanai T, Kawachi K, Kumamoto Y, Ogawa H, Yamada T and Matsuzawa H 1980 Spot scanning system for radiotherapy *Med. Phys.* **7** (4) 365–369

Kantor G, Destembert B, Breteau N and Schlienger M 1985 *Les protons en radiothérapy' Gazette Médicale* **92** 101–103

Kawachi K, Kanai T, Matsuzawa H and Inada T 1983 Three dimensional spot beam scanning method for proton conformation radiation therapy *Acta Radiol.* Supp 364

Khoroshkov V S, Barabash L Z, Barkhudaryan A V, Goldin L L, Lomanov M F, Onosovsky K K and Plyashkevich L N 1969 Proton beam of ITEP accelerator for radium therapy *Med. Radiol-Mosk.* **14** 58–62 (in Russian)

Kitigawa T 1988 Proton beam therapy *J. Rad. Res.* **29** 6

Koehler A M and Preston W M 1972 Protons in radiation therapy *Radiology* **104** 191–195

Koehler A M, Schneider R J and Sisterson J M 1975 Range modulators for protons and heavy ions *Nucl. Instrum. Methods in Phys. Res.* **131** 437–440

—— 1977 Flattening of proton dose distributions for large field radiotherapy *Med. Phys.* **4** 297–301

Larsson B, Leksell L and Rexed B 1963 The use of high energy protons for cerebral surgery in man *Acta Chirurg. Scand.* **125** 1–7

Larsson B, Leksell L, Rexed B, Sourander P, Mair W and Andersson B 1958 The high-energy proton beam as a neurosurgical tool *Nature* **182** 1222–1223

Larsson B, Liden K and Sarby B 1974 Irradiation of small structures through the intact skull *Acta Radiol. Ther. Phys. Biol.* **13** 512–534

Lennox A J 1991 Hospital-based proton linear accelerator for particle therapy and radioisotope production *Nucl. Instrum. Methods in Phys. Res.* B56/57 1197–1200

Miller D W, Slater J M, Awschalom M W and Allen M A 1987 Design of a hospital-based proton beam therapy centre *Med. Phys.* **14** 468

Montelius A, Grusell E, Russell K, Blomquist E, Pellettieri L and Lilja A 1991 Proton irradiation of arteriovenous malformations in the brain using a positioning system with titanium markers *Proc. 1st biennial ESTRO meeting on physics in clinical radiotherapy (Budapest, 1991)* p 31

Munzenrider J E, Austin-Seymour M, Blitzer P J, Gentry R, Goitein M, Gragoudas E S, Johnson K, Koehler A M, McNulty P, Moulton G, Osborne E, Seddon J M, Suit H D, Urie M, Verhey L J and Wagner M 1985 Proton therapy at Harvard *Strahlentherapie* **161** 756–763

Petti P 1992 Differential-pencil-beam dose calculations for charged particles *Med. Phys.* **19** (1) 137–149

Phillips M H, Pedroni E, Blattmann H, Boehringer T, Coray A and Scheib S 1992 Effects of respiratory motion on dose uniformity with a charged particle scanning method *Phys. Med. Biol.* **37** 223–234

Potts T M, Miller D, Prechter R, Prichard B and Slater J M 1988 The conceptual design of a patient handling system for isocentric proton beam therapy *Med. Phys.* **15** 798

Raju M R 1980 *Heavy particle radiotherapy* (New York: Academic) pp 188–251

Rossi B 1956 *High energy particles* (Englewood Cliffs: Prentice Hall) 2nd edn p 64

Sharma J, Bedi B S and Mukhopadhyay S 1991 Linear accelerator for cancer therapy—an indigenous effort *Electronics Information and Planning* **18** 167–177

Sisterson J M 1990 Overview of proton beam applications in therapy *Nucl. Instrum. Methods in Phys. Res.* **B45** 718–723

—— (ed) 1991 The proton therapy program at Louvain la Neuve *Particles Newsletter* no 8, June 1991

—— (ed) 1992 *Particles Newsletter* no 9, Jan 1992

Sisterson J M, Cascio E, Koehler A M and Johnson K N 1991 Proton beam therapy: reliability of the synchrocyclotron at the Harvard Cyclotron Laboratory *Phys. Med. Biol.* **36** 285–290

Sisterson J M and Johnson K N 1985 As we approach 3000: proton radiation therapy at the Harvard Cyclotron Laboratory *Med. Phys.* **12** 546

—— 1986 As we approach 3000: proton radiation therapy at the Harvard Cyclotron Laboratory *AAMD Journal* **11** 21–30

—— 1987 Proton radiation therapy: the Harvard cyclotron experience *Med. Phys.* **14** 469

Sisterson J M, Urie M M and Koehler A M 1987 Factors affecting the precise shaping of range in proton beams *Med. Phys.* **14** 468

Sisterson J M, Urie M M, Koehler A M and Goitein M 1989 Distal penetration of proton beams: the effects of air gaps between compensating bolus and patient *Phys. Med. Biol.* **34** 1309–1315

Slater J M, Archambeau J O, Miller D W, Notarus M I, Preston W and Slater J D 1992 The proton treatment center at Loma Linda university medical center: rationale for and description of its development *Int. J. Rad. Oncol. Biol. Phys.* **22** 383–389

Slater J M, Miller D W and Archambeau J O 1988 Development of a hospital-based proton beam treatment centre *Int. J. Rad. Oncol. Biol. Phys.* **14** 761–775

Slater J M, Slater J D, Archambeau J O and Miller D 1990 Proton therapy: a modality for expanding the parameters of precision radiation therapy *Loma Linda Radiology Medical Group Inc: Rad. Sci. report* **2** (1) Jan 1990

Suit H D, Goitein M, Munzenrider J, Verhey L, Urie M, Gragoudas E, Koehler A, Gottschalk B, Sisterson J, Tatsuzaki H and Miralbell R 1990) 'Increased efficacy of radiation therapy by use of proton beam *Strahlentherapie und Onkologie* **166** (1) 40–44

Suit H D, Goitein M, Tepper J, Koehler A M, Schmidt R A and Schneider R 1975 Exploratory study of proton radiation therapy using large field techniques and fractionated dose schedules *Cancer* **35** 1646–1657

Suit H D and Verhey L J 1988 Precision in radiotherapy *Brit. J. Radiol.* (Suppl 22) pp 17–24

Tobias C A, Lawrence J H, Lyman J, Born J L, Gottschalk A, Linfoot J, McDonald J 1964 Progress report on pituitary irradiation *Response of the nervous system to ionising irradiation* ed T J Haley and R S Snider (New York: Little, Brown and Co)

Tobias C A, Roberts J E, Lawrence J H, Low-Beer B V A, Anger H O, Born J L, McCombs R, Huggins C 1956 Irradiation hypophysectomy and related studies using 340 MeV protons and 190 MeV deuterons *Peaceful Uses of Atomic Energy* **10** 95–96

Tsunemoto H 1988 Radiation therapy with accelerated particles *J. Rad. Res.* **29** 6

Tsunemoto H, Morita S, Ishikawa T, Furukawa S, Kawachi K, Kanai T, Ohara H, Kitagawa T and Inada T 1985 Proton therapy in Japan *Rad. Res.* **8** (Suppl) S235–S243

Urie M, Goitein M, Holley W R and Chen G T Y 1986a Degradation of the Bragg peak due to inhomogeneities *Phys. Med. Biol.* **31** 1–15

Urie M, Goitein M, Sisterson J M and Koehler A M 1985 Proton dose distribution influenced by changes in drift space between patient and beam modifying devices *Med. Phys.* **12** 546

Urie M, Goitein M and Wagner M 1984 Compensating for heterogeneities in proton radiation therapy *Phys. Med. Biol.* **29** 553–566

Urie M, Sisterson J M, Koehler A M, Goitein M and Zoesman J 1986b Proton beam penumbra: effects of separation between patient and beam modifying devices *Med. Phys.* **13** (5) 734–740

Verhey L J and Munzenrider J E 1982 Proton beam therapy *Ann. Rev. Biophys. Bioeng.* **11** 331–357

Wagner M S 1982 Automated range compensation for proton therapy *Med. Phys.* **9** (5) 749–752

Webb S 1989 Optimisation of conformal radiotherapy dose distributions by simulated annealing *Phys. Med. Biol.* **34** 1349–1369

Wilson R R 1946 Radiological use of fast protons *Radiology* **47** 487–491

CHAPTER 5

CONFORMAL RADIOTHERAPY WITH A MULTILEAF COLLIMATOR

5.1. INTRODUCTION

Multileaf collimators (MLC) are regarded presently as the state-of-the-art method for generating arbitrary (and generally irregularly) shaped fields for x-ray therapy. They are still research tools in a number of specialized centres interested in conformation radiotherapy and are not widely available. They are also used for shaping neutron beams in view of the ease of varying the field shapes compared with moving very heavy volumes of ordinary block collimators (Eenmaa *et al* 1985, Chu and Bloch 1986,1987, Brahme 1988, Wambersie 1990, Bewley 1989). This chapter reviews the technology. Planning aspects of using a multileaf collimator, including the exciting possibility of tailoring the intensity of the radiation spatially across the field, are discussed in chapter 2. The multileaf collimator is sufficiently important to conformal radiotherapy to justify separate consideration here from other features of treatment machines which are discussed in chapter 7. The idea of moving leaves for radiation collimation is in fact rather an old one (Robinsohn 1906).

5.1.1. Early history

Two patents for means of collimating radiation may be seen as landmarks of some importance. The first (McGunnigle 1926/29) described apparatus for collimating not an ionizing radiation but an optical field. It is worthy of attention in that it describes how to collimate an irregular shape. The second (Green and McColm 1952/54) was a method for collimating ionizing radiation to square or rectangular fields, but with a novel approach which bears on MLCs.

The essence of McGunnigle's collimator may be grasped from figure 5.1. Four plates, shown as '9', are able to move relative to a groundplate '5', which is set up such that the direction of radiation is at right angles to it. Two of the plates '9' are on one side of the groundplate and two (shown dotted) on the other. Each plate is equipped with a pin '9' which is able to slide and rotate in slots '7'. Each plate

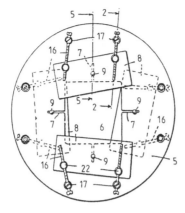

Figure 5.1. *McGunnigle's collimator for irregular fields. Four collimating plates '8' can be translated and rotated by pairs of screws to define an aperture with straight sides, but not necessarily rectangular. The application was not to collimate radiation fields, but apparatus of this type could be made to perform this task. (From McGunnigle (1924–26).)*

also carries two traveller posts '22' which are able to rotate relative to the plates. The groundplate is equipped with two similar posts '17' for each moving plate. Threaded rods '16' connect the posts '22' to the other posts '17', the unthreaded parts of the rods being held firmly at '17' so the rods can rotate but not slide. Conversely, the threaded parts of the rods pass through the posts '22' so that as the rods rotate the posts are moved along the length of the rods.

It should now be clear that by turning the two rods controlling each plate, the plate may be moved. If both screws turn the same way, the plate moves bodily forwards or backwards; if one screw is turned one way and the other turned the other way, then the plate can be made to rotate. Each of the four plates is controlled independently in the same way.

McGunnigle intended his apparatus for collimating projected light to specific shapes (for example illuminating a painting without the frame). If the plates were of appropriate thickness and material one could imagine a collimator for ionizing radiation built along the same lines. A mechanism similar to this idea has in fact been used very recently for a collimator for stereotactic radiosurgery (see chapter 3) (Leavitt *et al* 1991).

Green and McColm (1952/54) developed a collimator for a cobalt or therapeutic x-ray unit in which four lead blocks were able to move to create either a square or rectangular field. What was novel about this development was that the four blocks all lay in the same plane with the inside face of each one abutting the end face of its nearest neighbour in a clockwise direction (figure 5.2). As each block moved it carried with it that nearest neighbour, which clearly would then simply slide along the end face of the opposing block. The abutting faces were provided with tongue and grooving to eliminate stray radiation between the blocks. The confining of the four blocks to a single plane was important. This is essentially

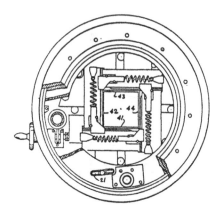

Figure 5.2. *Showing the arrangement used by Green and McColm for coplanar movement of four collimator blocks. (see text for description). (From Green and McColm (1952–54).)*

the same feature seen in MLCs; of course Green and McColm's apparatus could not provide for irregular fields.

Robinsohn (1906) patented an attenuating screen comprising multiple leaves to define an irregular field for a diagnostic x-ray set. The diagram of the apparatus (figure 5.3) does not show any method of automating the movement of the leaves. The idea resurfaced (figure 5.4) with the German patent by Maas and Alexandrescu (1982).

The idea of the multileaf collimator for *radiation treatment machines* appears to have originated with Gscheidlen (1959), who patented a device in which four sets of orthogonally disposed leaves could be moved to create an irregular field shape. The device is shown in figures 5.5 and 5.6 where, for simplicity, an arrangement with just five leaves per set is shown. The central leaf appears thicker than the other four in the set.

The leaves '10' were positioned upright in the direction of the central ray of the beam and the beam was laterally limited by the edges or walls of the upright collimator leaves. The leaves were disposed so as to move in the plane perpendicular to the central ray and supported by the plane surface '9'. The leaves were guided by the angle bars '12–15', and prism-shaped blocks '12a–15a' and screws '16, 17' were provided whereby the pressure on the leaves could be varied.

The leaves were moved by push rods '19' actuated by cranks '18', which in turn were driven by friction disks '21' carried by a shaft '20' by means of grooves and tongues and pressed against the cranks '18' by springs '23'. By turning the knob '26' the movement was actuated. The final resting position of each leaf was determined by a setting template '28' in which pins '30' could be positioned. The leaves would come to rest at these positions (which in general were different for each leaf) after which any further turning of the screw '26' would simply result in slipping at the friction rings. By means of endless pulleys '27', opposing pairs of leaves were simultaneously adjusted. In this way the shape of the irregular field

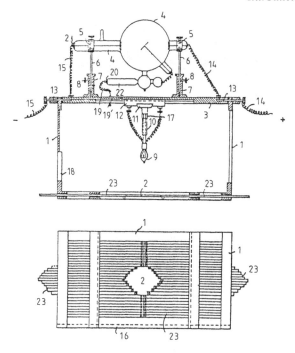

Figure 5.3. *The multi-element collimator for diagnostic radiology, patented by Robinsohn. The leaves are shown at '23' and were positioned by hand to give an irregularly shaped field '2'. (From Robinsohn (1906).)*

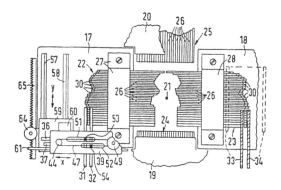

Figure 5.4. *Showing the collimator patented by Maas and Alexandrescu. This was more advanced than that of Robinsohn, in that it had the mechanism shown for moving the leaves '26'. Once again the application was in diagnostic radiology. (From Maas and Alexandrescu (1982).)*

was only limited by the quantization introduced by the width of the leaves and separation between the pins '30'.

The patent showed variations on this theme whereby several collimator systems

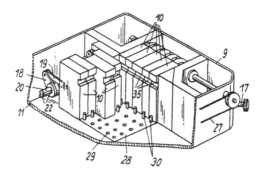

Figure 5.5.   *A schematic view of Gscheidlen's multileaf collimator patented in 1959, showing just two of the four sets of leaves. The actuating mechanism and the template for setting the field shape may be seen, as well as some of the guide bars (see text for description). (From Gscheidlen (1959).)*

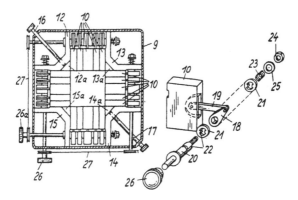

Figure 5.6.   *A plan view of Gscheidlen's multileaf collimator and, alongside, the details of the actuating mechanism for moving the leaves (see text for description). (From Gscheidlen (1959).)*

of this type might be arranged one above the other to provide greater collimation, and whereby arcuate leaves moving in arcuate guides (and multiple versions thereof) could be constructed to limit the field in such a way that the edge of the field lay tangential to the edges of the leaves.

These collimators were shown attached to a ^{60}Co treatment head. They embody most of the ideas germane to modern multileaf collimators with motor drives. However, the use of orthogonal pairs has not been carried over to the modern multileaf collimator. This orthogonal arrangement is possible in a single plane (unlike the conventional pairs of single lead-block collimators, one pair of which has to reside above the other), because there is no possibility of collisions between the orthogonally disposed collimators, since they comprise multiple leaves. This collimator does not appear to have been developed commercially, and indeed was technology ahead of its time.

5.2. MODERN DEVELOPMENTS

The concept of multileaf collimation is an attractive one since it dispenses with the cumbersome business of casting and positioning individual blocks. Under computer control the setting of arbitrary shapes can be very fast. There can be a high degree of conformity with the beam's-eye-view of the treatment volume with reduced irradiation of normal tissue. The shapes can be fine-tuned, and even varied during the course of fractionated delivery. The arrangement lends itself to image-driven treatment planning. A multileaf collimator opens up the possibility of using certain beam orientations which would not be useable with blocked fields because of collisions between the blocking tray and the patient. This is one of the exciting new possibilities with the multileaf collimator (Williams 1992).

Multileaf collimators can be characterized by:

1. number of leaf pairs,
2. width of leaves (or projected width at the isocentre),
3. compatability with accelerator type,
4. maximum field size,
5. the 'over-run distance' (see section 5.2.2).

A number of collimators are catalogued this way in table 5.1.

5.2.1. Brahme's patent

Brahme (1984, 1985/87, 1987, 1988) developed an MLC with 32 pairs of tungsten leaves which moved on a circular orbit. This so-called 'double-focused MLC' was characterized by a very small (5 mm) penumbra, comparable to that obtained with diverging beam blocks of low melting-point alloys. The high performance was also due to the helium atmosphere inside the head, reducing multiple electron scatter and the production of Compton electrons. The collimator accomodated x-rays between 2.5 and 50 MeV and also 50 MeV electrons. The projected leaf width at a 1 m isocentre was 1.5 cm, but the 'staircase' irregularity in the field shape due to the finite width of the leaves was smoothed in the resulting isodose contours. Similar observations were made for another collimator by Boyer *et al* (1991).

Brahme's (1985/87) patent contains details of a number of features which are desirable for a multileaf collimator and which improve on the rather simpler collimator patented by Gscheidlen (1959). These include:

- The leaves were curved such that both their upper and lower surfaces form part of an imagined circle having its centre at the source of radiation.
- The leaves were tapered in cross section and the group of 32 pairs, when viewed end on, have a fan-like appearance again centred at the source. This ensures that, whatever the positioning of the leaves, the tangents to all the edge surfaces converge at the source, ensuring a minimal penumbra. These two features were termed 'double focusing' (figure 5.7).

Table 5.1. *Details of the multileaf collimators developed to date for photon therapy.*

Author	Number of leaf pairs	Width of leaves (and/or * projected width at isocentre)	Accelerator	Over-run distance	Maximum field at 1m
Aoki (Tokyo)	9	10 mm / 20 mm *	Mitsubishi ML-20M	0	18 cm
Boesecke (Heidelberg)	28	4 mm / 7 mm *	Siemens Mevatron 77	15 mm	16 cm
Schlegel (Heidelberg)	27	3 mm / 5 mm *	Siemens Mevatron 77	15 mm	13.5 by 13.5 cm
Schlegel (Heidelberg)	40	1 mm / 1.7 mm *	Siemens Mevatron 77	20 mm	6.8 cm
Brahme (Stockholm)	32	7 mm / 15 mm *	Scanditronix	50 mm	40 by 32 cm
Ishigaki (Nagoya)	11	30 mm * (central pair) / 20 mm * (4 pairs) / 30 mm * (outer 6 pairs)	Mitsubishi ML-15M 3	50 mm	29 cm
Kobayashi (Nagoya)	11	25 mm (inner 9 pairs) / 87.5 mm (outer 2 pairs)	Toshiba LMR-4C TOKU	75 mm	
Mantel (Detroit)	12	14 mm	Toshiba LMR-16	0	30 cm
Bauer (Freiburg)	13	16 mm *	Philips SL 75/20	30 mm	12 by 20.8 cm
Uchiyama (Aichi)	11	20 mm *		50 mm	22 cm
Biggs (Mass. Gen.)	44	8 mm *		20 mm	35 by 33 cm
Varian	26	10 mm *	Clinac 2100C	200 mm †	40 by 26 cm
Philips	40	10 mm *	Philips SL	125 mm	40 by 40 cm ‡

† But no two leaves can differ in position by more than 16 cm.
‡ Only by completely ignoring the MLC and using secondary collimation.

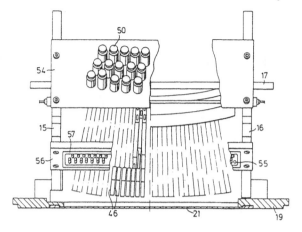

Figure 5.7. *Detailed schematic of the arrangement of the leaves in the double-focused mode. This is the 'end-on' view of the apparatus shown in figure 5.8. The leaves are focused to the source. '50' are the ends of the motors driving the leaves '46'. (From Brahme (1985/87).)*

- The leaves were supported on roller bearings (shown as '36', '37' and '38' in figure 5.8) and were driven by motors (such as at '50') driving screw threads '53' into attachments '51'.
- Each pair of leaves was provided with a step at their inner faces so that when completely closed there is little leakage of radiation (see '45' and '60' and figure 5.8).
- Adjacent leaves on the same side of the collimator were also provided with a step along the long edges for the same purpose.
- The field can be verified by a TV arrangement, and is also seen projected onto the skin surface in an analogous manner to conventional rectangular fields.

The collimator is shown in figure 5.9

5.2.2. *Ishigaki's MLC*

Multileaf collimators can be divided into two groups; those in which the leaves cannot pass beyond the line joining the source to the isocentre, and those in which they can. The latter are known as 'over-running' MLCs. Over-running MLCs allow the field to be tailored to a target volume (or part of a target volume) through which the axis of rotation does not pass. A collimator of this type has been described by Ishigaki *et al* (1988, 1990a,b) from Nagoya University, Japan (figure 5.10). Table 5.1 summarizes its features.

This MLC was attached to a Mitsubishi ML-15M 3 10 MV linear accelerator and used for both rotation therapy and multiple-fixed-port irradiation with irregularly shaped fields and a maximum of 16 ports. The planning was CT guided and the dose rate at each port could be varied. These authors did not state if this planning was guided by optimization theory (see chapter 2). They constructed a phantom,

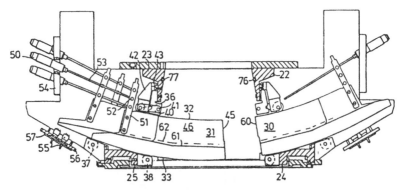

Figure 5.8. *Detailed schematic of the method of controlling the positions of opposing leaves in each pair for the multileaf collimator designed by Brahme. The leaves are shown at '30', '32', with faces '32', '45' moving on bearings '36'–'38' and driven by screw threads '53' controlled by motors '50' acting on connecting rods '51'. The rest of the diagram is supporting structure and the spaces were filled with helium gas at atmospheric pressure. The leaves are double-focused to the source (see figure 5.7). (From Brahme (1985/87).)*

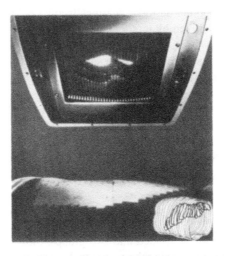

Figure 5.9. *The multileaf collimator head and part of the light-field projected on the patient using a 32 leaf-pair double-focused computer-controlled collimator of a medical microtron. The insert shows the patient contour and the shape of the target volume in each CT slice. (From Brahme (1988).)*

within which lay an irregular target volume. The planning problem demanded the over-running leaf facility. By experimenting with different placements for the rotation axis they found that the shape of each target area (defined as slices of the volume) could be fitted well by the 90% isodose contour when a full 360° rotation therapy was performed, and also by nine ports spaced at 40°.

Ishigaki *et al* (1990a,b) gave the accuracy of leaf setting as 2 mm. They

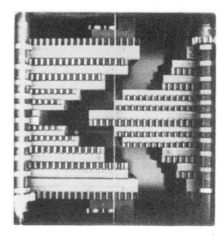

Figure 5.10. *View of the multileaf collimator developed by Ishigaki et al . There are 11 pairs of leaves and each leaf can over-run by 5 cm. An example of over-running is seen here. (From Ishigaki et al (1988).)*

used the overrunning collimators to delete a partial region within a field with multiple-fixed-field irradiation, claiming the result was better than using a gravity-assisted hanging-block method (see chapter 2). The computer-assisted fixed-field irradiation, with irregular fields shaped using the overrunning collimators, can produce more homogeneous target dose by changing the dose rate and deleting partial regions within a field, thus shielding critical organs. There is no need for the isocentre of rotation to be within the target volume.

5.2.3. Heidelberg's MLCs

Boesecke *et al* (1988) at the German Cancer Centre at Heidelberg have developed an MLC with 28 pairs of tungsten leaves, each 4 mm wide, arranged to provide a maximum field size of 16 cm at a distance of 1 m from the radiation source (figure 5.11). The collimator was constructed by the mechanical workshop of the Institute of Nuclear Medicine at the German Cancer Research Centre (DKFZ). The aim was to attach the MLC to a Siemens Mevatron 77 linear accelerator. By way of contrast with some other MLCs, in which the MLC is an integral undetatchable component, this one is accommodated in the accessory tray of the accelerator and is therefore somewhat limited in size. The surface of each blade has no step to prevent leakage radiation, but the leaves were double-focused to limit penumbra. Two clinical versions of the collimator were made with the number of leaves reduced to 27 pairs, each 3 mm wide (Schlegel 1992). One version of the collimator (the manual version) uses a template and clamps to set the field shape. In the motorized version there are six stepper motors: four control the movements of four leaves, and the leaves to be moved are selected by translating the leaf-moving assembly by two further stepper motors (Nahum and Rosenbloom 1989, Bortfeld 1992). Mechanical stresses, coupled with difficulties of coupling

Figure 5.11. *The tungsten multileaf collimator with 28 leaves per side developed at DKFZ Heidelberg, pictured in the mechanical workshop. This multileaf collimator is not motor driven; the leaves are set manually. (From Boesecke et al (1988).)*

powerful stepper motors, have been problems to overcome with the motorized version (Bortfeld 1992).

The manual Heidelberg collimator was used in a fixed-field mode for treatments with nine adapted field shapes spread over an angle of 330°. The field shapes were determined so as to match the beam's-eye-view of the target volume. This volume comprised not only the tumour, but a margin to accommodate potental tumour extensions and also positioning errors (figure 5.12). In turn, the beam's-eye-views were determined in a 3D treatment-planning package (also developed at this centre) using the contours of the volumes of interest determined from x-ray CT data. After planning, the isodose contours were displayed superposed on slices of interest, which could be transaxial, sagittal or coronal. Additionally, the facility to view three-dimensional isodose 'ribbons' superposed on three-dimensional shaded-surface anatomy was possible (figure 5.13). Other clinical applications were presented by Ésik *et al* (1991). The treatment-planning software (known as VOXELPLAN-Heidelberg) was described by Boesecke *et al* (1991).

The group at Heidelberg are currently developing an MLC with 1 mm wide leaves for accurate conformation in the brain (Bortfeld 1992).

5.2.4. *Kobayashi's MLC*

Kobayashi *et al* (1989) developed a multileaf collimator for a Toshiba LMR-4C TOKU linear accelerator operating at 4 MV. The MLC comprised 11 pairs of leaves, the inner nine being 25 mm wide and the outer two pairs being 75 mm wide. The leaf-pairs were able to over-run the mid-line to the rotation centre by some 75 mm. In addition, on the source side of the MLC, primary collimation was provided,

Figure 5.12. *The shape of the field generated for a multileaf collimator for an irregularly shaped target volume. The projection of the target volume can be seen as an array of black dots. The aperture of the collimator includes a 'safety zone' to accomodate potential tumour extensions and possible movement or positioning error. (From Boesecke et al (1988).)*

Figure 5.13. *A three-dimensional representation of a target volume and spinal chord with an 80% isodose (delivered by multiple-static-MLC-shaped fields) presented as ribbons. In the original picture the surface and the ribbons were shown in different colours. Provided the surface does not protrude through the ribbons, one can be certain that the dose to the target has reached the value labelling the isodose, or is above that. (From Boesecke et al (1988).)*

and on the patient side a wedge could be arranged if required. The unusual feature of this development was the incorporation of a movable wedge-shaped compensating filter (figure 5.14). Each pair of MLC leaves was accompanied by

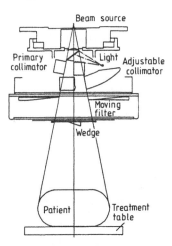

Figure 5.14. *Showing the position of the* MLC *in relation to the additional collimating elements for a Toshiba LMR-4C TOKU 4 MV linear accelerator. The moving filter is on the patient side of the assembly. (From Kobayashi et al (1989).) (Reprinted with permission from Pergamon Press Ltd, Oxford, UK.)*

a corresponding pair of wedge filters, constructed from aluminium with a wedge angle of $\tan\theta = 0.1$. As the MLC leaves varied in position, the location of the corresponding wedge filters was adjusted so that the tip of each filter just touched the beam edge defined by the ipsilateral leaf (figure 5.15). This geometrical arrangement ensured that the filter thickness varied linearly with increasing width of the field defined by each leaf pair. For example, when the width of the x-ray beam at the rotation centre was 200 mm, the filter was 76mm thick, but when the width of the beam was 100 mm, the filter was only 38 mm thick. Since the MLC leaves were able to traverse the mid-line to the isocentre, the movable wedges followed suit.

Kobayashi *et al* (1989) made measurements with film dosimetry for a number of irregularly shaped target distributions (the MLC leaves being adjusted to match the geometry) in round and oval phantoms. These experiments mimicked the way the leaves (and wedge filters) would be adjusted in clinical practice to accomodate irregular target volumes determined from multiple CT images. It was observed that the dose uniformity within the volume of the target was better when the wedge-filters were in use than when the x-ray field was shaped only by use of the MLC leaves alone. This was particularly the case when the calculated dose distributions were compared with the measured dose distributions for those parts of the target volume which were smaller than the widest part of the field and/or off-centre. The dose profiles were also sharper, showing that using the wedge-shaped filters protected the surrounding normal tissues more effectively from a high dose of irradiation.

The rationale for use of a variable-thickness compensator may lie in the well known observation that the penetration of a beam decreases as the field size

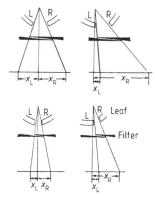

Figure 5.15. *Showing the relationships between the leaf movement and the wedge-filter movement for the collimation shown in figure 5.14. The tip of each wedge just touches the ipsilateral field edge. Four different positions of the collimator–wedge combination are shown. (From Kobayashi et al (1989).) (Reprinted with permission from Pergamon Press Ltd, Oxford, UK.)*

increases for a given megavoltage radiation (Green and Williams 1983). This is partly due to increased scatter as the field size increases.

5.2.5. *Mantel and Perry's MLC*

Mantel *et al* (1984) and Perry *et al* (1984) have also described the construction of an MLC with 12 leaf pairs of width 14 mm, attachable to a Toshiba LMR-16 linear accelerator. One somewhat unusual feature of their system was that the control motors and circuitry for activating the leaf movements were external to the collimator head. The system was used for dynamic rotation therapy.

5.2.6. *Aoki's MLC*

Aoki *et al* (1987) created an integrated radiotherapy treatment system in which imaging for planning and radiotherapy treatment were performed in the same room with the patient on a swivelled couch. A Mitsubishi multileaf collimator under computer control was installed. This was used for dynamic conformation therapy (which the authors called multileaf offset rotation—MOR) and fixed-port conformation (multileaf offset multiport—MOMP), but they also allowed the use of several axes of rotation when the treatment-planning problem called for it. The dynamic and static multi-axis techniques were dubbed multileaf offset multiaxis rotation—MOMAR and multileaf offset multiaxis multiport—MOMAMP. The development is also discussed in chapter 7 (section 7.3)

5.2.7. *Uchiyama's MLC*

An MLC with leaves 2 cm wide at the axis of rotation was also included by

Uchiyama *et al* (1986, 1990) in a system for conformal therapy based on the beam's-eye-view of the tumour. The 11 leaves could cross the mid-line by 5 cm.

5.2.8. Bauer's MLC

An MLC with 13 leaves was constructed at the University of Freiburg and used for multiport irradiation (10 fields arranged over 360° degrees) of oesophageal carcinoma (Bauer 1992). The leaves diverge in the longitudinal but not in the lateral field direction.

5.2.9. Biggs' MLC

Another home-made MLC has been constructed at the Massachussetts General Hospital (Biggs 1991). It has 44 pairs of tungsten leaves, independently motorized and capable of taking up position in 15 s. The leaves are focused as a sheaf lengthwise, but not in the orthogonal direction. Each leaf width projects to 8 mm at the isocentre; the penetration through the interleaf space is 8%. The leaves can only move a cm or two across the mid-line. There is no verification system as yet. The drive shafts are rigid and make the housing for the collimator quite bulky. The largest field size is 35 by 33 cm. The aim was to use the MLC to cut down on the number of custom-cast blocks and possibly, in time, achieve dynamic therapy.

Biggs *et al* (1991) gave their reasons for deciding on a single- rather than a double-focusing set of leaves (figure 5.16). To ensure the penumbra was acceptably small when very large fields were set, the leaves were focused in the plane orthogonal to the plane of the leaves (i.e. viewed end on, the tangents to the long edges of the leaves converged to the focus.) However, focusing was eliminated in the plane of the leaves. Biggs *et al* (1991) conducted a series of experiments comparing the penumbra produced by a lead block with a straight edge, a lead collimator with an edge focused to the source, a lead block having a straight edge with a single step of 0.5 mm, and finally with a lead block having a straight edge with a single step of 1 mm. Comparison was also made with the penumbra from a tungsten block with a straight edge. The width of several penumbra (95%–50%, 90%–50%, 80%–50%) was measured for different field sizes, at different depths in water for a 10 MV x-ray beam on a Clinac 18 accelerator. No statistically significant differences in penumbra were found at depth, whichever geometrical arrangements were compared. Some differences were noted at small depths in water.

This enabled Biggs *et al* (1991) to conclude that there was no need for the leaves to move on the arc of a circle, nor to have rounded ends, simplifying the construction of their multileaf collimator. The same conclusion (for fields less than 15 by 15 cm) was reached by Dries (1991).

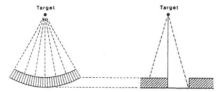

Figure 5.16. *A so-called 'single-focused' multileaf collimator. The leaves are focused to the source when viewed 'end-on' (left), but move in a plane and are not focused in the orthogonal direction (right). This arrangement should be contrasted with that shown in figures 5.7 and 5.8. (From Biggs et al (1991).)*

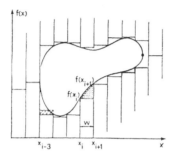

Figure 5.17. *Showing the fitting of a planar target area with a multileaf collimator. The dotted area is the excess region treated by the ith leaf. The best orientation of the leaves relative to the area is that which minimizes the sum of such dotted areas of excess. (From Brahme (1988).)*

5.3. BRAHME'S THEORY OF ORIENTATION

Brahme (1988) provided the arguments to answer the question of the optimal angulation of the MLC leaves (at some particular static orientation relative to the target volume). This led to the result that for a simple convex target volume the direction of motion of the leaves should be in the direction in which the volume presents its smallest cross section, i.e. the longest diameter of the target should be covered with as many leaves as possible. The argument can be best understood from figure 5.17.

Let the solid line represent the boundary of the target region and the leaves be of width w. Let $f(x)$ describe the boundary. For leaf i the overdosage is

$$O_i = \int_{x_i}^{x_{i+1}} (f(x) - f(x_i))\, dx. \tag{5.1}$$

If the boundary is monotonic and reasonably slowly varying, then it may be Taylor expanded about x_i with only the first three terms retained. This gives

$$O_i = \int_{x_i}^{x_{i+1}} \left[f'(x_i)(x - x_i) + f''(x_i)(x - x_i)^2/2 \right] dx. \tag{5.2}$$

Integrating, we obtain

$$O_i = w \left(f'(x_i) \, w/2 + f''(x_i) \, w^2/6 \right). \tag{5.3}$$

To obtain the total overdosage, integrate the expression

$$O = \int O_i = \int_{x_0}^{x_n} \left(f'(x) \, w/2 + f''(x) \, w^2/6 \right) dx \tag{5.4}$$

from one side of the target region to the other to obtain

$$O = \left(f(x_n) - f(x_0) \right) w/2 + \left(f'(x_n) - f'(x_0) \right) w^2/6. \tag{5.5}$$

From this we see that:

1.   the error is proportional to leaf width to first order, as one may expect,
2.   the error is proportional to the variation in the target area in the direction of the leaf movement,
3.   the leaves should be aligned to minimize the opening of the collimator from the fully closed position.

The problem reduces to finding the optimum way of arranging the leaves so as to minimize the volume (represented by an area 'seen' in the beam's-eye-view) of normal tissue outside the target volume. This unwanted area arises because target regions do not have convenient stepped outlines and cannot be precisely matched by MLCs with leaves of finite width. Indeed, by considering a large number of possible ways of fitting an elliptical area with an MLC of leaf width 1 cm, Brahme (1988) showed that poor arrangements could lead to up to 20% more area being treated than necessary. However, even this poorer arrangement was not as bad as making use of a rectangular field which would treat some 35% extra area. The reader could draw other geometrical target areas and go through the same type of investigation of possibilities leading to different figures for the overdosage, but the principle would be the same: using an MLC reduces the dose to normal tissue outside the target area. When several MLC-defined fields are combined, the advantages are even greater.

The problem becomes worse if the target area is not convex. Then (see figure 5.18) the range of directions which the leaf motion can take is restricted to less than 360°. The figure shows that as the concavity becomes deeper the range of angles becomes smaller. Indeed, if the opening to the cavity is smaller than its interior, it is no longer possible to fit the target area however small the leaf width is. Such cases as these could only be accommodated by dividing the field into sub-fields, collimated one after the other. The problem becomes even more difficult if there are multiple concavities. Using sub-fields may create matching difficulties and is inherently less attractive.

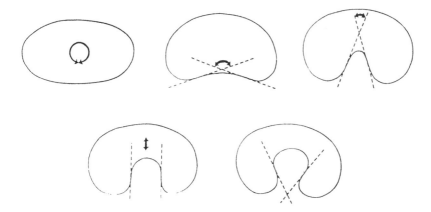

Figure 5.18. *The possible range of orientation (solid arrows) of the direction of movement of collimator leaves for convex and convex/concave outlines. As the concavity increases, so the range of possible angles falls. The dashed lines are the tangents at the inflection points of the target outline. The region in the bottom right-hand figure cannot be treated by a single setting of the leaves of a multileaf collimator because the opening is narrower than the region to be treated within the concavity. Two adjacent fields could overcome this, each collimated separately by the multileaf collimator. (From Brahme (1988).)*

5.4. OPTIMIZED BEAM PROFILES

Källman *et al* (1988) developed a method for delivering a dose by scanning an elementary slit beam from an MLC. The density of application of such a slit for the collimator is computed from a prescription of the (2D) target area, based on an iterative method. Let us define the three-dimensional dose distribution from an elementary slit beam by $d(r)$. This is the most elementary dose distribution which can be generated from a multileaf collimator—that corresponding to all the leaves shut with just one leaf marginally open. Such a dose distribution may be calculated from the basic point-spread functions for monoenergetic photon interactions (Ahnesjö *et al* 1987). The dose distribution may then be approximated by a continuum of collimated elementary slit beams, each turned on for a different time, which is called the density of application. Suppose the density of application of such slit beams is $\Phi(r)$, then the dose is given by

$$D(r) = \int \int \int \Phi(r') \, d(r - r') \, \mathrm{d}^3 r'. \tag{5.6}$$

Here we assume that the elementary distribution is independent of position and then this is a convolution integral (in three dimensions). Because D and d are known, the density Φ can be computed by deconvolution. Källman *et al* (1988) perform this iteratively via:

$$\Phi_0(r) = aD(r) \tag{5.7}$$

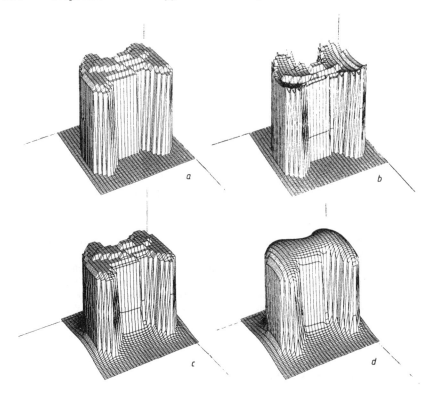

Figure 5.19. *Showing (a) a target volume (cancer of the cervix) with graded doses to the involved regional lymph nodes, (b) the irradiation density computed by the method of Källman et al , (c) the resulting dose distribution according to the irradiation density in (b) for* ^{60}Co *irradiation, and (d) the resultant dose distribution for a flat field. (From Källman et al (1988).)*

$$\Phi_{k+1}(r) = C\left(\Phi_k(r) + a\left(D(r) - \Phi_k(r)\bigotimes d(r)\right)\right) \qquad (5.8)$$

where the subscript k refers to the kth iteration, C is a positivity constraint operator and a controls the speed of convergence. This is formally identical to iterative methods to remove blurring from medical images (see e.g. Webb 1988). It is also formally similar to the iterative methods introduced by the same authors (Lind and Källman 1990) used to perform 'inverse computed tomography' (see chapter 2), but note that in that case the kernel d has a completely different meaning, being the *point irradiation density* from rotation therapy.

The opening density Φ is a measure of the dose monitor units that should be delivered, or alternatively the length of time the collimator should be opened locally, assuming constant accelerator dose rate. The irradiation could also be accomplished by a moving elementary slit beam with the dwelling time at each location given by Φ. An example of the results of this method is in figure 5.19.

Similar iterative techniques were used by Lind and Brahme (1985, 1987) to

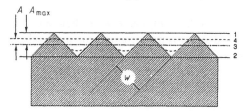

Figure 5.20. *Schematic diagram of the projection of an MLC field edge with the direction of travel of the leaves at 45° to the desired field edge. The area blocked by the MLC is indicated by shading. The leaf width is 'w'. Line 1 represents the location of the field edge for 'exterior field blocking', line 2 represents the edge of the field for 'interior field blocking', line 3 represents the field edge for 'leaf centre insertion', and line 4 represents the location of the field edge for 'variable insertion' blocking strategy. The sinusoid represents the approximate location of the 50% isodose contour. The strategies are discussed in section 5.5. (From Zhu et al (1992).)*

determine the optimum scanning density of pencil beams from the desired fluence profile.

The use of multiple geometrically shaped multileaf-collimated radiation fields for optimizing 3D planning is discussed in chapter 2 (Morrill *et al* 1991, Mohan 1990, Webb 1990, 1991, 1992).

5.5.   HOW THE LEAF POSITIONS MAY BE DETERMINED

Since each MLC has leaves of a finite width, the shape of any collimated field has irregular staircase-like edges. Imagine the MLC leaves approaching a straight-line field edge at 45°, as shown in figure 5.20. The question arises of how to place the leaves in relation to the field edge, and Zhu *et al* (1992) have provided the data to answer this question. If the leaf width is 'w', the saw-tooth irregularity will translate to a sinusoidal variation in dose at depth with the maximum amplitude of the variation being

$$A_{max} = w/\sqrt{(8)}. \tag{5.9}$$

In practice, measurements of the dose variation, with film dosimetry for cast cerrobend blocks, double-focused to simulate an MLC, showed that amplitude A of the 50% isodose contour (see figure 5.20) was some 1.2 mm smaller than A_{max} for 6 MV radiation and was some 1.4 mm smaller than A_{max} for 18 MV radiation. The difference between this amplitude and the value of A_{max} was *independent of the width 'w' of the leaves*, a particularly convenient and useful result. The sharp serrations of the leaves were thus 'rounded out' in the dose at depth (as also discussed by Brahme).

Several options for placing the MLC leaves naturally arise:

1. The inner corners of the leaves are tangent to the field edge. No leaves project

into the field. Zhu *et al* (1992) call this 'exterior field blocking'; the entire target volume is thus exposed, but this may irradiate too much adjacent normal tissue.

2. The outer corners of the leaves are tangent to the field edge. Zhu *et al* (1992) call this 'interior field blocking'; no normal tissue is exposed to the target dose, but, since the leaves project into the field, some target locations may be underdosed.

3. The edge of the field bisects the ends of the leaves. Zhu *et al* (1992) call this 'leaf-centre insertion'; the 50% contour wanders in and out of the field.

4. The edge of the field is set so the leaves protrude into the field by A_{max} minus 1.2 mm (6 MV) or minus 1.4 mm (18 MV). Zhu *et al* (1992) call this 'variable insertion' and this guarantees that the 50% isodose contour will undulate up to and outside the field margin.

Zhu *et al* (1992) investigated these four possibilities, confirming the predictions by film dosimetry. They also showed that convolution dosimetry techniques (see chapter 2) would predict the dose distributions to within 3 mm accuracy and take account of variable field shape, scatter and electron transport.

5.6.   SYSTEMS COMMERCIALLY AVAILABLE IN 1992

The system described in section 5.2.1, due to Brahme, is available commercially via Scanditronix. That described by Aoki *et al* (1987) is available from Mitsubishi. Two other systems are commercially available, although only a small number are in service worldwide in 1992. These are from Varian and Philips.

5.6.1.   *The Varian MLC*

The Varian MLC (figure 5.21) has 26 leaves, each of width 0.5 cm projecting to 1 cm at a 1 m isocentre (Varian 1992). The leaves can retract to 20 cm either side of the mid-line, giving a maximum field of 40 cm by 26 cm. The leaves may over-run by the full 20 cm, but, when they do so, no two leaves may differ in translated position by more than 16 cm because of the limitations of the secondary collimation which allows this large over-run. All leaves move in a plane (rather than on the arc of a circle) and have rounded ends to limit penumbra. No two leaves can close to more than within 2 mm at isocentre. The collimator is detachable from the accelerator, which retains the conventional block tray and wedge support mechanisms below it. Experiments with the shape of the leaf ends were performed in collaboration with the University of Pennsylvania and the final design (partially rounded, partially flat) gave a penumbra of some 4 mm at 6 and 15 MV for various leaf positions (Galvin 1990). The average transmission of the leaves themselves was 4% of the open field (the leaves were 5 cm thick) and the transmission through the gaps between the leaves added a further 1%. The leaf sides were double-stepped to minimize this extra penetration. All leaves were focused to the source in the direction orthogonal to the leaf movement. The collimator weighs 200 lb.

Figure 5.21. *The Varian multileaf collimator. (From Varian product literature.)*

The leaves can be positioned to an accuracy better than 1 mm with a repeatability of 0.1 mm and presently the speed of leaf movement is 1.5 cm s^{-1}. The leaf position is verified by measuring potentiometers, rather than positive identification with a TV system. At present the leaves cannot be moved when the gantry is rotating, so, when calculating the times to reset fields, these have to be the sum of sequential gantry rotation and leaf movement times. At present the interaction with the planning process is under development. The MLC is an integral part of the accelerator head, but is controlled by an independent computer. There is a facility to enter the required leaf positions by digitizer tablet. An outline can be traced and then one of three options exercised: either the leaves completely expose the field within the contour, or completely shield the region outside the contour, or an option in-between (Kay 1992). The user interface is not yet standardized and is specific to one site. In time, the MLC interface will have access to the same data-base as the linac control. The wedge can be orthogonal to the direction of movement of the leaves.

Boyer *et al* (1991) have reported measurements of penumbra and transmission, and also the stair-step shaping of the isodoses at depth.

5.6.2. *The Philips MLC*

The Philips MLC has 40 leaves, each projecting to 1 cm at an isocentre of 1 m. The blades are 7 cm thick, giving a mean transmission of some 1.8% at 6 MV and 2% at 20 MV. The leaves may over-run by 125 mm and can be positioned to within an accuracy of 0.25 mm in 10 s. The leaves have rounded ends and move in a plane. There is a single step between adjacent leaves and the transmission through the step is a further 2.8% at 6 MV and 3.1% at 20 MV (Williams 1990, Williams and Jordan 1991). In addition, there are four conventional collimators;

two for x-collimation and two for y-collimation (the direction of movement of the MLC leaves). The y-collimator has transmission 12.3% at 6 MV and 15% at 20 MV, and the corresponding figures for the x-collimator are 1.2% and 1.4%. When an irregular field is collimated, the total leakage will therefore depend on which combinations of collimators come into play, and can be deduced by producting the appropriate fractional leakages. In addition, scatter from the primary radiation in the open field will of course contribute into these areas. Williams and Jordan (1991) found that the 80% isodose contour was about 4 mm inside the leaf tips and the 50% contour went through the mid-points of the ends of the leaves.

The leaf positions are verified by a TV system. The leaves diffusely reflect light onto the face of a CCD TV camera. The source of light is a halogen bulb. The apex of the wedge is parallel to the direction of movement of the leaves.

The Manchester group work out the dosimetry for this collimator using integrations over experimentally determined empirical scatter functions (Hounsell and Wilkinson 1990). The Philips collimator presently comes with three methods of setting the field shapes (Jordan and Williams 1991):-

1. regular rectangular fields are the simplest to set,
2. predefined shapes can be set via eight parameters,
3. irregular fields require the input of some 80 parameters, essentially two for each leaf.

Williams and Kane (1991) found that the third method, time consuming though it was, was needed for many of the fields set at Manchester.

Kirby and Williams (1991) used the MLC in conjunction with a Philips SRI-100 MVI system to set up the fields for conservative breast irradiation. They made use of the ability to set fields where one edge corresponded to the central beam (so-called 'half-fields') so that match-line problems were minimized between the tangential and the AP fields (Kirby and Kane 1991).

Hounsell *et al* (1992) have described how the treatment field is determined from either digitized simulator radiographs or a digitized image-intensifier video signal or from a digitally reconstructed radiograph produced from CT data. The Manchester group have created a method of image display which allows the locations of the MLC leaves to be overlaid on the digital image-display system. They describe how the leaves may be positioned to satisfy one or other of the alternative strategies discussed by Zhu *et al* (1992). An additional use of these digital data is the means to compare with digital portal images for position verification (Moore 1992).

The leaf positions are held in a 'prescription file', which is passed to the computer of the treatment unit. Hounsell *et al* (1992) have found that, with a typical treatment preparation time of only some 30 minutes, the use of the MLC has led to a high through-put of patients with conformal fields.

The second Philips MLC to be installed in the UK has recently arrived at the Royal Marsden Hospital, Sutton and a third is at London's Royal Free Hospital.

5.7.   MULTILEAF COLLIMATORS FOR PROTON THERAPY

In chapter 4 we discussed the use of high-energy protons for radiation therapy. Proton fields may be shaped geometrically by sweeping elementary pencils of protons with scanning magnets, or by expanding a small-area field with a combination of scattering foils and beam-stopper. In the former case, a multileaf collimator can be used as a back-up collimator, as in the irradiation facility at Chiba, Japan (Tsunemoto *et al* 1985, Kanai *et al* 1991). In the latter case, the multileaf collimator is the only beam-shaping device, as in the facility at Tsukuba, Japan (Tsunemoto *et al* 1985).

5.8.   SUMMARY

We should be somewhat surprised that it has taken so long to engineer multileaf collimators. It is self-evident that avoiding primary radiation at non-target-volume sites is desirable and that custom-built blocking is tedious (Spelbring *et al* 1987). However, the engineering difficulties should not be underestimated and those centres who have chosen to build their own collimators have reported that it has not been an easy task. The main engineering problems are machining tungsten accurately (including the steps at the interfaces), arranging the motorization of the leaf movements and verifying the leaf positions. The 'interface' to the planning computer and to the treatment machine itself also has to be engineered. At the time of writing, MLCs are still not widely available, so it is too early to make authoritative statements concerning their utility. With an eye of faith, one expects to see them become the standard treatment accessory to the machine head.

REFERENCES

Ahnesjø A, Andreo P and Brahme A 1987 Calculation and application of point spread functions for treatment planning with high energy photon beams *Acta Oncol.* **26** 49–56

Aoki Y, Akanuma A, Karasawa K, Sakata K, Nakagawa K, Muta N, Onogi Y and Iio M 1987 An integrated radiotherapy treatment system and its clinical application *Rad. Med.* **5** 131–141

Bauer J, Hodapp N and Frommhold H 1992 A comparison of dose distributions with and without multileaf-collimator for the treatment of oesophageal carcinoma *Proc. ART91 (Munich)* (abstract book p 49) *Advanced Radiation Therapy: Tumour Response Monitoring and Treatment Planning* ed A Breit (Berlin: Springer) pp 649–653

Bewley D K 1989 *The physics and radiobiology of fast neutron beams* (Bristol: Adam Hilger) pp 69–71

Biggs P 1991 private communication (AAPM travelling fellow visit, May 21st 1991)

Biggs P, Capalucci J and Russell M 1991 Comparison of the penumbra between focused and nondivergent blocks—implications for multileaf collimators *Med. Phys.* **18** (4) 753–758

Boesecke R, Becker G, Alandt K, Pastyr O, Doll J, Schlegel W and Lorenz W J 1991 Modification of a three-dimensional treatment planning system for the use of multi-leaf collimators in conformation radiotherapy *Radiother. Oncol.* **21** 261–268

Boesecke R, Doll J, Bauer B, Schlegel W, Pastyr O and Lorenz W J 1988 Treatment planning for conformation therapy using a multileaf collimator *Strahlentherapie und Onkologie* **164** 151–154

Bortfeld T 1992 private communication

Boyer A L, Nyerick C E and Ochran T G 1991 Field margin characteristics of a multileaf collimator *Med. Phys.* **18** 606

Brahme A 1984 Multileaf collimation of neutron, photon and electron beams *Proc. 3rd Ann. ESTRO meeting (Jerusalem, 1984)* p 218

—— 1985/87 Multileaf collimator *US Patent* no 4672212

—— 1987 Design principles and clinical possibilities with a new generation of radiation therapy equipment *Acta Oncologica* **26** 403–412

—— 1988 Optimal setting of multileaf collimators in stationary beam radiation therapy *Strahlentherapie und Onkologie* **164** 343–350

Chu J C H and Bloch P 1986 Static multileaf collimator for fast neutron therapy *Med. Phys.* **13** 578

—— 1987 Static multileaf collimator for fast neutron therapy *Med. Phys.* **14** 289–290

Dries W J F 1991 Comparison between the penumbra of focused and straight standard shielding blocks *Proc. 1st biennial ESTRO meeting on physics in clinical radiotherapy (Budapest, 1991)* p 67

Eenmaa J, Kalet I and Wootton P 1985 Dosimetric characteristics of the University-of-Washington clinical neutron therapy system (CNTS) multileaf variable collimator *Med. Phys.* **12** 545

Ésik O, Bürkelbach J, Boesecke R, Schlegel W, Nṁeth G and Lorenz W J 1991 3-dimensional photon radiotherapy planning for laryngeal and hypopharyngeal cancers: 2 Conformation treatment planning using a multileaf collimator *Radiother. Oncol.* **20** 238–244

Galvin J 1990 private communication

Green D T and McColm C S 1952–54 Beam therapy collimating unit *US Patent no* 2675486

Green D and Williams P C 1983 *Brit. J. Radiol.* (Suppl 17) 61–86

Gscheidlen W 1959 Device for collimation of a ray beam *US Patent no* 2904692

Hounsell A R and Wilkinson J M 1990 Data for dose computation in treatments with multileaf collimators *The use of computers in radiation therapy: Proceedings of the 10th Conference* ed S Hukku and P S Iyer (Lucknow) pp 40–43

Hounsell A R, Sharrock P J, Moore C J, Shaw A J, Wilkinson J M and Williams

P C 1992 Computer-assisted generation of multi-leaf collimator settings for conformation therapy *Brit. J. Radiol.* **65** 321–326

Ishigaki T, Itoh Y, Horikawa Y, Kobayashi H, Obata Y and Sakuma S 1990a Computer assisted conformation radiotherapy *The use of computers in radiation therapy: Proceedings of the 10th Conference* ed S Hukku and P S Iyer (Lucknow) pp 76–79

—— 1990b Computer assisted conformation radiotherapy *Med. Phys. Bulletin of the Association of Medical Physicists of India* **15** (3,4) 185–189

Ishigaki T, Sakuma S and Watanabe M 1988 Computer-assisted rotation and multiple stationary irradiation technique *Europ. J. Radiol.* **8** 76–81

Jordan T J and Williams P C 1991 Commissioning and use of multi-leaf collimator system *Proc. 1st biennial ESTRO meeting on physics in clinical radiotherapy (Budapest, 1991)* p 3

Källman P, Lind B, Eklof A and Brahme A 1988 Shaping of arbitrary dose distributions by dynamic multileaf collimation. *Phys. Med. Biol.* **33** 1291–1300

Kanai T, Kawachi K and Hiraoka T 1991 An irradiation facility and a beam simulation program for proton radiation therapy *Nucl. Instrum. Methods in Phys. Res.* A **302** 158–164

Kay A 1992 Varian (Crawley) Open Day Lecture. February 21st 1992

Kirby M C and Kane B 1991 Megavoltage imaging—its role in breast treatment verification *Rad. Magazine* (Sept 1991) **17** (196) 22–23

Kirby M C and Williams P C 1991 Initial experience of megavoltage imaging for breast treatment verification *Proc. Radiology and Oncology 91 (Brighton, 1991)* (London: BIR) p 82

Kobayashi H, Sakuma S, Kaii O and Yogo H 1989 Computer-assisted conformation radiotherapy with a variable thickness multileaf filter *Int. J. Rad Oncol. Biol. Phys.* **16** 1631–1635

Leavitt D D, Gibbs F A, Heilbrun M P, Moeller J H and Takach G A 1991 Dynamic field shaping to optimise stereotactic radiosurgery *Int. J. Rad. Oncol. Biol. Phys.* **21** 1247–1255

Lind B and Brahme A 1985 Generation of desired dose distributions with scanned elementary beams by deconvolution methods *Med. Bio. Eng. and Comp.* **23** (Suppl 2) 953

—— 1987 Optimisation of radiation therapy dose distributions using scanned electron and photon beams and multileaf collimators *The use of computers in radiation therapy* ed Bruinvis *et al* (Proc. 9th ICCRT) pp 235–239

Lind B K and Källman P 1990 Experimental verification of an algorithm for inverse radiation therapy planning *Radiother. Oncol.* **17** 359–368

Maas W and Alexandrescu M 1982 Röntgenuntersuchungsgerät mit einer fernsteuerbaren Primärstrahlenblende *German patent app* DE 3030332 A1

Mantel J, Lefkofsky M A, Higden D and Perry H 1984 Use of a multileaf collimator for dynamic rotation therapy *Proc. 3rd Ann. ESTRO meeting (Jerusalem, 1984)* p 219

McGunnigle H A 1926–29 Apparatus for altering the shape and size of a projected beam of light *US Patent no* 1709626

Mohan R 1990 Clinically relevant optimisation of 3D conformal treatments *The use of computers in radiation therapy: Proceedings of the 10th Conference* ed S Hukku and P S Iyer (Lucknow) pp 36–39

Moore C J 1992 Correlation of megavoltage and simulator images *Proc. BIR Conf. on Megavoltage Portal Imaging (London, 1992)* (London: BIR)

Morrill S M, Lane R G and Rosen I 1991 Constrained simulated annealing for optimised radiation therapy treatment planning *Comp. Meth. Progr. Biomed.* **33** 135–144

Nahum A and Rosenbloom M E 1989 private communication (notes on a visit to Heidelberg, July 1989)

Perry H, Mantel J, Vieau D M, Higden D L and Lefkofsky M M 1984 Dynamic treatment: automation of a multileaf collimator *Proc. 8th Int. Conf. on the Use of Computers in Radiation Therapy (Toronto, 1984)* (IEEE Computer Society, Toronto) pp 479–481

Robinsohn I 1906 Apparat zur radiologischen Belichtung lebender und lebloser Objekte *German patent no* 192300

Schlegel W 1992 private communication to S Webb (visit to Heidelberg, October 1992)

Spelbring D R, Chen G T Y, Pelizzari C A, Awan A and Sutton H 1987 A computer simulated evaluation of multivane collimator beam delivery systems *The use of computers in radiation therapy* ed Bruinvis et al (Proc. 9th ICCRT) pp 153–156

Tsunemoto H, Morita S, Ishikawa T, Furukawa S, Kawachi K, Kanai T, Ohara H, Kitagawa T and Inada T 1985 Proton therapy in Japan *Rad. Res.* **8** (Suppl) S235–S243

Uchiyama Y, Kimura C, Ueda T, Morita K and Watanabe M 1986 Systematization of radiotherapy units by computer system *Rad. Syst. Res. Suppl.* (Suppl. 3) pp 217–220

Uchiyama Y, Morita K, Kimura C, Mizutani T and Fuwa N 1990 Integrated system of the computer-controlled conformation radiotherapy *The use of computers in radiation therapy: Proceedings of the 10th Conference* ed S Hukku and P S Iyer (Lucknow) pp 52–55

Varian 1992 *Varian's multileaf collimator* Varian Associates product literature

Wambersie A 1990 Present and future work at the radiotherapy and radiobiology departments of the UCL-St Luc University Hospital, Brussels *ICRU News* **2** (90) (Dec 1990) 16–18

Webb S 1988 The mathematics of image formation and image processing *The physics of medical imaging* ed S Webb (Bristol: Adam Hilger) pp 534–566

—— 1990 A new dose optimising technique using simulated annealing *Proc. 9th Ann. Meeting of ESTRO (Montecatini, 1990)* p 263

—— 1991 Optimisation by simulated annealing of three-dimensional conformal treatment planning for radiation fields defined by a multileaf collimator *Phys. Med. Biol.* **36** 1201–1226

—— 1992 Optimisation by simulated annealing of three-dimensional conformal treatment planning for radiation fields defined by a multileaf collimator. 2:

Inclusion of two-dimensional modulation of the x-ray intensity *Phys. Med. Biol.* **37** 1689–1704

Williams P C 1990 private communication

—— 1992 Shaping the treatment volume *Proc. 50th Ann. Congress of the British Institute of Radiology (Birmingham, 1992)* (London: BIR) p 15

Williams P C and Jordan T J 1991 Performance characteristics of the Philips multileaf collimator *Proc. Radiology and Oncology 91 (Brighton, 1991)* (London: BIR) p 83

Williams P C and Kane B 1991 Use of predefined shapes for producing irregular fields with a multileaf collimator *Proc. Radiology and Oncology 91 (Brighton, 1991)* (London: BIR) p 82

Zhu Y, Boyer A L and Desobry G E 1992 Dose distributions of x-ray fields shaped with multileaf collimators *Phys. Med. Biol.* **37** 163–173

CHAPTER 6

MEGAVOLTAGE PORTAL IMAGING

6.1. THE NEED FOR HIGH-QUALITY PORTAL IMAGING

It must be continually emphasized that precision in radiotherapy depends on the accuracy of a sequence of stages in the planning and delivery phases; since a chain is only as strong as its weakest link it is important that no aspect of the process receives less attention than the others. Hendee (1978) prophetically wrote: 'With improved dose-perturbation correction techniques, will not the accuracy of dose distributions be determined primarily by how well the patient geometry used to collect CT data simulates the patient geometry employed for radiation therapy?' Until comparatively recently, portal imaging for ensuring the accuracy of positioning *at the time of treatment* was not a very active focus of attention. Despite many obvious deficiencies, film was the widely favoured detector of megavoltage photons. Although improving portal imaging does not solely benefit conformal radiotherapy, it has received greater attention as conformal therapy becomes a more realistic prospect. This is possibly because of the statement that conformal therapy needs to be more accurate, in the sense that the tighter the dose distribution to the target, the more serious patient mispositioning in the treatment fields will become (Swindell and Gildersleve 1991). This is discussed philosophically by Smith (1990).

The main limitations of film are:

- It has low contrast for megavoltage radiation. Scattered radiation contributes to this low contrast.
- The images are not taken in 'real-time'. Development times are relatively long. This prohibits using the images to adjust the position of the patient prior to treatment.
- The data are analogue. This precludes more than qualitative comparisons with simulator and DRR images.
- The cost of the film is largely irrecoverable.

A number of alternative detectors have been developed which each overcome some or all of these limitations. For example, digital images not only open up possibilities for quantitative comparison (e.g. registration) with simulator images or digital projections from CT data, but can also be processed (e.g. by windowing,

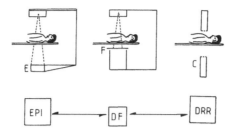

Figure 6.1. *An electronic portal image (EPI) made by a digital portal imager (E) on a linear accelerator (left) may be compared with digitized film (DF) taken on a radiotherapy simulator (centre) or with a digitally reconstructed radiograph (DRR) created from x-ray CT (C) data (right). Each image may have a different magnification and pixel size which must be corrected before quantitative comparisons are made. These comparisons assess how well the patient positioning is maintained through imaging, simulation and treatment.*

histogram equalization etc) to overcome poor contrast (Smith 1987) (figure 6.1). Imaging can be performed in real-time, allowing intervention and intelligent use of the data. The cost is transferred to the development phase and/or the capital outlay, each image being essentially 'free'. An argument that film is at least hard-copy can now be easily overcome as computer-driven hard-copy devices become readily available.

A variety of technologies have emerged and are naturally somewhat competitive. There are pioneering 'home-made' systems, as well as commercially available products. There is plenty of debate over which is the 'best' system (figure 6.2). The main developments have been: fluoroscopic area detectors, scanning-photodiode and scintillator-plus-photodiode arrays, matrix ionization chambers, photostimulable phosphors, xeroradiography and silicon area detectors (Boyer *et al* 1992). The main features of each will now be described.

6.2. FLUOROSCOPIC DETECTORS

Fluoroscopic detectors were the earliest developments, being the natural analogue of the fluoroscopic imager so common on radiotherapy simulators. Benner *et al* (1962) claimed to be the first to image very high-energy (30 MV) radiation with a television fluoroscopic technique. The energetic photons were generated by a Betatron and imaged on to a 1.5 mm thick lead sheet bonded to a layer of large ZnCdS crystals. The lead sheet converted photons into electrons, which generated light in the fluorescent material with a very small, but not negligible, efficiency. The low-luminance photon emission was sensed by an experimental prototype orthocon television camera.

The very characteristics which make high-energy photons desirable for radiotherapy, namely their lack of differential absorption between bone and soft tissue and their small attenuation, minimizing the effects of tissue inhomogeneities

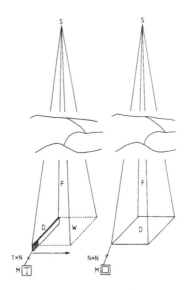

Figure 6.2. *The two main classes of electronic portal imager. Left—a one-dimensional array of radiation detectors (D) scans across the radiation field (F) and builds up the image on a monitor (M) from a series of* $(1 \times N)$ *vectors of pixels. The radiation (W) not falling on the detector is wasted, although each part of the radiation field is of course sampled in turn as the detector scans (arrow). A detector of this type must have a high quantum efficiency. The detector is shown schematically and in practice has many elements and may have two rows of elements, the rows offset by half a detector spacing to improve spatial resolution. Right—a two-dimensional detector (D) imaging the whole field (F). The image is displayed as an* $(N \times N)$ *matrix of pixels on a monitor (M). Such area detectors generally have a much poorer quantum efficiency. Both detectors are shown at a large distance from the isocentre, which reduces the contribution from scattered radiation.*

and improving depth dose, are the enemies of megavoltage imaging. To overcome this Benner *et al* (1962) had the patient swallow a plastic tube containing lead, gold or tungsten to act as a visible landmark with some contrast! Gold surgical clips served the same purpose for other anatomical sites. Modern portal imaging shows artificial-limb implants effectively for the same reasons.

To put these developments in perspective, it should be recalled that it was common practice to set up a patient using images with diagnostic x-rays, followed by wheeling the patient (supposedly without moving on the couch) to the therapy set in the same room (figure 6.3).

Baily *et al* (1980) (at the University of California) and Leong (1986) (at the Massachusetts General Hospital, Boston) describe systems with similar characteristic features. In Baily's system, megavoltage x-rays are incident on an area detector comprising a $\frac{1}{16}''$ stainless-steel plate cemented to an E-2 fluorescent screen (Dupont). Baily's detector was 17" × 17". The light from the scintillator

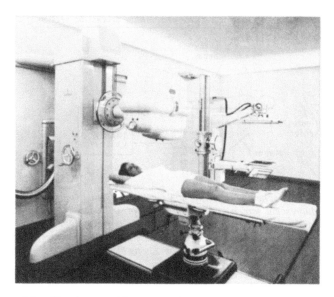

Figure 6.3. *Showing an old practice of having a diagnostic x-ray set for arranging the patient position in the same room as the treatment machine. The patient was wheeled unmoving between the two machines. (From Siemens Reineger Werke News, July 1957.)*

was bent through 90° by a front-surfaced mirror on to a video system via a 13 mm focal-length lens, a silicon-intensified target (SIT) Vidicon tube and a video camera. The 90° bend is necessary to avoid irradiating the camera. The limiting resolution (for 6 MV radiation) in Baily's system was 0.8 lp mm^{-1}, measured at the surface of the fluorescent screen. The system was used to image 6 MV and ^{60}Co radiation. Leong passed the video output into a real-time digital frame store capable of storing 512 × 512 images, either 8 or 16 bits deep at 30 frames per second (see figure 6.4). This, combined with a pipelined image processor, enabled numerical operations on this data at the same rate as frames were acquired. For example, time averaging (integration) for noise reduction, discrete convolutions, decalibration, grey-scale manipulation etc, were all possible.

The system, in addition to creating digital images, also acted as a differential dosemeter (single frames digitized to 8 bits) or integral dosemeter (summed frames digitized to 16 bits). In the latter mode a theoretical dynamic range of 65 000 was possible. Leong compared the output of the detector as a differential dosemeter with that of a Farmer ionization chamber showing linearity over doserates from 100 to 500 cGy min^{-1}. He also compared the output of the detector as an integrating dosemeter with the monitor ionization chamber of the treatment machine, also showing linearity with dose. Both these comparisons were made with only 2500 detector pixels, to avoid the need to consider detector spatial non-uniformity. Furthermore, the system compared favourably with film when measuring wedged fields (figure 6.5). There were no saturation effects. The

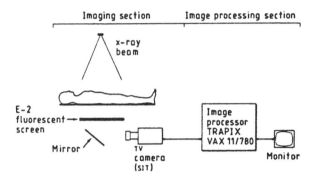

Figure 6.4. *A schematic showing how electronic portal imaging is achieved with a fluorescent screen, a 45° angled mirror, a TV camera and a frame store/image-processing computer system. This is the system developed by Leong and colleagues, but all such systems have similar features, although they may use different equipment. (From Leong (1986).)*

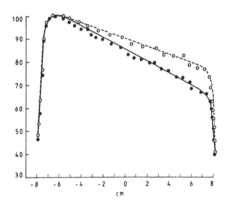

Figure 6.5. *The curves show the performance of Leong's portal imaging system acting as a dosemeter compared with film (Kodak XV2) for two wedged fields. The solid lines show the film response and the single points are from the digital fluoroscopy system. (From Leong (1986).)*

system could thus be used with confidence to map the exit dose from a treatment portal. In integrating mode it could be used to become a total-dose monitor and act as a 'verify' system. For example, it could be coupled to switch off the radiation beam if too many monitor units had been set, and could detect incorrect wedges. If there were a need for extensive image processing, e.g. frame averaging, the images themselves took some 2–4 s to be presented on a TV monitor.

Leong provided an early indication that the human observer was possibly the new 'weak link in the chain'. Could such an observer detect errors rapidly? Could these be rapidly corrected? These questions are still debated.

Shalev *et al* (1989), in Winnipeg, have also developed on-line portal imaging using a metal-plus-phosphor detector viewed via a 45° mirror by a TV camera. They used an SIT video camera and obtained images by averaging 128 or 256 frames (4–8 s acquisition time) and post-processing the data (Leszczynski *et al* 1988, 1990). A major consideration with video-based MVI systems is the need to optimise the transfer characteristics of each stage of the system; for example there are very large light losses in the optical coupling between scintillator and TV. If the magnification is m, the refractive index of the scintillator is n, the F-number of the lens is F and its transmission is T, then the optical transmission factor is

$$P_3 = \frac{T}{16F^2n^2} \left[\frac{m}{(1+m)} \right]^2 \tag{6.1}$$

which is very small for practical values of the variables (Swindell 1991). Shalev has a factor '4' not '$16n^2$' in the above equation. Others (e.g. Munro *et al* 1990a) have '8'. The correct factor is '$16n^2$', because the 'object' has a refractive index and is self-luminant, rather than reflecting incident light in a Lambertian manner (for which the 'usual' formula with the '4' was developed. Under these different circumstances the lens transmission factor is simply the fractional geometrical solid angle collecting the light produced with the transmission of the glass, and the '4' is appropriate.) Swindell (1991) has clearly distinguished between these conditions.

The fluoroscopic imaging system can be considered as a 'cascade or chain of events'. The output signal-to-noise ratio SNR_{out} is related to the input signal-to-noise ratio SNR_{in} via (Barrett and Swindell 1981) the equation:

$$SNR_{out}^2 = \frac{SNR_{in}^2 P_1}{(1 + (1/P_3 P_4 g_2))} \tag{6.2}$$

where

- P_1 is the probability of an x-ray interacting with the metal converter plate and producing a high-energy electron which enters the phosphor (the quantum efficiency of the metal plate),
- g_2 is the number of photons emitted from the screen per absorbed electron,
- P_3 is the probability that a photon generated within the screen will pass through the lens and be incident on the photocathode of the TV camera,
- P_4 is the probability that a light photon will create a photoelectron in the target of the TV camera.

This analysis assumes the conversion of high-energy electrons to light photons is a Poisson process and all other stages in the cascade are binary selections. The product $P_3 P_4 g_2$ is all important, being the number of light photons detected by the TV camera for every x-ray photon interacting in the metal plate. It is necessary for this figure to be as high as possible so that the noise is not amplified by the imaging system. g_2 is determined by the type of screen and its thickness. Unfortunately P_3 is usually very small.

Set against this large loss factor due to P_3, an area detector has some advantages over a scanning linear array (see section 6.3). The area detector has a very poor quantum efficiency (P_1 is only about 1% (Shalev 1991, Shalev *et al* 1988)) but of course captures data over the whole field of view simultaneously. The scanning linear array, to compensate, needs a very high quantum efficiency of the order 50% or more, and, to this end, thick high-density crystals are used. The area detector can make real-time movies (Shalev et al 1988); the scanning linear array can only make movies with a time periodicity of the scan.

Munro *et al* (1990a,b) (in Toronto) have constructed a digital fluoroscopic megavoltage imaging system for radiotherapy localization, paying careful attention to optimizing the features of the design which lead to noise propagation. Munro *et al* (1990b) estimate $P_3 = 2.8 \times 10^{-4}$ for a F=0.95 lens and a demagnification of 20 and lens transmission coefficient of 0.9. Munro *et al* (1990b) have an '8' rather than the correct '$16n^2$' in the denominator of the expression (see equation (6.1)) for P_3 and the correct value is therefore at least a factor 2 smaller than stated. To counter the small P_3, Munro *et al* (1990b) use a thick phosphor screen (400 mg cm^{-2}) which gives a light output measured to be $g_2 = 2.5 \times 10^4$ light photons per high-energy electron. With P_4 estimated to be 0.75, the product $P_3 P_4 g_2$ turns out to be about five. It might be a factor of five smaller, however, due to other uncertainties. Munro *et al* (1990b) considered six different thicknesses of phosphor, showing the light output was roughly linear with thickness (there is a small non-linearity due to optical attenuation in the phosphor) and the use of thick gadolinium oxysulphide phosphor bonded to a 1 mm thick copper plate was one of two important improvements they made to this type of system.

The poor quantum efficiency of the area detector and the potential degradation of SNR_{out} by a small $P_3 P_4 g_2$ product has to be compensated by video frame averaging. In the Winnipeg system (Leszczynski *et al* 1988) static images have signal-to-noise ratio improved by a factor of 16 by averaging 256 frames. However, for making movies the acquisition time has to be shortened to some $\frac{1}{2}$ s so averaging over only 16 frames is possible, which improves the signal-to-noise ratio by only a factor of four. Additional non-uniformity corrections and contrast enhancement by adaptive histogram equalization were also needed. This system is being evaluated clinically (Shalev *et al* 1991).

The second major improvement made by Munro *et al* (1990b) to overcome the effects of noise was to arrange for some of the integration to be performed on the TV target rather than in the frame grabber. Their electronics allowed two modes of operation:

1. Frames grabbed at 30 frames per s continuously for 256 frames (total imaging time 8.5s).

2. Charge integrated for up to 2 s on the TV target with sequential frames integrated in the framestore of the framegrabber. As the latter could hold 256 frames, this second mode was capable of imaging over a period as long as 8.5 minutes, i.e. throughout an irradiation.

In the second mode it can be shown that the noise contributed by the TV camera itself is minimized (since it only contributes on data readout and the number of readouts is $\frac{1}{60}$ of that in the first mode.) The shot noise (included in P_4) is proportional to the square root of the signal; thermal and amplifier noise can be ignored by comparison. A potential disadvantage of mode 2 is that the dark current increases proportionally to the length of time the signal is accumulated, but Munro *et al* (1990b) show this is always less than 5% of the real signal and can essentially be ignored.

Experimentally, they showed that the use of these two improvements (optimized phosphor screen and integration on the TV target) lead to much better imaging in practice, with contrast as low as 0.7% visible with a 7 cGy irradiation. Further improvements might arise if P_1 can be optimized. There are no reliable estimates of the electron production from high-energy x-rays interacting in thin metal plates, although studies are in progress (Wowk *et al* 1991).

Visser *et al* (1990) have developed a fluoroscopic system that is marketed by Philips. The prototype has been in use at the Daniel den Hoed Cancer Centre since July 1988. A demountable 30 × 40 cm^2 rare-earth fluorescent screen (411 mg cm^{-2}) is viewed via a 45° mirror by a CCD camera permanently installed on the linac. The field at the isocentre is 19 × 25 cm^2 maximum. Integration is performed in the CCD camera for between 0.1 and 1 s with further integration in the frame store. The system is calibrated daily with an open-field measurement. Images may be filtered for edge enhancement and histogram equalization and also grey-scale windowed. With a 7 s exposure and a 25 cm absorber present, a dose of 42 cGy allows a signal-to-noise ratio of 105. The spatial resolution was measured as 1.5 mm FWHM (compared with 0.87 mm FWHM for film). Measurements with a Lutz and Bjarngard (1985) test phantom gave 0.74% contrast visible.

Fluoroscopic portal-imaging systems have been developed by several commercial manufacturers. The system of Shalev and colleagues evolved into the 'BeamView Plus' product from Siemens, which creates a 26 cm by 33 cm image at the isocentre with some 0.5–3 cGy dose depending on the site. A feature of the system is that the imager can be telescoped into the gantry for 'parking'. The system of Munro and colleagues evolved into the 'Theraview' product from S and S Inficon. This also has a retractable bellows for parking the imager. Images with a spatial resolution of 1 lp mm^{-1} and contrast resolution of 0.5% are formed with 1–2 cGy dose. The imager has a 40 cm by 40 cm metal plate and the image size at isocentre obviously depends on the geometry; the detector can move vertically as well as be retracted. The Philips system is known as 'SRI-100 MVI' and has a rigid frame without the option to retract into the gantry.

One of the major disadvantages of all the systems discussed so far is their physical bulk. A group at the Mallinckrodt Institute of Radiology have developed an optical coupling which does away with the mirror and is inherently more compact. Wong *et al* (1990a, 1991) linked a 40 cm by 40 cm input screen to a 3 cm by 3 cm output plane via bundles of fibre-optic image reducers; 256^2 plastic fibres were used. At one end these were coupled to a fluorescent screen with 300

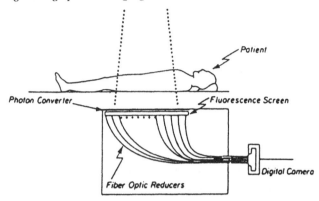

Figure 6.6. *The prototype optical coupling in the fibre-optic imager. The fluorescent screen is coupled to the front face of a digital camera by bundles of fibres. This dispenses with the need for a 45° mirror and reduces the bulk of the detector. (From Wong et al (1990).) (Reprinted with permission from Pergamon Press Ltd, Oxford, UK.)*

mg cm^{-2} of gadolinium oxysulphide beneath a 3 mm copper sheet. At the other end they coupled to the input of a TV camera with a digital 512^2 pixel output (figure 6.6). The fibres were in bundles of 16^2, each bundle optically isolated from its neighbours to reduce inter-bundle cross-talk. The bundles turned through a right angle within 10 cm of the input screen, leading to a shallow device. The efficiency of the fibre coupling system is much the same as for an efficient lens system (Boyer *et al* 1992). This is because nearly all the optical photons are turned around in the tapering light-pipe due to multiple reflection (Swindell 1992b).

The raw images suffered considerable spatial distortion and non-uniformity of more than 60%, both effects needing software correction. The usual digital windowing and histogramming facilities were available. Typically 8 to 128 frames of video data were averaged to generate each image. Despite these corrections, the prototype system was judged poorer than optimal with a spatial resolution of 0.3 line pairs per mm and a density resolution of 1% for an object 1.5 cm in diameter. The non-uniformity of the light-pipes could have been reduced by pre-selecting only those fibres with common optical properties. Nevertheless, useful phantom and clinical images were obtained. Indeed Graham *et al* (1991) and Jones and Boyer (1991) have reported using the system (marketed as 'Fibervision') to assess patient movement during therapy. The latter developed a Fourier-based correlation technique to compare digital portal images with simulator images.

6.3.   SCANNING LINEAR ARRAY OF DIODES AND CRYSTAL DETECTORS

The scanning-diode array for megavoltage radiation imaging originated with Taborsky *et al* (1982) and was further developed by Lam *et al* (1986). Taborsky *et al* (1982) performed a design study to show the feasibility of using a scanning linear array of photodiodes as a portal detector. In the design study the diode array was fixed in space and the object to be imaged was translated through the field. The detectors were silicon ($Z=14$) photodiodes of radius 1 mm embedded in 2 mm thick lead. About 11% of the (4 MeV) x-rays interacted with the lead/diode detector, the rest passing straight through undetected. A Monte Carlo calculation of the self absorption by the lead of the secondary Compton electrons showed that some 30% only were detected by the photodiode. Thus the quantum efficiency of the detector was some 3%. They estimated that each accelerator pulse delivered 1.77×10^7 photons cm^{-2}. Thus for each pulse some $1.77 \times 10^7 \pi (0.1)^2 0.03 = 1.83 \times 10^4$ primary photons were detected, from which the fractional noise from the photon statistics alone was some 0.7%. They estimated electronic noise as some 0.5% and so the total signal-to-noise ratio was 116. More complicated calculations, taking into account the variable output per accelerator pulse, showed the signal-to-noise ratio was 119 for an open field. The effect of this variability was calibrated out by mounting a detector of five diodes close to the accelerator collimator. Taborsky *et al* (1982) built a small test prototype with 62 diode pairs (mounted one above the other and connected in parallel to double the signal per 2 mm element of detector) and used it to make small images with a 62×67 matrix of sample points, displayed with just eight grey levels. They also noted that other room-temperature detectors with higher conversion efficiencies could have been used, such as cadmium telluride ($Z=50$) and mercuric iodide ($Z=62$).

Lam *et al* (1986) (at Johns Hopkins Medical Institute, Baltimore) constructed a linear array of 255 silicon diodes, with the centre-to-centre separation between diodes of 2 mm. A layer of lead 1.1 mm thick was placed above each diode for electronic build-up. The line of diodes was scanned under computer control, the step size between scan lines being 2 mm. The maximum number of scan lines of 224 gave a maximum field size of 44.8 cm (in the scan direction) by 51 cm orthogonal to this. The cathodes of all diodes were grounded and the anode of each diode was connected to a charge amplifier and thence to a sample-and-hold unit. The output voltage of each assembly was proportional to the charge generated at the diode by the impinging x-rays. The use of sample-and-hold ensured the signals detected at all diodes were sampled simultaneously and held at the peak voltage. The computer which controlled the stepper motor governing scanning also accepted the signals which were multiplexed and passed to a single analogue-to-digital converter, working at 12-bit resolution (figures 6.7 and 6.8).

After scanning, the signal level of each diode was corrected for variable detection efficiency and background noise using prestored data obtained by scanning an open field. A further single photodiode placed near the collimators of the accelerator was used to calibrate the image for intensity fluctuations in the x-ray beam. The final image was video-reversed (so the image resembled a film image with dark corresponding to minimum photon absorption) and displayed as a 256^2 or 512^2 image with 8 bits (256 grey levels). The detector was 50 cm below the isocentre (at 1 m) giving a spatial resolution of 1.3 mm and measurements

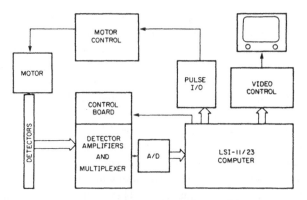

Figure 6.7. *Block diagram of Lam's electronic portal imaging scanner. The one-dimensional line of detectors move under motor control. The output from the detectors is amplified and multiplexed, finally being displayed on a video monitor. (From Lam et al (1986).)*

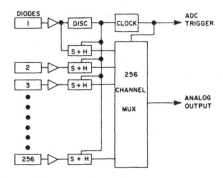

Figure 6.8. *Schematic of the signal-processing circuitry for Lam's scanner. The outputs from the diodes are held in a sample and hold (S + H) before being multiplexed and output. (From Lam et al (1986).)*

showed that, at doses comparable to those which would be received at treatment, the contrast resolution (of 1 cm diameter disks) was better than 0.8% on a Varian Clinac 4/100 4 MV accelerator.

The image was constructed throughout the *full course* of a treatment fraction. The accelerator was run at a pulse repetition frequency of 200 pulses per second. A typical treatment fraction required 5000 pulses, and 20 pulses represented 0.33 cGy. The need to image throughout a treatment fraction is a severe disadvantage for a scanning system of this type and demands that the quantum efficiency of the detector be increased to completely offset the disadvantage of viewing only a small part of the radiation field at any one moment in the scan. Scanning silicon diodes have much the same quantum efficiency as a TV-based system, but loose out because the latter is an area detector visualizing the whole radiation field. A scanning system with some 50% efficiency has been developed

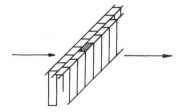

Figure 6.9. *Illustrates the arrangement of individual detectors in the Royal Marsden Hospital portal scanner. Each crystal presents a square face (shown hatched) to the radiation beam. One line of crystals is offset by half the square size relative to the other. Samples are collected from both lines by integrating for the time it takes one line to move half a square. This arrangement satisfies the Shannon sampling theorem. The arrow indicates the direction of scanning. Only a few detecting elements are drawn.*

by Swindell's group (Morton 1988, Swindell *et al* 1988, Morton and Swindell 1988, Morton *et al* 1988, 1991b) making use of zinc tungstate detectors.

Lam *et al* (1986) also digitized simulator radiographs for comparison with the digital images at treatment. They commented that portal images were 'comparable' in quality to film images, and more readily compared with simulator images.

The system of Swindell at the Royal Marsden Hospital, Sutton (Morton 1988, Morton and Swindell 1988, Morton *et al* 1988, 1991b) has a scanning array of 128 zinc tungstate crystals arranged in two rows of 64, with the crystals of each row offset by half a crystal size relative to the other row (figure 6.9). The crystals are 5×5 mm in area and 25 mm deep, arranged for practical reasons in 6×6 mm spacing. Morton (1988) determined that the light output of these crystals was some twice that of bismuth germanate (BGO) crystals, resulting in some 2000 optical photons reaching the photodiode per x-ray interaction, a favourable situation (see section 6.11) for noise suppression. These crystals are non-hygroscopic, almost resistant to radiation damage and have a high (2.2) optical refractive index.

The output from this continuously scanning detector was integrated for the time taken to traverse half the detector spacing (i.e. for 3 mm). This constructed a 128 $\times$ 128 matrix image with the output from one row of 64 detectors filling the odd columns and the output from the other row of detectors filling the even columns. The data were low-pass filtered to remove machine variations. The detector was 2 m from the radiation source to reduce the contribution of scattered radiation to the image, giving (with a magnification of 2) an image of size 19×19 cm at the isocentre. The image took 3.8 s to construct, requiring a dose of some 0.55 *c*Gy (the dose to the patient was greater—25 *c*Gy—because the beam irradiated the full portal whilst the detector sampled only a part of the portal at any one time). The experimentally determined spatial resolution was 2.5 to 3 mm and the density resolution was 0.5%, in line with the design specification. This system (figure 6.10) is now being extensively used for clinical studies of the accuracy of patient

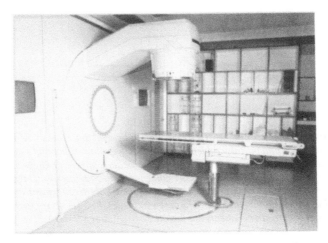

Figure 6.10. *The Royal Marsden Hospital electronic portal imaging instrument attached to a Philips SL25 accelerator. The scanner slots into the triangular mounting plate, which is firmly bolted to the rotating gantry of the linac. The acquisition electronics are located within the arm holding the scanning detector. The figure shows the prototype detector; an improved version is under development. (Courtesy of W Swindell, E Morton and P Evans.)*

positioning (Gildersleve *et al* 1992c). Images produced by the portal scanner are compared with digital images captured by the image intensifier of the radiotherapy simulator or simulator-film digitizer (figure 6.11).

Before this comparison can be done, three spatial corrections are made to the simulator images (Morton *et al* 1991a) which were captured by a frame-store coupled to a personal computer. (Incidentally, this provided the additional advantage that workers in the simulator suite could view an image captured in this way after the x-rays had been switched off, thus saving on patient and staff radiation exposure). Firstly, the TV pixel aspect ratio of 4:3 is corrected to a square image. Secondly, pin-cushion distortion is removed, observing that the displacement of the image of a point from its true position can be expressed as a Taylor series of the distance of that point from the origin, with just two terms needed. Rudin *et al* (1991) have also reported distortion removal with a polynomial fit. Thirdly, the simulator images were re-sampled to the same pixel size as the megavoltage portal images. Tools for comparing simulator images with megavoltage images (Evans *et al* 1992a) and for correcting patient mispositioning (Evans *et al* 1992b) have been developed.

Gildersleve (1992) and Gildersleve *et al* (1992b,c) used the Royal Marsden Hospital portal scanner to quantitate the random and systematic errors in patient positioning during *routine* radiotherapy, retrospectively analysing data for brain and pelvic irradiations. Being compact, the equipment was unobtrusive and could be used for all but a few geometries without any danger of collisions. Random error was defined to be the departure of the field positioning from the mean.

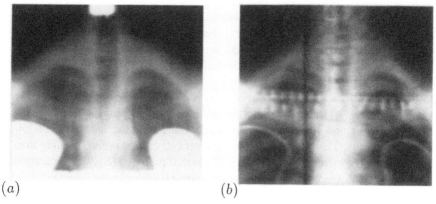

(*a*)   (*b*)

Figure 6.11. *(a) Portal image of a mantle field captured with the Royal Marsden Hospital portal imager shown in figure 6.10. The lung shields are seen as large white areas to the bottom left and right. The important anatomical details (lungs, ribs and soft tissue) are clearly visible and the images can be windowed as appropriate to visualize these more clearly on a display monitor. (b) The corresponding simulator image. The image is made up of two films to cover the large field. The location of the desired shielding has been drawn by the clinician on to the films. (Courtesy of W Swindell, E Morton and P Evans.)*

Systematic error was the departure of this mean position from the 'gold standard' at simulation. For the brain, 95% of treatments had random positioning error less than 3 mm. The mean systematic positioning error was about 1%, although (and this was an important conclusion of the study) *individual patients* had systematic errors as large as 6 mm. For the pelvis, 95% of treatments had random positioning error less than 4.5 mm. The mean systematic positioning error was close to zero, although *individual patients* had systematic errors as large as 6 mm. There was a trend for patients with large systematic errors to also exhibit large random errors. Whilst this study showed that positioning errors were within the margin which many radiotherapists added 'for safety' in determining the target volume, the existence of some 'out-liers' in the distribution emphasized the need to verify patient positioning with portal imaging. Following this retrospective study, an 'intervention' study was commenced late in 1991 (Gildersleve *et al* 1992a).

A scanning diode array has also been constructed at Massachusetts General Hospital, comprising cadmium telluride detectors (Biggs 1991). The development is at an early stage.

6.4. MATRIX IONIZATION-CHAMBER DETECTORS

Research at the Antoni van Leeuwenhoekhuis in Amsterdam has resulted in the development of three prototype liquid-filled matrix ionization chambers for portal imaging. The first was a small prototype of 32^2 chambers (Meertens *et al* 1985) and the clinical prototype had 128^2 chambers (van Herk and Meertens 1987,1988). The present version has 256^2 chambers (Meertens 1989, Meertens *et al* 1990b).

The small prototype matrix ionization chamber of Meertens *et al* (1985) comprised only 32^2 cells, giving a small field of view of 78 × 78 mm, operating in much the same way as described below for the clinical prototype. The data collecting electronics were, however, very slow, 120 s being needed for the complete image. The small prototype allowed variable spacing between the electrodes (1, 3 and 10 mm were tried) and could be filled with air or 2,2,4-trimethylpentane. The liquid gave much greater cell current at each spacing than the air, with little increase in the noise and was thus preferred (e.g. at a separation of 1 mm the SNR was 200 (density resolution some 0.5%) with the liquid medium and 15 with the air.) The smallest spacing was also preferred at a fixed voltage of 300 V since, although more ions were produced with a wider gap, the lower field strength allowed greater recombination, an effect dominating the increased production. With a liquid-filled chamber operating at 300 V and a separation of 1 mm, the spatial resolution was 3.8 mm with a ^{60}Co source. The spatial resolution (measured by differentiating an edge-response function) was the same in orthogonal directions and the point-spread function was deconvolved from the measured images to improve their appearance. The point-spread function with the liquid medium was narrower (at corresponding separations) for the liquid medium than for the air. Hence 1 mm spacing with the liquid medium became the preferred conditions for operation and were adopted for the larger clinical prototype.

In the clinical prototype the ionization chambers comprised a pair of 1.5 mm thick glass-fibre printed circuit boards etched with a 2.5 mm pitch to give 128 rows on the front plate, serving as high-voltage electrodes, and 128 columns on the rear plate, serving as low-current-collecting electrodes. Each crossing point serves as an ionization chamber for measuring the local radiation intensity. The electrodes were separated by a 1 mm film of the liquid iso-octane as ionization medium. There was no additional radiation build-up material. Thus the field of view at the detector was 320 × 320 mm. The ionization current in each chamber was measured with a current-to-voltage converting electrometer. The ionization matrix was scanned row by row, by switching the high voltage to different rows sequentially and reading out the 128 column electrodes. The data were taken through a 12 bit analogue-to-digital converter, stored on computer and the whole image took 3.1 s to construct. Images could be displayed as a 256 × 384 array with 8-bit accuracy. This detector had the advantage of no moving parts, all 'scanning' being achieved electronically (figure 6.12). The advantage of a detector based on ion transport with low mobility is that the ion lifetime is very long, resulting in an inherent charge integration in the liquid. The Netherland's group have also considered a chamber based on electron transport (Boyer *et al* 1992).

The raw images required correction for electrometer offset (a variation of some 10%) and leakage current (a variation of some 1%). Corrections were also made for changes in the ionization-chamber sensitivity (as great as 40%). This arises because the response is very sensitive to the local thickness of the liquid

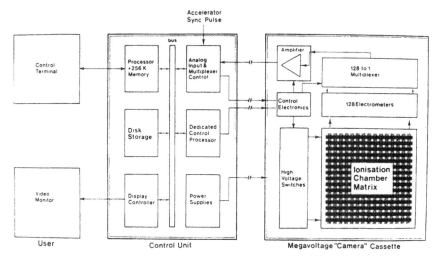

Figure 6.12. *Schematic of the liquid-ionization chamber megavoltage imaging system. The camera cassette is connected with the control unit (placed outside the treatment room) by a 15 m length of cable. The figure shows the 128^2 detector. (From van Herk and Meertens (1988).)*

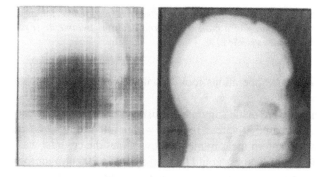

Figure 6.13. *(a) Raw image data of the Rando/Alderson phantom imaged with the 128^2 matrix ionization-chamber portal imager in figure 6.12, showing the electrometer offset and sensitivity variations, electrode shape variations and variations due to the changing thickness of the chamber. (b) The same image after the application of all correction steps. (From van Herk and Meertens (1988).)*

layer, which can vary with orientation of the detector on the gantry as deforming pressures come into play. However, these corrections can be made during the data capture itself. Figure 6.13 shows an image of a head before and after these corrections.

Additionally (as with most megavoltage imaging systems) a number of fast image processing algorithms were used, including linear contrast enhancement, global histogram equalization, various 3 × 3 spatial filters, as well as image

Figure 6.14. *Van Herk and Meertens' electronic portal detector, as used to verify the positioning of a patient for a radiotherapy treatment. (From van Herk and Meertens (1988).)*

arithmetic. All these operations took less than 0.6 s and were thus essentially 'semi-real-time'.

Measurements were made (figure 6.14) using a Philips SL 75-10 accelerator at 8 MV and showed a point-spread function with a full-width at half-maximum of some 1.5 mm. Phantom measurements showed that the contrast was such that a 2.5 mm extra water thickness could be seen against a 200 mm uniform thickness of water. The magnitude of the PSF was taken to indicate the absence of cross-talk between ionization cells, despite the absence of chamber walls.

Van Herk and Meertens (1988) discuss the need for pattern recognition software, so that the portal images which can be generated quickly can be used to intervene in the treatment fraction. They also call for studies to evaluate the use of electronic portal imagers in clinical practice and suggest they may also be used as dosimeters. The authors of other systems make similar points. One special claim for the matrix ionization chamber is that being 'thin' (the overall thickness of the whole system is no more than 5 cm) it can easily take the place of a film and be used with the same convenience.

The clinical prototype has been superceded by a detector with 256^2 chambers of size 1.27 mm by 1.27 mm by 1 mm (Meertens *et al* 1990b, van Herk 1991, 1992). The efficiency is some 1%. A detailed analysis of the ion-chamber performance, comparing with theoretical predictions, has been published by van Herk (1991).

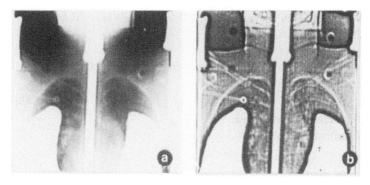

Figure 6.15. *Portal images in the thoracic region made with the high-resolution 256² imaging device during one treatment session: (a) mantle field, (b) same image with additional high-pass filtering. (From Meertens et al (1990b).) (Reprinted with permission from Pergamon Press Ltd, Oxford, UK.)*

The 256^2 detector could also be operated as a 128^2 detector with a lower noise level. Figure 6.15 shows clinical images from this high-resolution device.

The Netherlands Cancer Institute system is available commercially as Portalvision from Varian Inc, Palo Alto, California (formerly the Dynaray ID from Asea Brown Boveri). With the system 20 cm below the isocentre, the image field is 27×27 cm². Data may be acquired in either 1.5, 2.8 or 5.5 s. In the commercial system the fluid space is only 0.8 mm thick. The contrast resolution is quoted as better than 0.5% (Varian 1992). The detector weighs 7 kg. Wade and Nicholas (1991) report favourable use of the prototype system for a wide variety of patient treatment sites. They note the need to recalibrate for oblique angles. The prototype digital cassette was used free-standing, but the latest Varian version 3.1 is attached to the treatment machine via a motorized support which enables a variable source-to-detector distance, as well as lateral and longitudinal movements. This is also of more rigid construction and does not need to be recalibrated at each gantry angle.

Meertens *et al* (1990a) describe methods for measuring field-placement errors using digital portal images. Bijhold *et al* (1991a) describe the edge-detection algorithms developed for use with this system and Bijhold *et al* (1991b) show how using features visible in simulator and portal images, assessment of patient set-up error may be made. Bijhold *et al* (1991c) report clinical usage of the system, showing how intervention became possible. Bucciolini *et al* (1991) have used the liquid-filled portal system (which was developed in Holland) in Florence.

Hoogervorst *et al* (1991) have reported a development of a 32×32 air-filled ionization chamber, similar to the portal imager, used for measuring the flatness of fields.

6.5. IMPROVEMENTS TO PORTAL IMAGING WITH FILM AS DETECTOR

Three improvements to portal imaging have been made, retaining the use of film as detector:

- use of a film digiter,
- use of a metal converter plate,
- use of gamma multiplication.

6.5.1. Film digitization

A film digitization system was constructed by Leong (1984).   A trans-illuminated film was viewed by a TV camera with the different optical densities creating signals which were then digitized.   The recorded digital images were directly viewable on a monitor (Lexidata System 3400) attached to the controlling computer (Vax11/780) or were input to several digital-image processing techniques and displayed to show a modified or enhanced version of the image. Contrast re-scaling, histogram equalization and filters applied in the Fourier domain were implemented. It was claimed that the resulting images were more easier to interpret than peering at low-contrast radiographs on a light box showing information *present* but not easily *perceived* with the naked eye.

The images were corrected for the non-uniform illumination of the light box; no spatial distortion was observed. The spatial resolution was limited by the discrete sampling which, in principle, leads to aliasing, since (with such a small grain size) the image is essentially not band-limited. However, it was thought that all the useful information was within the band limit of the digitization (the images were described as 'pseudo-band-limited') and that any aliasing would only introduce noise which was removed by the Fourier filters.

The system was shown to have a useful dynamic range of about 80, not a limitation because of the inherent lack of contrast in portal films. When 16 or more frames were averaged, the signal-to-noise ratio (defined as the ratio of the mean to the standard deviation in the response to a uniform field) was about 60, indicating that 6 bits only were required to digitize the image. The principal source of noise was thermal noise in the TV pre-amplifier; film-grain noise and photodetector noise could be ignored.

It is thus clear that there is a much smaller dynamic range and signal-to-noise ratio in digitized portal films than in some on-line electronic portal imagers.

Meertens (1985) digitized film differently using a densitometer with a spatial resolution of 1 mm and density resolution of 0.2% displaying images as a 256^2 pixel matrix on the console of a CT scanner.   The films had been exposed sandwiched between two copper screens of thickness 0.6 mm (front) and 0.4 mm (rear). Thus he brought into play all the computational facilities available for inspecting CT tomograms, such as grey-scale windowing, filtering, distance measurement and inspection of regions of interest. A method was also developed to superimpose data from the simulator film for comparison purposes. The main flaw with these procedures was the time factor, too long to allow real-time intervention at the point of treatment, a limitation noted by all workers (including Meertens) proposing real-time portal imaging.

Digitization of port films has been widely applied. Amols and Lowinger (1987) used a black-and-white video camera to digitize to 512^2, 8 bits deep, whilst Grimm *et al* (1987) used a Dupont laser scanner to create images at 1024^2. In both cases the digitized images were further processed by contrast enhancement, windowing, edge enhancement and filtering to render structures of interest more clearly. Grimm *et al* (1987) reported significant improvements in head and neck imaging. Crooks and Fallone (1991) built a system around a personal computer.

6.5.2. *Influence of metal screens*

Metal screens are generally used with film. Without the screen both primary and scattered photons largely pass undetected through the film emulsion, so the film can only respond to electrons *generated in the patient* and specifically in the last few cm of the patient (since electrons of N MeV travel only some $N/2$ cm in tissue). This is undesirable because of the long range of such electrons. If there is an air gap between the exit surface of the patient and the detector, the image will be swamped by the electrons from scattered photons and will be blurred. Even those electrons resulting from primary interactions will not deposit their energy at the correct location because of the finite air space across which they are transported. The purpose of the metal screen is to totally remove the contribution of electrons *from the patient* (provided the screen is thicker than the maximum electron range in the metal—some $N/22$ cm for a lead screen) and instead generate new electrons by photon interactions in the lead. These travel a distance of some 11 times less than they would in tissue and so are sharply imaged. The screen provides a *flat* build-up layer directly next to the film (figure 6.16). The electron production in the screen reflects the scatter-to-primary ratio of photon fluxes, provided both have built up to electronic equilibrium. The front screen does *not* act as an intensifier since it does not provide any extra electronic contribution to the film. It simply replaces the patient-generated electrons by a new set generated in the film and more localized to the points of interaction. (By contrast a rear screen does have an intensifying effect, since it provides backscattered electrons to increase the electronic contribution to the film—see later; this effect is by a factor of about 1.8.)

The effect on the image contrast of screens of different material and thickness can easily be related to their effect on the scatter-to-primary ratio of dose at the film. Droege and Bjarngard (1979) studied this experimentally with some very clear conclusions. Suppose two parts of a film receive primary doses P_1 and P_2 and the same scatter dose S, then the dose contrast is

$$C = (P_2 + S) / (P_1 + S). \qquad (6.3)$$

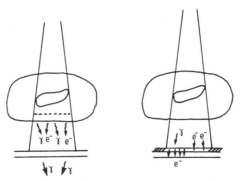

Figure 6.16. *Showing the effect of a metal screen in front of a portal film. The patient is shown in cross section and the fan lines represent a beam irradiating the target area shown. In the left-hand figure, both photons (primary and scattered) and electrons exit the patient. The electrons have come from the last few cm of the patient nearest the film (shown dotted). Most photons pass straight through the film, which responds to the electrons. In the right-hand figure, the same particles exit the patient. The electrons from the patient are stopped by the metal screen, in which the photons interact to produce a new set of electrons to which the film responds. This is a more favourable situation.*

Writing $\Delta P = P_2 - P_1$ and $P = P_1$ we have

$$C = 1 + \frac{\Delta P}{P}\left(\frac{1}{1 + S/P}\right). \tag{6.4}$$

This gives a film-density contrast

$$\Delta D = \gamma \log (C). \tag{6.5}$$

where γ is the slope of the Hurter–Driffield curve.

It therefore follows that the only factors which can affect the contrast in density (which is the important physical measure of whether a small object can be perceived) are γ and the scatter-to-primary ratio S/P (figure 6.17). Droege and Bjarngard (1979) found experimentally that whatever the source energy (^{60}Co, 4 MV and 8 MV), whatever the screen metal and whatever its thickness, the Hurter–Driffield curve of film density versus film dose (corrected for the transmission factor of the screen) was *the same* and so the film *gamma* was unaffected by the presence of the screen (figure 6.18). This simply reflects that the film responds to electrons generated in the screen and, provided these are in equilibrium, it is irrelevant what material they come from or how thick it is.

To consider further image contrast on portal films, attention therefore turns to the S/P ratio. This was investigated for contact geometry (screen and film touching the exit surface of a phantom) and for a geometry with an air gap (screen and film some 25 cm from the exit surface of the phantom). In contact geometry

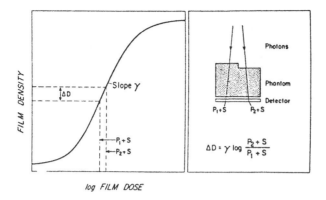

Figure 6.17. *Illustrating the factors which can affect radiographic contrast. These factors are the film γ and the scatter-to-primary ratio. (From Droege and Bjarngard (1979).)*

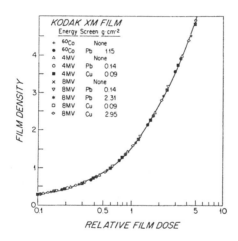

Figure 6.18. *Showing the measurements which indicate that the film contrast is independent of x-ray energy and metal screen composition. All the measurements lie on a single curve. (From Droege and Bjarngard (1979).)*

the S/P ratio initially increased as the thickness of the screen is increased, but after a certain thickness settled to the same value as without the screen at all. This optimum thickness is 1.5 g cm^{-2} for copper and 2.5 g cm^{-2} for lead, the result largely independent of beam energy (figure 6.19). In a geometry with an air gap, the S/P ratio decreased monotonically as the screen thickness is increased, from a high value with no screen thickness to a roughly constant value after 1.5 g cm^{-2} of either material. The use of an air gap decreased the S/P ratio considerably from the corresponding value in contact geometry (as also shown by Swindell *et al* 1991), thus improving contrast (excepting the extra blurring introduced by finite source size).

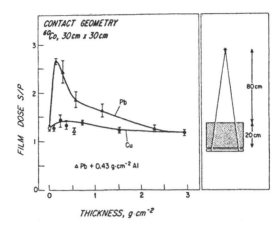

Figure 6.19. *Showing the effect on the scatter-to-primary ratio S/P of front metal screens of different thicknesses and two materials used in megavoltage contact radiography at ^{60}Co energy. When the thickness reaches a certain value, the S/P ratio is the same as when there is no metal at all. This is the best thickness to use. (From Droege and Bjarngard (1979).)*

To understand the behaviour of the S/P curve for contact geometry as a function of screen thickness, it is necessary to consider carefully the photon–electron transport in phantom, screen and film. Firstly, with no screen the dose to the film is provided almost entirely by the *electrons* which are generated from the layer of phantom or patient nearest the film; the film itself stops negligible *photons* at these energies. Hence the scatter-to-primary ratio simply reflects the ratio of electrons from scattered-to-primary photons at the exit of the phantom, which in turn mirrors the ratio of scattered-to-primary photons at the exit surface. Droege and Bjarngard (1979) checked this was so using published depth-dose data.

A thin screen has a dramatic effect on the mostly forward-travelling electrons liberated in the phantom by primary photons, turning a large fraction away from the film. As the thickness of the screen is increased, both scattered and primary photons liberate electrons *in the screen*, the build-up depth being smaller for scatter than for primary. Once electron equilibrium is achieved, the film dose becomes proportional to the screen dose from such photons. However, since the energetics are different, these doses will decrease at a different rate with increasing screen thickness. The overall result is the form of the curve shown in figure 6.19. Although the thick screen produces almost exactly the same scatter-to-primary ratio as no screen, it is preferred because of the resolution-degrading effect of long-range electrons generated in the patient or phantom, all of which have been excluded from the film by the screen (figure 6.16). In summary *the main purpose of the screen is to remove patient-generated electrons, replacing them by a new set generated in the screen and more localized to the points of interaction.* The scatter-to-primary ratio of a screen of build-up thickness or greater has to be the same as with no screen, because one can think of a lead screen as acting just like 11 times its thickness of 'extra patient' as far as producing electronic fluxes proportional to the two photon fluxes; primary and scatter. All that changes is the location of the production sites in relation to the film (i.e. right next to it).

If we write

$$\Delta P/P = e^{\mu t} - 1 \simeq \mu t \qquad (6.6)$$

for the ratio of the change in primary dose to the primary dose from a small extra attenuating thickness t of water-equivalent material of linear attenuation coefficient μ and put this in equations (6.3) and (6.5), we find the minimum perceptible tissue-step thickness t_{min} given by

$$t_{min} = \ln(10)\frac{\Delta D_{min}}{\gamma}\left(\frac{1 + S/P}{\mu}\right). \qquad (6.7)$$

Using Droege and Bjarngard's data for S/P and the known values of μ at the three energies studied, it turns out that t_{min} is remarkably the same at all three energies. Yet it is known that the contrast in portal images with ^{60}Co radiation is worse than with 8 MV; the difference is attributed to changes in spatial resolution.

Galkin *et al* (1978) also investigated the dependence of film γ on the presence of metal screens. From measurements with Kodak Rapid-Process therapy film (RP/TL) and the radiations from ^{60}Co (Picker C-3000), 4 MV (Clinac-4 accelerator) and 45 MV Betatron (Brown Boveri), they determined that the presence of a single lead screen made no difference to the film gamma, in agreement with Droege and Bjarngard (1979). Sandwiching the film betwen two screens made small increases in γ. However, using just a single light-emitting screen (US Radium Radelin Ultra-detail screen) increased the film γ considerably, an effect even greater with a pair of light-emitting screens. This is because the film responds differently to electrons (from the metal screen) and to optical photons (from the light-intensifying screen).

In Sweden, Jevbratt *et al* (1971) used low-sensitivity industrial film with a γ of 6.5 to record portal images from ^{60}Co radiation, a 6 MeV linear accelerator and a 42 MeV betatron, improving the sharpness by employing a sandwich of 2 mm thick lead screens to remove the scattered electrons from the patient and replace them by a new set of electrons generated in the screens. They also investigated graphic film with a γ of 12 and very slim latitude.

Hammoudah and Henschke (1977) showed experimentally that the effect of a front lead screen was to decrease the effect of scattered electrons and thereby increase contrast with no intensifying effect, whereas the rear screen produced an intensifying effect. A regular diagnostic film sandwiched between two lead layers was some ten times more sensitive than supervoltage localization films. They found films used with fluorescent screens were too sensitive, being over-exposed at the minimum monitor setting. Similar observations on the different physical effects of front and back metal intensifying screens were made by Halmshaw (1966) in his large text on industrial radiography.

6.5.3. Film enhancement by analogue processing

Instead of numerical methods for enhancing digitized films, an analogue technique using purely photographic processes has been developed by Reinstein and Orton (1978,1979). This method is called 'gamma multiplication' for reasons which will soon be obvious. The original poor-contrast portal radiograph is copied by a contact process onto a second previously unexposed film. This print has reversed contrast, of course, being exposed more where the density of the portal film was less. The contrast on this reversal film is the product of the gamma of the portal film and of this copy film. Next, the process is repeated by making a contact print of the second film onto a third film, at which stage an image with the same polarity as the original is recovered. The contrast on this third film is the product of all three gammas of the individual films. If the first is a Kodak XV with γ = 2.3, the second is Kodak XTL with $\gamma = 3.5$ and the third is also Kodak XTL, the gamma multiplication is 28.

The method is rapid and inexpensive. Good contact is needed between films and this is ensured by placing both under a heavy glass screen. The contact prints are made by normal darkroom techniques and require no special apparatus. Reinstein and Orton (1978, 1979) verified the gamma multiplication experimentally by imaging a thin step wedge, providing initially very small contrasts, and observing the contrast magnification at each stage.

There are trade-offs, of course. Increased contrast comes at the expense of decreased latitude. Indeed, contrast enhancement may be so great with two XTL films that a third-stage film with lower gamma may be preferable. A second negligible disadvantage is that copying decreases spatial resolution, but this is not a limiting factor with portal radiographs at high energy.

With reference to figure 6.20, the gamma multiplication may be shown mathematically: Figure 6.20(a) shows the formation of the original film with an incident intensity of radiation X_0 producing primary output intensities X_1 and X_2 corresponding to passage through two regions in the brain with contrast. The corresponding film densities are D_1 and D_2. By definition of the film gamma,

$$\gamma_1 = \frac{D_1 - D_2}{\log X_1 - \log X_2}. \tag{6.8}$$

Figure 6.20(b) shows the first copying stage. Light of intensity I_0 is shone onto the film giving the two light output intensities I_1 and I_2, which create the new densities D'_1 and D'_2 on the copy film. The copy-film gamma γ_2 is then given by

$$\gamma_2 = \frac{D'_1 - D'_2}{\log I_1 - \log I_2}. \tag{6.9}$$

Since, by definition of optical density $D_1 = \log I_0 - \log I_1$, this reduces to

$$\gamma_2 = \frac{D'_1 - D'_2}{D_2 - D_1} \tag{6.10}$$

which, by equation (6.8), gives

$$\gamma_2 = \frac{D_1' - D_2'}{(\log X_1 - \log X_2)(-\gamma_1)}. \tag{6.11}$$

Since $(D_1' - D_2')/(\log X_1 - \log X_2)$ is the definition of the combined gamma, which we can write $\gamma_{1,2}$, we have

$$\gamma_{1,2} = -\gamma_1\gamma_2. \tag{6.12}$$

This shows the gamma multiplication at the first stage. The minus sign shows the reversal process. The argument is repeated at the third stage. We have (see figure 6.20(c)) light of intensity I_0' irradiating the copy film and giving rise to two intensities I_1' and I_2', which in turn give densities D_1'' and D_2'' on the third film. Again by definition

$$\gamma_3 = \frac{D_1'' - D_2''}{\log I_1' - \log I_2'} \tag{6.13}$$

and, from the definition of optical density (by analogy with equations (6.9) and (6.10) above),

$$\gamma_3 = \frac{D_1'' - D_2''}{D_2' - D_1'}. \tag{6.14}$$

Since $(D_1' - D_2')/(\log X_1 - \log X_2) = -\gamma_1\gamma_2$, we have

$$\gamma_3\gamma_1\gamma_2 = \frac{D_1'' - D_2''}{\log X_1 - \log X_2} \tag{6.15}$$

the right-hand side of which is the overall effective gamma, γ_{eff}, and so finally

$$\gamma_{\text{eff}} = \gamma_1\gamma_2\gamma_3. \tag{6.16}$$

The phase reversal is reversed back to a positive image and this proves the 'gamma multiplication'.

6.6. IMAGING WITH A PHOTOSTIMULABLE PHOSPHOR PLATE (FUJI SYSTEM)

Computed radiographic imaging has been developed by the Fuji Photo Film Co (Sonoda *et al* 1983) for imaging at diagnostic radiology energies. The same equipment has been used for megavoltage portal imaging by Wilenzick *et al* (1987). The image receptor comprises a flexible plate, some 1 mm thick, coated with europium-activated barium fluorohalide compounds in crystal form in an organic binder. The photostimulable phosphors act as energy traps when exposed to ionizing radiation, producing a stored or latent image. When illuminated by red

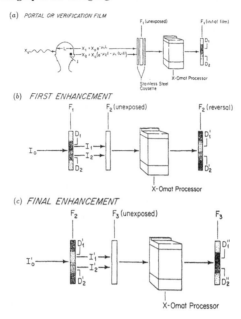

Figure 6.20. *The method of 'gamma multiplication'. This is a three-stage process: (a) showing the production of a portal radiographic film, (b) showing the production of an enhancement reversal film, (c) showing the production of the third and final enhancement film. The symbols are defined in the text. (From Reinstein and Orton (1979).)*

laser light at 633 nm, the stored energy is emitted as light. The usual technique to do this is to shine the laser light onto a scanning mirror, which illuminates one small part of the plate, the emitted light being taken into a light guide to a photomultiplier tube and thence converted to electrical energy.

The imaging sequence (figure 6.21) starts by releasing any traps by flooding the plate with light. After exposure to radiation, an initial low-resolution (256^2) image is 'pre-read' with a low laser intensity to set the sensitivity. Then the image is read at high resolution (2000^2) with 8 bits per pixel at a higher laser power. The images can be stored as digital computer files and processed in the usual ways, for example by windowing, contrast adjustment (by simulating some Hurter–Driffield response curve), by unsharp masking, etc.

This imaging modality is sensitive over four decades of response, exhibiting a linear dose-photoluminescent response function. For this reason it can cope with the overexposure by some two orders of magnitude compared with how it is used at diagnostic energies.

Wilenzick and Merritt (1987) used the system to record images at 6 and 10 MV and showed, using the Lutz–Bjarngard portal-film test object (Lutz and Bjarngard 1985), that the computed radiography system outperformed conventional portal imaging with film. They discussed the interesting possibility of using the portal

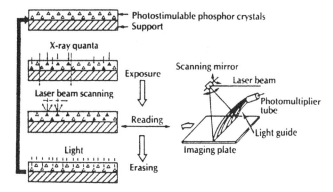

Figure 6.21.   *The imaging method with a photostimulable phosphor screen showing exposure of the plate, release of latent energy by scanning laser, and erasure of the image to prepare the plate for re-use. The laser-stimulated fluorescence is detected by a photomultiplier tube from which the digitized signal is obtained. The mirror is used to scan the laser beam over the plate. This technique is also used in diagnostic radiology. (From Wilenzick and Merritt (1987).)*

data for transit dosimetry (see section 6.12).

A commercial version of the portal imager, based on the use of a photostimulable phosphor, is manufactured by Siemens (the 'FCR 7000 Digiscan'). This can give five image formats with pixel sizes varying between 100 and 200 μm, giving a spatial resolution of between 2.5 and 5 lp mm^{-1} (Scheck *et al* 1992). The images are impressive, but of course this is not a 'real-time' method of megavoltage imaging.

6.7.   PORTAL IMAGING BY RECONSTRUCTING FROM PROJECTIONS

Bova *et al* (1987) described a novel detector comprising a rotating and translating one-dimensional, strip, ion chamber, collecting projection data from which the portal image could be *reconstructed* by filtered backprojection. Several types of ion chamber were tested, with a liquid-filled chamber emerging as favourite. The images are digital and exhibit good scatter rejection. A spatial resolution of 4.1 mm was achieved when reconstructing from 180 angular samples (Boyer *et al* 1992) (figure 6.22).

Now one may imagine a set of long, thin ionization chambers positioned parallel to each other so as to form a detection *plane*. At one orientation of the detector a 1D projection of the portal image is recorded. Now imagine the detector rotates, collecting a 1D projection at each of a large number of orientations in 180°. The 2D portal image may now be reconstructed from these 1D projections. There is no need for the detector to translate. This is the same principle as engineered by Keyes (1975) and by Webb *et al* (1992) to make a novel gamma-camera collimator which improves on the performance of a matrix of parallel holes with direct capture of

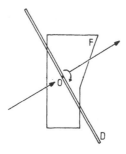

Figure 6.22. *Illustrating the principles of how a 2D image can be made from a one-dimensional detector D which integrates all the signal along its length. Imagine this detector translates in the direction of the arrow without rotating across the field F to be imaged; a one-dimensional projection of the 2D image would be formed. Now if this scanning movement is repeated at a set of orientations over 180°, the 2D image could be reconstructed from the set of projections.*

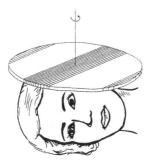

Figure 6.23. *A rotating parallel-walled slit collimator for a gamma camera. The collimator rotates through 180°, collecting a projection of the 2D planar image at each orientation from which the 2D image may be reconstructed.*

the 2D image (figure 6.23). The equivalent megavoltage imaging system has not been engineered.

6.8.   PORTAL IMAGING BY XERORADIOGRAPHY

Another diagnostic imaging technique which has been pressed into use for high-energy portal imaging is xeroradiography (Fingerhut and Fountinell 1974, Wolfe *et al* 1973). Xeroradiography dates back to the 1890's and was extensively developed for diagnostic mammography from the 1960's (Boag 1973). The same commercial equipment has been used for megavoltage imaging. The xeroradiographic imaging process produces images with strong edge enhancement and little broad-area contrast (Dance 1988).

The xeroradiographic process exploits the photoconductive properties of amorphous selenium. The receptor is a thin layer of selenium (125μm) deposited on an aluminium backing plate. The detector is charged prior to irradiation. During irradiation the energy absorbed creates electrons and holes. These charge carriers migrate to the surface of the selenium under the action of the electric field, forming a latent image by subtraction from the original uniform charge distribution. The charge image is developed by being sprayed with a fine aerosol of blue powder and a powder pattern builds, reflecting the charge image. The powder pattern is transferred to paper by contact and fused by heating.

Fingerhut and Fountinell (1974) used xeroradiography to make images using a 4 MeV accelerator with a dose of 4 cGy. The wide latitude of the process allowed good visualization of anatomical detail. No details were published. Development took 90 s and they claim the method completely replaced the use of film at their clinic. It might be observed, however, that the method has not really caught on elsewhere. In principle a digital image could be produced.

At much the same time, Wolfe *et al* (1973) used xeroradiography to capture portal images from a ^{60}Co source, presenting images of a variety of treatment sites in the body, showing the versatility of xeroradiography consequent on its wide latitude. For example, air spaces and bones were simultaneously visualized in one single exposure (radiation dose 6 to 9 cGy) of a large treatment field (for bronchogenic carcinoma) designed to include regional lymph nodes. Wolfe *et al* (1973) claimed the method was 'quick and convenient'; they quoted times of some 2 minutes for development, claiming this was sufficiently short for the therapist to verify the accuracy of positioning the radiation field and readjust if necessary, repeating the imaging. Today this statement would find opponents amongst those striving for near-real-time portal imaging for truly interventional studies.

6.9.   SOLID-STATE IMAGERS

Antonuk *et al* (1990a,b, 1991a,b,c) have reported the development of a solid-state detector consisting of amorphous silicon photodiodes and thin-film transistors. A prototype detector of 11.5 cm by 10.8 cm was constructed with a 450 μm pixel pitch. This is a very new development; the eventual aim is a 30 cm by 30 cm system or even bigger.

The detector comprises a matrix of hydrogenated amorphous silicon (a-Si:H) photodiodes each coupled to a single a-Si:H field-effect transistor (FET). This matrix is constructed by techniques which are similar to photolithographic and etching methods used to create conventional crystalline silicon circuits. The silicon detector is typically 1 to 2 μm thick, deposited on a chrome metal substrate, comprises three layers of a-Si:H with different doping, capped by an indium tin oxide conducting layer. X-ray photons are converted to optical photons by a suitable metal + phosphor screeen. Antonuk *et al* (1991a) have studied the effects on signal and noise of different screens with different thicknesses, together with the variations with thickness of the a-Si:H detector. Other important considerations are the effect of the 'dead space' taken up by the data lines, FET lines and the bias lines.

The development is known as 'MASDA', multi-element amorphous silicon detector array. The potential benefit for radiotherapy portal imaging is the compactness of the detector; comparable to a film cassette. The major advantage of MASDA is that, although the optical photon production efficiency is much the same as for all systems using a metal plate and a phosphor, the detecting plane is right next to the source of optical photons, resulting in very small losses. Hence some 50% of the light signal can be measured, unlike the $\leqslant 1\%$ for a fluoroscopic camera or fibre-optic system (Boyer *et al* 1992). Amorphous silicon materials are also inherently radiation resistant. Research is progressing towards a detector of comparable dimensions to a film; very recently the first radiotherapy images have been obtained (Antonuk *et al* 1992).

6.10.   DIAGNOSTIC IMAGING ON A LINEAR ACCELERATOR

There have been attempts to take diagnostic-energy radiographs with a diagnostic x-ray tube attached to a high-energy linear accelerator. Since these are not true 'megavoltage images' they are included here more for completeness. Diagnostic-quality x-ray films were obtained many years ago by mounting an angularly offset x-ray source and detector on the side of a cobalt treatment unit (Holloway 1958, Shorvon *et al* 1966). Ideally the diagnostic energy source should coincide exactly in position with the high-energy treatment source and, according to Biggs *et al* (1985), this has been engineered at Stanford University by Varian Associates. However, the next-best solution is to mount an x-ray source on the therapy machine, offset from the treatment head by a precise angle (figure 6.24). Biggs *et al* (1985) report a development of this type at Boston. The important feature of the development was that the diagnostic imaging assembly included an exact replica of the support system for holding the blocking tray. The axes of the diagnostic and therapy beams coincided at the isocentre of the machine and were 45° apart. The low- and high- energy x-ray sources were exactly the same distance from the isocentre. The extra weight of the diagnostic tube was counterbalanced. Great care was taken to provide devices for precisely engineering the position of the diagnostic x-ray tube, and which allowed a good programme of quality assurance thereafter. A 'two-bubble' spirit-level-like device was used when moving from the diagnostic-imaging mode to the therapy mode. Two spirit levels were mounted on the gantry at the same angular separation as the two x-ray axes, and movement between the two modes was such that one or other bubble was centred for each mode. This ensured that the diagnostic-energy tube during imaging was exactly in the same place as occupied by the therapy source during therapy. Possibly the reason these ideas were not taken further is that they removed the ability to rotate the collimator (i.e. to 'put on head twist').

Figure 6.24. *View of a diagnostic x-ray unit attached to the head of a linear accelerator. The central axis of the diagnostic beam intersects that of the therapy beam at the isocentre and the two lines are at a fixed orientation to each other whatever the gantry angle. There is a complete replica of the accessories found on the linac (e.g. the wedge tray) incorporated into the diagnostic x-ray unit. This arrangement allows 'diagnostic-quality' radiographs to be obtained with the patient in the treatment position. (From Biggs et al (1985).) (Reprinted with permission from Pergamon Press Ltd, Oxford, UK.)*

6.11. THEORETICAL CONSIDERATIONS OF DOSE AND IMAGE SIGNAL-TO-NOISE RATIO

Viewed as a radiological imaging system, albeit at an unusually high energy, many of the phenomena, well understood in relation to diagnostic-energy x-radiology, can be investigated by similar mathematical arguments. Surprisingly, this had not been done in detail until the recent paper by Swindell *et al* (1991).

They established that the minimum dose $D_{\min}$ at the build-up depth $d_{\max}$ required to see an object of area l^2 and thickness x in a slab of tissue of thickness T and linear attenuation coefficient μ with monoenergetic photons of energy E_p, a detector of quantum efficiency η, a scatter-to-primary ratio SPR with a differential signal-to-noise ratio $DSNR$, was

$$D_{\min} = E_p \left(\frac{\mu_{en}}{\rho}\right) \frac{DSNR^2}{(\Delta\mu x l)^2} \left(\frac{L_1}{L_1 - t + d_{\max}}\right)^2 \left(1 + \frac{1}{g}\right) \frac{e^{\mu T}}{\eta} (1 + SPR)$$

(6.17)

where L_1 is the distance from the source to the centre of the 'object' to be viewed, which is itself a distance t below the surface of the slab, and $\Delta\mu$ is the change in linear attenuation coefficient between the object and its surroundings (figure 6.25). μ_{en}/ρ is the mass-energy absorption coefficient of the soft tissue. The factor g accounts for the Poisson noise in the random detection process (it is the same

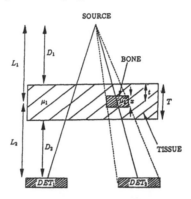

Figure 6.25. *The geometry used to evaluate equation (6.17). A tissue slab of thickness T contains a piece of bone of thickness x at a depth t below the surface. Two lines of radiation are shown, one passing through the bone onto a detector element DET_2 and the other passing through just soft tissue to a detector element DET_1. The bone and tissue have linear attenuation coefficients μ_2 and μ_1, respectively. (From Swindell et al (1991).)*

as the factor $P_3 P_4 g_2$ in section 6.2). The factor $\left(1 + \frac{1}{g}\right)$ is almost exactly unity for scintillation-detection portal imagers, but can be as large as three for TV-based systems. Swindell *et al* (1991) present detailed calculations of the SPR in different geometries which, together with the nomogram (figure 6.26) for bone 'object' in soft tissue (derived from this equation when $SPR = 0$), can be used to predict the dose-size trade-off under various circumstances. Even small air gaps (say 40 cm) were found to reduce the SPR to less than 10%. Following Rose's theory (Rose 1948, 1973, Webb and Johnson 1990) the $DSNR$ was set to 10 (i.e. a conservative estimate of double the usual Rose factor of 5 for the smallest $DSNR$ that can be perceived).

The 'diagnostic equivalent' of the above equation was derived by Dance (1988). When comparing TV-based systems with scanning-detector arrays, the different quantum efficiency of the two systems must be considered, as well as the obvious fact that one system involves scanning and the other does not. Because the x-ray beam cannot be steered in synchrony, the dose to the patient is higher than the dose forming the image with a scanning system.

The number of bits required to represent the image faithfully works out to be about a minimum of 10 bits in typical conditions.

6.12. PORTAL DOSE IMAGES; TRANSIT DOSIMETRY

Portal images, whether accrued in analogue fashion on film or by electronic imagers, have to date largely been correlated with simulator films, or radiographs digitally reconstructed from CT (DRR) to assess position uncertainty. The use of DRR is not yet well established, perhaps because of the limited spatial resolution

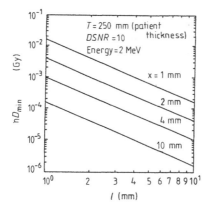

Figure 6.26. *Nomogram showing the minimum dose required to see a bone of square shape ($l \times l$ mm with $\mu = 0.0083$ mm^{-1} embedded in the surface ($t = 0$) of a 25 cm thickness of tissue ($\mu = 0.0049$ mm^{-1}). The curves are parametrized in terms of the bone insert thickness, x. The effect of scatter is not included here, as equation (6.17) shows how the nomogram changes with different scatter-to-primary ratios and these can be worked out from Monte Carlo calculations for different geometries. (From Swindell et al (1991).)*

compared with simulator films, arising because of the finite slice width of the CT data from which they are generated (Sontag and Purdy 1991). Whilst these comparisons are valuable and can give indications of patient mispositioning, they cannot, for example, determine whether a wedge has been inadvertently inverted or an incorrect wedge used. Ideally the comparison should be between measured portal *dose* and the predicted portal *dose*.

Measurements of portal dose are called portal dose images or PDIs. Portal dose can be predicted if a 3D CT data-set is known and the model for calculating the exit portal dose is accurate. The latter must have a full description of scattered radiation. It is not adequate to simply ray-trace the primary radiation through the patient.

Wong *et al* (1990b) have compared the predictions of the 'Delta Volume' calculation method with measurements using TLD chips and an ionization chamber for a phantom study of known composition and for a Rando phantom where CT data were available. They report that for these well controlled conditions agreement was better than 1%. The same could not be said for an attempt to match a PDI for a patient with a calculation, when the portal dose was found for a small number of spot measurements. Errors as large as 10% arose. However, when a film was digitized to make an areal PDI, more informative results were obtained. Wong *et al* (1990b) were able to demonstrate that adjusting the position of the patient could bring the calculated PDI into good alignment with the measured PDI. This is a very difficult area of research and there is still much to be done (Swindell 1992a).

Ying *et al* (1990) described a very complicated way of using portal dose images. A portal dose image was calculated from 3D CT data and compared with the measured PDI. Then the ratio of the two measurements was distributed back along fan lines through the CT volume to adjust the CT data. From these data, a new calculation of PDI was made and the process cycled until the calculated and measured PDI were close enough in agreement. At that stage the adjusted CT data were used to compute patient dose. All this took place of course after the patient had been treated, since the forward dose calculations were lengthy.

6.13.   MEGAVOLTAGE COMPUTED TOMOGRAPHY

In Tucson, Arizona, and later at the Royal Marsden Hospital, Sutton, Swindell and colleagues have developed megavoltage computed tomography (MVCT) (Simpson *et al* 1982, Swindell *et al* 1983, Lewis and Swindell 1987, Lewis *et al* 1988, 1992). A Swedish development has also been reported (Brahme *et al* 1987, Källman *et al* 1989) as well as a replication of Swindell's system in Japan (Nakagawa *et al* 1991). Megavoltage CT scanning has been reported in the literature of non-destructive testing, but is rare with medical applications (Kanamori *et al* 1989).

Strictly MVCT does not fit into the scope of a review of portal imaging, but is included here because the work is complementary. The developments achieve several goals. Firstly, it may be noted that x-ray linear attenuation coefficients derived from a machine operating at megavoltage energies would be immediately applicable to making tissue inhomogeneity corrections in planning (whereas for a diagnostic machine a conversion is needed). Secondly, images taken on a linac would help to ensure the patient was in the treatment position. The aim is to take images just prior to treatment to verify the positioning of the patient by comparison with CT images at treatment simulation. This provides a connection with CT imaging on a simulator, since these latter images might well be obtained for certain treatments (e.g. breast radiotherapy) with a simulator non-diagnostic computed tomography (NDCT) scanner (Webb 1990). Periodic checks of the patient anatomy (to determine, for example, whether a tumour has changed its size) become feasible during the course of fractionated radiotherapy.

Swindell's work has progressed in several phases. Using the well known relationship between image spatial and density resolutions, slice width, beam energy, delivered dose, incident x-ray energy, detective quantum efficiency and object size (Barrett *et al* 1976), they calculated that, by relaxing the spatial resolution requirement to some 2 mm, images with a signal-to-noise ratio greater than 200 could be calculated with a dose of about 8 *c*Gy. When the signal-to-noise ratio was relaxed to 100 the dose could be as low as 2 *c*Gy. The first system constructed had a multi-element detector made of Pilot-B plastic scintillator coupled to photodiodes. The 80 individual elements were arranged on the arc of a circle of radius 140 cm centred on the x-ray source. 110 fan-beam projections (12 bits deep) were collected in a 220° arc and images were reconstructed by a convolution and backprojection (CBP) method with a Shepp–Logan filter into 128^2 matrices using a PDP 11/34 computer. Simpson *et al* (1982) showed that the relationship between reconstructed CT numbers and tissue electron density was precisely linear. The measured resolution was some 4 mm and a signal-to-noise ratio greater than 100 was achieved.

The plastic scintillator was soon abandoned in favour of a detector fabricated from 96 bismuth germanate crystals. These had a superior physical density (7.13 g cm^{-3}, instead of 1.1 g cm^{-3} for the plastic) and a stopping power of 83% instead of 25%. The distance of the detector from the source was also increased from 140 cm to 180 cm, decreasing the scatter contribution to the data. The reconstructions showed once again a linearity of better than 1 % between CT number and electron density, a spatial resolution of 3 mm and signal-to-noise ratio in excess of 100. The dose delivered was some 10 cGy and indicated that there was still room for improvement in the data gathering and handling. Also, annoying circular artefacts were seen, caused by electronic drift. The data were, however, clearly adequate for treatment planning and verification purposes and Swindell *et al* (1983) showed a variety of *in vivo* images, mainly of head structure in view of the relatively small field of view (28–32 cm). They concluded that the limit on spatial resolution in such a system would probably be set by the finite spot size of the x-ray source and they felt it was still an open question as to whether megavoltage CT would allow tumour visualization.

Lewis and Swindell (1987) and Lewis *et al* (1988, 1992) report setting up a very similar system to the second system (above) at the Royal Marsden Hospital, Sutton (figure 6.27). The major improvements over the Tucson system included: (i) image formation with less dose (5 cGy) but with no reduction of image quality, (ii) image reconstruction during little more than the time it takes to scan, (iii) shorter reconstruction time, (iv) the detector takes just a few minutes to attach instead of the half an hour for the prototype, making working with the system much more convenient, (v) the system has been put into clinical use. An image from this system is shown in figure 6.28.

6.14. SUMMARY

Megavoltage portal imaging, after years of being restricted to the use of film, is now being approached seriously as a field for improvement. Electronic portal imaging has become the state of the art, but has not yet quite reached maturity, in that real-time portal imaging is not yet a routine reality in all centres. Consequently, studies of the effectiveness of improved portal imaging are still to be done. Several groups are beginning to use portal imagers to assess the importance of field mispositioning with respect to the patient (Griffiths 1992, Shalev 1992).

There is a variety of quite distinct technical instrumentation and it is difficult to make quantitative comparisons with which the authors of each system would agree. Swindell *et al* (1991) have provided a very detailed study of the underlying physics which brings out many of the important considerations, and this paper

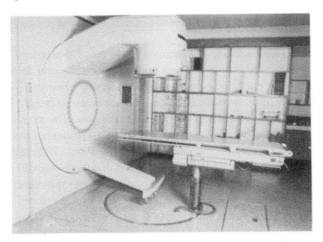

Figure 6.27. *The Royal Marsden Hospital megavoltage CT system. The curved detector is attached to a Philips SL25 accelerator. This slots into the triangular mounting plate which is firmly bolted to the rotating gantry of the linac. (Courtesy of W Swindell, E Morton, D G Lewis and P Evans.)*

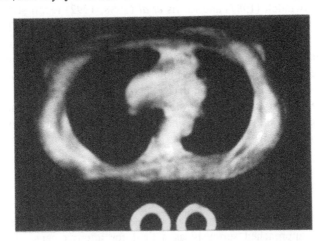

Figure 6.28. *A megavoltage CT scan taken on the system at the Royal Marsden Hospital. The scan shows the external contour; the lungs and the bones and soft-tissue detail can be resolved when viewing the image on a monitor with a windowing facility. The two circular structures below the patient are part of the patient support system. (Courtesy of W Swindell, E Morton, D G Lewis and P Evans.)*

should be appraised by the serious researcher. In this chapter we have discussed some but not all of these aspects, whilst trying to present the operating features of each system as simply and objectively as possible. This is a field which will expand and indeed is only now beginning to attract serious commercial interest.

In summary, the need is for a detector which can operate in real-time, produce a high resolution digital radiograph, has a high quantum efficiency, be of convenient size, reject scattered radiation and possess a high signal-to-noise ratio so that the low contrast inherent in portal radiology is challenged. At present, good detectors exist, together with means to make accurate measurements. Attention should now be turned towards quantifying positional errors and using portal measurements to compute the dose actually delivered to the patient.

REFERENCES

Amols H I and Lowinger T 1987 An inexpensive microcomputer based port film image enhancement system *Med. Phys.* **14** 483

Antonuk L E, Berry J, Boudry J, Huang W, Longo M J, Mody P, Morton E J, Schick H, Seraji N, Yorkston J and Street R A 1992 Flat-panel imagers for radiotherapy and diagnostic radiographic and fluoroscopic imaging *Proc. Megavoltage Imaging Meeting (Newport Beach, CA, 1992)* (New York: American Association of Physicists in Medicine)

Antonuk L E, Boudry J, Kim C W, Longo M, Morton E J and Yorkston J 1991a Signal, noise, and readout considerations in the development of amorphous silicon photodiode arrays for radiotherapy and diagnostic x-ray imaging *Proc. SPIE* **1443** (Medical Imaging V: Image Physics) pp 108–119

Antonuk L E, Boudry J, Yorkston J, Wild C F, Longo M J and Street R A 1990a Radiation damage studies of amorphous silicon photodiode sensors for applications in radiotherapy x-ray imaging *Nucl. Instrum. Methods in Phys. Res.* A **299** 143–146

Antonuk L E, Yorkston J, Boudry J, Longo M J, Jimenez J and Street R A 1990b Development of hydrogenated amorphous silicon sensors for high energy photon radiotherapy imaging *IEEE Trans. Nucl. Sci.* **37** (2) 165–170

Antonuk L E, Yorkston J, Boudry J, Longo M J and Street R A 1991b Large area amorphous silicon photodiode arrays for radiotherapy and diagnostic imaging *Nucl. Instrum. Methods in Phys. Res.* A **310** 460–464

Antonuk L E, Yorkston J, Morton E J, Boudry J, Kim C W and Longo M J 1991c Flat-panel, self-scanning, solid state imagers for real-time radiotherapy and digital fluoroscopy *Med. Phys.* **18** 610

Baily N A, Horn R A and Kampp T D 1980 Fluoroscopic visualisation of megavoltage therapeutic x ray beams *Int. J. Rad. Oncol. Biol. Phys.* **6** 935–939

Barrett H H, Gordon S K and Hershel R S 1976 Statistical limitations in transaxial tomography *Comp. Biol. Med.* **6** 307–323

Barrett H H and Swindell W 1981 *Radiological imaging: the theory of image formation, detection and processing* vol 1 pp 97–100, 287–289

Benner S, Rosengren B, Wallman H and Netteland O 1962 Television monitoring of a 30 MV X-ray beam *Phys. Med. Biol.* **7** 29–34

Bijhold J, Gilhuijs K G A, van Herk M and Meertens H 1991a Radiation field edge detection in portal images *Phys. Med. Biol.* **36** 1705–1710

Bijhold J, van Herk M, Vijlbrief R and Lebesque J V 1991b Fast evaluation of patient set-up during radiotherapy by aligning features in portal and simulator images *Phys. Med. Biol.* **36** 1665–1679

Bijhold J, Lebesque J V, Hart A A M and Vijlbrief R E 1991c Maximising setup accuracy using portal images as applied to a conformal boost technique for prostatic cancer *Proc. 1st biennial ESTRO meeting on physics in clinical radiotherapy (Budapest, 1991)* p 18

Biggs P J 1991 private communication (AAPM visiting fellow, May 21st 1991)

Biggs P J, Goitein M and Russell M D 1985 A diagnostic x-ray field verification device for a 10 MV linear accelerator *Int. J. Rad. Oncol. Biol Phys.* **11** 635–643

Boag J W 1973 Xeroradiography *Phys. Med. Biol.* **18** 3–37

Bova F J, Fitzgerald L T , Mauderli W M and Islam M K 1987 Real time megavoltage imaging *Med. Phys.* **14** 707

Boyer A L, Antonuk L, van Herk M, Meertens H, Munro P, Reinstein L E and Wong J 1992 A review of electronic portal imaging devices (EPIDs) *Med. Phys.* **19** (1) 1–16

Brahme A, Lind B and Näfstadius P 1987 Radiotherapeutic computed tomography with scanned photon beams *Int. J. Rad. Oncol. Biol. Phys.* **13** 95–101

Bucciolini M, Pallotta S, Cionini L, Milano F and Renzi R 1991 Control of patient setup during radiation treatments *Proc. 1st biennial ESTRO meeting on physics in clinical radiotherapy (Budapest, 1991)* p 22

Crooks I and Fallone B G 1991 PC based selective histogram equalisation for contrast enhancement of portal films *Med. Phys.* **18** 618

Dance D R 1988 Diagnostic radiology with x-rays *The physics of medical imaging* ed S Webb (Bristol: Adam Hilger) pp 20–73

Droege R T and Bjarngard B E 1979 Influence of metal screens on contrast in megavoltage x-ray imaging *Med. Phys.* **6** 487–493

Evans P M, Gildersleve J Q, Morton E J, Swindell W, Coles R, Ferraro M, Rawlings C, Xiao Z R and Dyer J 1992a Image comparison techniques for use with megavoltage imaging systems *Brit. J. Radiol.* **65** 701–709

Evans P M, Gildersleve J Q, Rawlings C and Swindell W 1992b The implementation of patient position correction using a megavoltage imaging device on a linear accelerator *Brit. J. Radiol.* in press

Fingerhut A G and Fountinell P M 1974 Xeroradiography in a radiation therapy department *Cancer* **34** 78–82

Galkin B M, Wu R K and Suntharalingam N 1978 Improved technique for obtaining teletherapy portal radiographs with high-energy photons *Radiology* **127** 828–830

Gildersleve J 1992 Clinical aspects of megavoltage imaging *MD thesis* University of London

Gildersleve J, Dearnaley D P, Evans P M, Law M, Rawlings C and Swindell W 1992a A randomised trial of patient repositioning during pelvic radiotherapy

Proc. BIR Conf. on Megavoltage Portal Imaging (London, 1992) (London: BIR)

Gildersleve J, Evans P M, Dearnaley D P, Rawlings C and Swindell W 1992b Reproducibility of patient positioning, as assessed by an Integrated Megavoltage Imaging system *Proc. BIR Conf. on Megavoltage Portal Imaging (London, 1992)* (London: BIR)

Gildersleve J, Swindell W, Evans P, Morton E, Rawlings C and Dearnaley D 1992c Verification of patient positioning during radiotherapy using an integrated megavoltage imaging system *Proc. ART91 (Munich)* (abstract book p 135) *Advanced Radiation Therapy: Tumour Response Monitoring and Treatment Planning* ed A Breit (Berlin: Springer) pp 693–695

Graham M L, Cheng A Y, Geer L Y, Binns W R, Vannier M W and Wong J W 1991 A method to analyze 2-dimensional daily radiotherapy portal images from an on-line fiber-optic imaging system *Int. J. Rad. Oncol. Phys.* **20** 613–619

Griffiths S 1992 Quality assurance—what is possible? *Proc. 50th Ann. Congress of the British Institute of Radiology (Birmingham, 1992)* (London: BIR) p 29

Grimm D F, Wilson C R and Gillin M T 1987 Digital processing of radiotherapy port films *Med. Phys.* **14** 483

Halmshaw R 1966 *Physics of industrial radiography* (London: Heywood) p 488

Hammoudah M M and Henschke U K 1977 Supervoltage beam films *Int. J. Rad. Oncol. Biol. Phys.* **2** 571–577

Hendee W R 1978 Computed tomography in radiation therapy treatment planning *Int. J. Rad. Oncol. Biol. Phys.* **4** 539–540

van Herk M 1991 Physical aspects of a liquid-filled ionization chamber with pulsed polarizing voltage *Med. Phys.* **18** (4) 692–702

—— 1992 An electronic portal imaging device: physics, development and application *PhD thesis* University of Amsterdam

van Herk M and Meertens H 1987 A digital imaging system for portal verification *The use of computers in radiation therapy* ed I A D Bruinvis *et al* (Proc. 9th ICCRT) pp 371–373

—— 1988 A matrix ionisation chamber imaging device for on-line patient setup verification during radiotherapy *Radiother. Oncol.* **11** 369–378

Holloway A F 1958 A localising device for a rotating cobalt therapy unit *Brit. J. Radiol.* **31** 227

Hoogervorst B R, van Herk M and Mijnheer B J 1991 Development of a two-dimensional array dosimeter *Proc. 1st biennial ESTRO meeting on physics in clinical radiotherapy (Budapest, 1991)* p 26

Jevbratt L, Lagergren C and Sarby B 1971 Direct beam control in radiotherapy with high energy photons *Acta Radiol. Ther. Phys. Biol.* **10** 433–442

Jones S M and Boyer A L 1991 Investigation of an FFT-based correlation technique for verification of radiation treatment setup *Med. Phys.* **18** (6) 1116–1125

Källman P, Lind B, Iacobeus C and Brahme A 1989 A new detector for radiotherapy computed tomography verification and transit dosimetry *Proc.*

17th Int. Congress on Radiology (Paris, 1989) p 83

Kanamori T, Kamata S and Ito S 1989 Cross-sectional imaging of large and dense materials by high energy x-ray CT using linear accelerator *J. Nucl. Sci. Tech.* **26** (9) 826–832

Keyes W I 1975 The fan-beam gamma camera *Phys. Med. Biol.* **20** 489–493

Lam K S, Partowmah M and Lam W C 1986 An on-line electronic portal imaging system for external beam radiotherapy *Brit. J. Radiol.* **59** 1007–1013

Leong J 1984 A digital image processing system for high energy x-ray portal images *Phys. Med. Biol.* **29** 1527–1535

Leong J 1986 Use of digital fluoroscopy as an on-line verification device in radiation therapy *Phys. Med. Biol.* **31** 985–992

Leszczynski K, Miskiewicz K, Cosby S, Shalev S, Reinstein L and Meek A 1988 On-line megavoltage imaging 2: Image processing *Electronic Imaging 88; International electronic imaging exposition and conference (World Trade Centre, Boston, 1988)* (Waltham: Institute for Graphic Communications) pp 1080–1083

Leszczynski K, Shalev S and Cosby N S 1990 An adaptive technique for digital noise suppression in on-line portal imaging *Phys. Med. Biol.* **35** 429–439

Lewis D G, Morton E J and Swindell W 1988 Precision in radiotherapy: a linear accelerator based CT system *Megavoltage Radiotherapy 1937–1987 Brit. J. Radiol.* (Suppl. **22**) p 24

Lewis D G and Swindell W 1987 A megavoltage CT scanner for radiotherapy verification *The use of computers in radiation therapy* ed I A D Bruinvis *et al* (Proc. 9th ICCRT) pp 339–340

Lewis D G, Swindell W, Morton E, Evans P and Xiao Z R 1992 A megavoltage CT scanner for radiotherapy verification *Phys. Med. Biol.* **37** 1985–1999

Lutz W and Bjarngard B 1985 A test object for evaluation of portal films *Int. J. Rad. Oncol. Biol. Phys.* **11** 631–634

Meertens H 1985 Digital processing of high-energy photon beam images *Med. Phys.* **12** 111–113

Meertens H 1989 On line acquisition and analysis of portal images *PhD thesis* University of Amsterdam

Meertens H, Bijhold J and Strackee J 1990a A method for the measurement of field placement errors in digital portal images *Phys. Med. Biol.* **35** 299–323

Meertens H, van Herk M, Bijhold J and Bartelink H 1990b First clinical experience with a newly developed electronic portal imaging device *Int. J. Rad. Oncol. Biol. Phys.* **18** 1173–1181

Meertens H, van Herk M and Weeda J 1985 A liquid ionisation detector for digital radiography of therapeutic megavoltage photon beams *Phys. Med. Biol.* **30** 313–321

Morton E J 1988 A digital system for the production of radiotherapy verification images *PhD thesis* University of London

Morton E J, Evans P M, Ferraro M, Young E F and Swindell W 1991a Development of video frame store and distortion correction facilities for an external-beam radiotherapy treatment simulator *Brit. J. Radiol.* **64** 747–750

Morton E J, Lewis D G and Swindell W 1988 A method for the assessment of radiotherapy treatment precision *Megavoltage Radiotherapy 1937–1987 Brit. J. Radiol.* (Suppl. **22**) p 25

Morton E J and Swindell W 1988 A digital system for the production of radiotherapy verification images *The use of computers in radiation therapy* ed I A D Bruinvis *et al* (Proc. 9th ICCRT) pp 375–377

Morton E J, Swindell W, Lewis D G and Evans P M 1991b A linear array scintillation crystal-photodiode detector for megavoltage imaging *Med. Phys.* **18** (4) 681–691

Munro P M, Rawlinson J A and Fenster A 1990a A digital fluoroscopic imaging device for radiotherapy localisation *Int. J. Rad. Oncol. Biol. Phys.* **18** 641–649

—— 1990b Therapy imaging: a signal-to-noise analysis of a fluoroscopic imaging system for radiotherapy localisation *Med. Phys.* **17** (5) 763–772

Nakagawa K, Aoki Y, Akanuma A, Onogi Y, Karasawa K, Terahara A, Hasezawa K and Sasaki Y 1991 Development of a megavoltage CT scanner using linear accelerator treatment beam *J. Japan. Soc. Therap. Radiol. Oncol.* **3** 265–276

Reinstein L E and Orton C G 1978 Contrast enhancement of high energy radiotherapy films *Med. Phys.* **5** 332

—— 1979 Contrast enhancement of high energy radiotherapy films *Brit. J. Radiol.* **52** 880–887

Rose A 1948 The sensitivity performance of the human eye on an absolute scale *J. Opt. Soc. Am.* **38** 196–208

—— 1973 *Vision—human and electronic* (New York: Plenum)

Rudin S, Bednarek D R and Wong R 1991 Accurate characterization of image intensifier distortion *Med. Phys.* **18** (6) 1145–1151

Scheck R J, Wendt Th and Panzer M 1992 Digital portal radiography in radiation therapy *Proc. BIR Conf. on Megavoltage Portal Imaging (London, 1992)* (London: BIR)

Shalev S 1991 The physical limitations of DQE in portal imaging systems *Med. Phys.* **18** 616

—— 1992 *Report on 2nd Annual Int. Workshop on Electronic Portal Imaging (Newport Beach, 1992)* (private communication)

Shalev S, Lee T, Leszczynski K, Cosby S, Chu T and Reinstein L 1989 Video techniques for on-line portal imaging *Comp. Med. Im. Graph.* **13** 217–226

Shalev S, Leszczynski K, Reinstein L and Meek A 1988 On-line megavoltage imaging 1: system design *Electronic Imaging 88; International electronic imaging exposition and conference (World Trade Centre, Boston, 1988)* (Waltham: Institute for Graphic Communications) pp 1074–1079

Shalev S, Ryder S, Cosby S and Miskiewicz K 1991 Clinical experience with on-line portal imaging *Proc. ART91 (Munich)* p 138

Shorvon L M, Robson N L K and Day M J 1966 A new technique for treatment planning and field verification *Clin. Radiol.* **17** 139–140

Simpson R G, Chen C T, Grubbs E A and Swindell W 1982 A 4MV CT scanner for radiation therapy: The prototype system *Med. Phys.* **9** (4) 574–579

Smith V 1987 Routines for enhancement of radiation therapy images *The use of computers in radiation therapy* ed I A D Bruinvis *et al* (Proc. 9th ICCRT) pp 37–40

Smith A R 1990 Evaluation of new radiation oncology technology *Int. J. Rad. Oncol. Biol. Phys.* **18** 701–703

Sonoda M, Takano M, Miyahara J and Kato H 1983 Computed radiography utilising scanning laser stimulated luminescence *Radiology* **148** 833–838

Sontag M and Purdy J A 1991 (Writing Chairs) State-of-the-art external photon beam radiation treatment planning *Int. J. Rad. Oncol. Biol. Phys.* **21** 9–23

Swindell W 1991 The lens coupling efficiency in megavoltage imaging *Med. Phys.* **18** 1152–1153

—— 1992a Transit dosimetry *Proc. BIR Conference on Megavoltage Portal Imaging (London, 1992)* (London: BIR)

—— 1992b Imaging and portal verification *Review paper at the 50th Ann. Congress of the British Institute of Radiology (Birmingham, 1992)* (London: BIR) p 29

Swindell W and Gildersleve J 1991 Megavoltage imaging in radiotherapy *Rad. Magazine* (Sept 1991) **17** (196) 18–20

Swindell W, Lewis D G and Morton E J 1988 Radiological imaging methods based on a linear accelerator: implementation and application *Brit. J. Radiol.* **61** 717–718

Swindell W, Morton E J, Evans P M and Lewis D G 1991 The design of megavoltage projection imaging systems: some theoretical aspects *Med. Phys.* **18** 855–866

Swindell W, Simpson R G and Oleson J R 1983 Computed tomography with a linear accelerator with radiotherapy applications *Med. Phys.* **10** (4) 416–420

Taborsky S C, Lam W C, Sterner R E and Skarda G M 1982 Digital imaging for radiation therapy verification *Optical Eng.* **21** 888–893

Varian 1992 *Varian Portalvision* Varian Associates product literature

Visser A G V, Huizenga H, Althof V G M and Swanenburg B N 1990 Performance of a prototype fluoroscopic radiotherapy imaging system *Int. J. Rad. Oncol. Biol. Phys.* **18** 43–50

Wade J P and Nicholas D 1991 Clinical application of an electronic imaging device for assisting patient set-up in radiotherapy *Brit. J. Radiol.* **64** 596–602

Webb S 1990 Non-standard CT scanners: their role in radiotherapy *Int. J. Rad. Oncol. Biol. Phys.* **19** 1589–1607

Webb S, Binnie D M, Flower M A and Ott R J 1992 Montecarlo modelling of the performance of a rotating slit-collimator for improved planar gamma-camera performance *Phys. Med. Biol.* **37** 1095–1108

Webb S and Johnson C V 1990 The perception of detail in radioisotope images with Gaussian noise *Phys. Med. Biol.* **35** 1145–1151

Wilenzick R M, Merritt C R B and Balter S 1987 Megavoltage portal films using computed radiographic imaging with photostimulable phosphors *Med. Phys.*

14 389–392

Wolfe J N, Kalisher L and Considine B 1973 Cobalt 60 treatment field verification by xeroradiography *Am. J. Ront.* **118** 916–918

Wong J W, Binns W R, Cheng A Y, Epstein J W and Klarmann J 1991 Development of a high resolution fiber-optic radiotherapy imaging system *Med. Phys.* **18** 609

Wong J W, Binns W R, Cheng A Y, Geer L Y, Epstein J W, Klarmann J and Purdy J A 1990a On-line radiotherapy imaging with an array of fibre-optic image reducers *Int. J. Rad. Oncol. Biol. Phys.* **18** 1477–1484

Wong J W, Slessinger E D, Hermes R E, Offutt C J, Roy T and Vannier M W 1990b Portal dose images I: Quantitative treatment plan verification *Int. J. Rad. Oncol. Biol. Phys.* **18** 1455–1463

Wowk B, Shalev S and Leszczynski K W 1991 Optimisation of metal/phosphor screens for on-line portal imaging *Med. Phys.* **18** 616

Ying X, Geer L Y and Wong J W 1990 Portal dose images II: Patient dose estimation *Int. J. Rad. Oncol. Biol. Phys.* **18** 1465–1475

CHAPTER 7

TREATMENT MACHINE FEATURES FOR CONFORMAL THERAPY

7.1. THE EARLIEST TREATMENT MACHINE FOR CONFORMAL THERAPY WITH A ^{137}CS SOURCE

In this chapter we briefly review some of the machine technology for conformal therapy and gather together some of the 'accessories' found on such machines. By way of introduction, probably the first radiotherapy treatment machine to make use of a high-activity ^{137}Cs source was designed to investigate the optimal geometrical movements for 'concentrating' radiation in a small volume of the patient (Brucer 1956). This early machine for conformal therapy, constructed for the Oak Ridge Institute for Nuclear Studies, Tennessee, by W F and John Barnes Co, Rockford, Illinois, was loaded with 1540 Ci of the reactor fission waste product ^{137}Cs. The treatment machine was never intended as a routine clinical tool, but designed to investigate the relative merits of conical, pendulum and rotational therapy (see appendix 2A). The source could make any movement over the surface of a sphere whose radius was variable between 50 and 90 cm, with the beam forever directed at a central fixed point in space. Meanwhile the patient couch, a highly modified surgical table, could also execute a variety of translations and rotations. The aim was to develop a test bed, whereby a variety of patterns of irradiation were studied independent of the practicality of delivery in the clinic. Brucer (1956) wrote 'The literature of radiation therapy is full of therapeutic decisions that actually were forced by the limitations of a piece of machinery. The purpose of this excedingly complex device is thus to study therapeutic procedures unhampered by the bias of mechanical restrictions.'

Movements were controlled by analogue computer with motor-driven cams controlling the beam trajectory. Movements of a wedge filter could be coordinated with the movement of the source to accommodate the change in depth of the tumour with changing beam trajectory. The beam could be switched on and off for parts of the orbit via further mechanical cams. In view of the potential danger of collisions, touch-sensitive motor cut-outs and beam shielding was provided. The machine maintained an accuracy, which would today be regarded as rather poor,

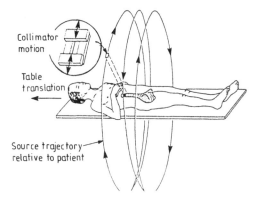

Figure 7.1. *An example of conformal therapy achieved by dynamic means. The linac gantry rotates and the couch translates. The jaws of the accelerator adopt different widths depending on the view of the target. The treatment shown is for retroperitoneal nodes and pelvic sidewalls. (From Levene et al (1978).)*

of less than 1 cm, largely set by the design of the cams.

7.2.   TRACKING UNITS

A high-dose volume can be made to conform with an irregularly shaped 3D target by dividing the irradiation into a large number of contiguous or slightly overlapping elemental fields. The isocentre is made to follow a complicated locus through the patient. This is the principle of 'tracking therapy units'. In such units the couch can move laterally and vertically, and of course the gantry can rotate. These movements, with the beam on, can be continuous. The method is also sometimes called 'dynamic therapy'. The subject has a long history. At some centres tracking units have been based on the linear accelerator (Chung-Bin and Wachtor 1979, Bjärngard *et al* 1977, Levene *et al* 1978, Chin *et al* 1981, Saunders and Chin 1986). Figure 7.1 shows how the high-dose volume could be conformed to an irregularly shaped volume by rotating the treatment head of a linear accelerator with the jaws adopting a varying width and the couch translating. The name of the Royal Free Hospital (RFH) in London is always associated in the UK with the 'tracking cobalt unit'. The first developments were made in the 1950's (Jennings 1985) and later versions of the machine are still being developed (Brace 1985, Davy 1987).

With a cobalt machine, the dose is proportional to the time spent irradiating a point, or alternatively inversely proportional to the tracking speed at which the point passes the beam. Thus exposure-time profiles (ETP) can specify the irradiation, leading to absorbed-dose profiles (ADP). For example, several non-uniform ETPs designed to shield organs at risk might be combined to give a uniform ADP in some target volume (Davy 1985). Tracking machines may be used in many different modes, depending on which components are actually in motion.

Tracking units have perhaps not become as popular as hoped by their pioneers. A number of reasons may be advanced. Certainly the treatment planning is complex, generally unique to each unit, and thus somewhat non-mainstream. For this reason (and also because excellent descriptions can be found elsewhere (Davy 1985)) this will not be discussed in detail here. Equally, verification, important for all radiotherapy, is particularly important for tracking therapy and is complicated.

Shentall and Brace (1987) have described how the treatment planning for the RFH tracking unit is performed for spinal axis irradiation. The planning process optimizes the velocity of the longitudinal couch speed and includes an end-of-track feature to avoid the large penumbra (equivalent to the field size) which would be associated with having a constant couch speed and fixed field size at the beginning and end of the track. This end-of-track method is, to start with, the minimum field length in the direction of tracking and increases the field size at twice the speed of the couch movement (and reverses the process at the cessation of irradiation). At the start, this means the trailing edge of the field stays stationary relative to the patient (small penumbra). The leading edge creates a wide penumbra, which is cancelled exactly by the subsequent movement of the full-width field.

Blake *et al* (1987) have shown how combining dynamic collimator motion with a static patient is another way to produce the same effect as moving the patient relative to a fixed small field. Computer-controlled dynamic collimation is a feature of the Philips SL-25 linear accelerator. One pair of collimator jaws can cross the centreline by 12.5 cm. Imagine they are set to a small separation and swept at constant velocity. The resulting dose profile is shown in figure 7.2. At either end, the dose ramps to the maximum dose. This large penumbra is equal to the collimator aperture and for practical reasons the end-of-track technique for the RFH cobalt unit cannot be used. The penumbra can be reduced to no more than obtained with a large single static field by appending extra field segments with just one jaw moving.

Richter *et al* (1987) achieve dynamic conformal therapy by a combination of computer control of the linac jaws, collimator and gantry rotation and four movements of the patient support system (couch translation—three orthogonal movements—and isocentric rotation.) The machine is a Philips SL 75/20.

Schlegel *et al* (1987) control the couch movements computationally and they also control the shape of the field by dynamically adjusting a multileaf collimator. The equipment is a Siemens Mevatron-77 linac. The planning system treats dynamic therapy as a series of multiple sub-beams.

7.3.   A TRACKING LINAC WITH MULTILEAF COLLIMATOR AND CT COMBINATION

Nakagawa *et al* (1987) have engineered to add computer-controlled longitudinal

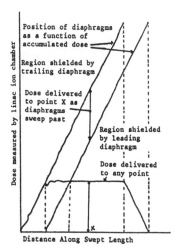

Figure 7.2. *Showing the dose profile resulting from a pair of scanned asymmetric jaws. The two approximately straight lines show the position of the two jaws as a function of accumulated dose. The lower line shows the dose delivered to each point as a function of distance. (From Blake et al (1987).)*

couch movement to a linear accelerator facility. The distinguishing feature of the development is that there is a CT scanner, in the same room as the linac, sharing the same treatment couch. The scanner and linac face each other (figure 7.3). The patient is scanned in the treatment position and the target volume is determined in each CT section. The linear accelerator (Mitsubishi ML-20M) is equipped with a multileaf collimator (Mitsubishi Multi-16). With the patient couch re-oriented and acting as the treatment couch, the width of two or three open pairs of leaves is dynamically adjusted to the projected width of the target as each slice passes through the beam, with the couch translating longitudinally. The gantry angle can also be dynamically controlled by computer.

7.4. THE UNIVERSAL WEDGE FOR THE LINEAR ACCELERATOR

Wedge filters are traditionally used in radiotherapy to assist shaping the dose distribution. For many accelerators a range of removable wedges are available. The wedge angle θ_w is defined (ICRU 1976) as the angle between the normal to the central axis and the line tangential to the isodose curve at a depth of 10 cm.

It is more convenient, and possibly avoids errors, if the accelerator is fitted with a universal wedge giving a fixed wedge angle θ_w. For example, the Philips SL/75 accelerator has a universal wedge of nominally $\theta_w = 60°$. The universal wedge avoids the need to enter a treatment room to set a removal wedge in place, means only one wedge field needs to be measured at commissioning and may be more accurately engineered. By treating a patient for part of each fraction with the

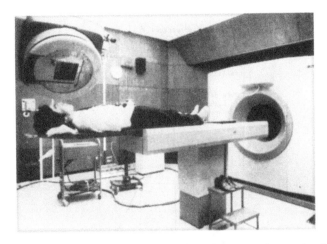

Figure 7.3. *The arrangement in Tokyo with a linear accelerator (equipped with a multileaf collimator and computer-controlled moving couch) in the same room as a CT scanner for target localization. (From Nakagawa et al (1987).)*

universal-wedged field and for part of each fraction with an open field, a wedged distribution with any wedge angle can be delivered.

Let D_w be the depth-dose distribution with such a wedge in place and D_0 be the depth-dose distribution with no wedge. A dose distribution D with a wedge angle θ_e, less than θ_w, may be constructed by combining a fraction B of the (universal) wedged field with a fraction $(1 - B)$ of the open unwedged field; i.e.

$$D = (1 - B)D_0 + BD_w. \tag{7.1}$$

Petti and Siddon (1985) showed that the fraction B should be determined from

$$B = f / [(\tan\theta_w / \tan\theta_e) + f - 1] \tag{7.2}$$

where

$$f = \frac{\partial D_0 / \partial z}{\partial D_w / \partial z} \tag{7.3}$$

is the ratio of the slopes of the central axis depth-dose curves for the open and universal-wedged fields (at 10 cm depth). Equation (7.2) is analytically correct; Petti and Siddon (1985) give the derivation.

They showed, from measured data, that for fields smaller than 15×15 cm^2, f was almost unity, but for large fields such as 20×20 cm^2, f departed from unity by as much as 20%, indicating the use of the full equation (7.2). Under conditions where $f = 1$, equation (7.2) reduces to

$$B = \tan\theta_e / \tan\theta_w \tag{7.4}$$

a result proposed by Philips Medical Systems Division (1983) and discussed by Zwicker *et al* (1985). Approximate equation (7.4) predicts θ_e to within 3° for fields up to 20 × 20 cm^2 and even better for smaller fields. A formula due to Tatcher (1970)

$$B = \theta_e/\theta_w \qquad (7.5)$$

is inadequate when the universal wedge angle is (as in Petti and Siddon's study) 60°, but accurately predicts θ_e to about a degree if θ_w is less than or equal to 30°. Mansfield *et al* (1974), however, appeared to verify Tatcher's equation for ^{60}Co radiation and a 60° universal wedge. In practice, the fraction B is usually determined so that θ_e matches the wedge angle for some particular removable wedge, allowing the planning experience gained with removable wedges to carry over to the universal wedge. However, the two situations may not be entirely identical because of changes in the scattered dose (McParland 1990).

Both McParland (1991) and Petti and Siddon (1985) showed the effective wedge angle θ_w itself increased with field size by about 0.5°–0.75° cm^{-1} for a Philips/ MEL SL75/5 accelerator equipped with a universal wedge of nominal angle 60°. McParland (1991) quoted the increase was from 49° at 5 × 5 cm^2 to 60° at 20 × 20 cm^2. This wedge was 81 cm from the isocentre and made of 96% lead and 4% antimony alloy.

McParland (1991) studied the effect of field size and also angulation of the universal-wedged fields on the dose to the build-up region. The former was a small effect, but the latter was noteworthy.

A problem arises in treatment planning in that it may not be at all clear ahead of time what would be the best wedge angle to choose for a particular set of circumstances. The problem is worse if a number of fields are to be combined to achieve a conformal plan. The problem can be specified: for N portals, with each portal comprising two fields—one open and one wedged with the fixed wedge angle—what are the $2N$ weights to be attached to each field? Cheng and Chin (1987) have shown that, by specifying the dose to a predefined set of points in the 3D volume and defining a suitable cost function, the problem can be cast as an optimization problem and so solved.

7.5. THE DYNAMIC WEDGE FOR THE LINEAR ACCELERATOR

Another way to produce a symmetric field with isodose contours angled relative to the axis of symmetry, without using either a fixed external wedge or a combination of a fixed wedge and an open field (the universal wedge), is the so-called 'dynamic wedge'. The dynamic wedge is the name given to the technique whereby a series of asymmetric fields of different sizes are combined. One jaw of the accelerator collimator remains fixed for all these sub-fields; the other moves to a series of locations varying from the full width of the required wedged field down to close to the first collimator jaw position. The static asymmetric sub-fields are delivered

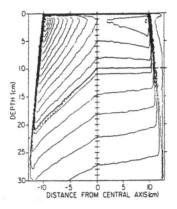

Figure 7.4. *Isodose distributions for a 20 cm wide mixed-wedge field (45°
over half the field, 0° over the other half). The wedge angle is defined at a
depth of 10 cm. Distributions of this kind can be obtained by using a dynamic
wedge. (From Leavitt et al (1990).)*

with a different number of monitor units per sub-field; the distribution of these
values determines the wedge angle.

Leavitt *et al* (1990) describe implementing this technique for a Varian Clinac
2100C accelerator. The full-field segment is delivered first, followed by each
smaller field in turn. In principle, the dose segments can be delivered for changes
in collimator position as small as 1 mm, but Leavitt *et al* (1990) found increments
of 5 mm were acceptable for routine use, giving the same dose distribution as
obtained with smaller increments.

The advantages claimed for the dynamic wedge are:

- The prescribed isodose line at the wedge angle extends over a greater fraction
  of the field than can be achieved using a standard fixed wedge.
- The wedge angle can be specified arbitrarily and at any depth.
- So-called 'customized wedge shapes' can be delivered with the wedge angle
  varying across the field, even being zero for part of the field, as shown in figure
  7.4.

Set against these are the potential disadvantage of increased time per field
because one jaw has to move and have its position verified.

The number of monitor units for each asymmetric field is computed as follows:
The computer is told the desired wedge angle and central axis depth at which this
is defined. Reference points are distributed along the line corresponding to this
idealized wedge isodose line, there being as many reference points as sub-fields,
i.e. the first reference point receives only primary dose from the full-width field;
the second receives primary dose from the first and second field and so on until the
last reference point receives dose from only the smallest asymmetric field. Hence
the number of monitor units for the first field must give the entire dose required
by the first reference point (actually reduced by 2% because some scatter from
other points will contribute). The number of monitor units for the second field has

to provide the difference between what the second reference point requires and what it has already received from the first sub-field... and so on. The weights so determined may be fine tuned by an optimization technique and it turns out that the monitor units per setting does not vary smoothly between fields, as would be the case if the jaw moved at constant velocity. The monitor unit weights for customized fields can be similarly found.

The set of asymmetric fields is delivered automatically under computer control.

A dynamic wedge is a standard feature of Varian linear accelerators, in which it may travel continously at constant velocity if required.

7.6. WEDGES WITH THE MULTILEAF COLLIMATOR ON A LINEAR ACCELERATOR

When a multileaf collimator is attached to a linear accelerator it is sometimes necessary to locate the field wedge on the patient side of the blocking tray. This minimizes the distance from the source to the blocking tray and can maintain this distance the same as without the MLC. Ochran *et al* (1992) studied this for a Varian linear accelerator equipped with a multileaf collimator. Their study came to the following conclusions:

1.   The shape of the wedged isodose distributions exhibited no essential differences from when the wedges were mounted above the blocking tray closer to the source.

2.   The average difference between the nominal wedge angle and the observed wedge angle was 2° at 18 MV and 5.5° at 6 MV for the four wedges of nominal angles 15°, 30°, 45° and 60°. The measured angle was made between the isodose line and the normal to the central axis at 10 cm depth.

3.   The wedge factor (defined as the output for the wedged field divided by the output for the open field) was measured at d_{max} and at 5 and 10 cm depth. The wedge factor varied with beam energy, field size and wedge angle, as expected, but had little depth dependence.

4.   The surface dose with the wedge below the blocking tray was larger than when it was above. The surface dose increased with decreasing SSD and as the field size increased, consistent with the theory that this was due to increasing low-energy scattered radiation.

7.7. LINEAR ACCELERATORS WITH INDEPENDENT COLLIMATORS

The modern multileaf collimator is sufficiently important to conformal therapy to have been discussed at length earlier in a separate chapter. Traditionally, however, linear accelerators have two sets of jaws, orthogonal to each other which can collimate square or rectangular fields symmetrical about the central beam axis. Recently some manufacturers have introduced independent collimators.

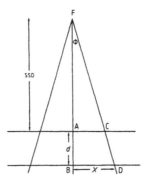

Figure 7.5. *Showing the nomenclature for beams. F is the source, SSD is the source-to-skin distance, d is the depth of a measurement point B on axis. Off-axis points may be specified by either the distance X or the angle ϕ. ϕ_0 would refer to the angle made by the central axis of an asymmetrical field. (From Thomas and Thomas (1990).)*

The jaws can be moved separately, cross the central axis and thus may define rectangular asymmetric fields. A number of studies have been made of the dosimetry, including Chui and Mohan (1986), Khan *et al* (1986) and Loshek and Keller (1988). The latter showed that an asymmetric field could be treated as a blocked symmetric field.

Alternatively, the dose at a point in such circumstances can be computed directly. Thomas and Thomas (1990) have shown this may be expressed as

$$d(S, d, \phi, \phi_0) = DD_{\text{sym}}(S, d) \, OF_{\text{sym}}(S) \, OCR_{\text{sym}}(\phi - \phi_0, d)$$
$$\times \frac{POCR(\phi, d)}{POCR(\phi - \phi_0, d)} \tag{7.6}$$

where $d(S, d, \phi, \phi_0)$ is the dose per monitor unit for a field of area S at depth d, at position labelled by angle ϕ, asymmetrically centred at angle ϕ_0 (figure 7.5). $DD_{\text{sym}}(S, d)$ is the on-central-axis depth dose (dose at depth d divided by dose at build-up depth d_{max}) for the same area field at the same depth, but symmetrically placed about the central ray of the machine. $OF_{\text{sym}}(S)$ is the output factor (dose per monitor unit at d_{max}) on the central axis for the symmetrically placed field. $OCR_{\text{sym}}(\phi, d)$ is the off-centre ratio (ratio of the dose at a point off the central ray to the dose at the same depth on the central ray) at angle ϕ from the central ray, at depth d for a *symmetrically placed* field. $POCR(\phi, d)$ is the primary off-centre ratio at angle ϕ and depth d (which is the ratio of the dose due to primary radiation at a point off-axis to the primary radiation at the corresponding depth on-axis). $POCR(\phi, d)$ is given in terms of the in-air POCR at the SSD, $POCR(\phi, 0)$ by

$$POCR(\phi, d) = POCR(\phi, 0) \exp(-\mu(\phi) \, d / \cos \phi) / \exp(-\mu_0 d) \tag{7.7}$$

and we note that μ is a function of angle ϕ, indicating the hardening of the beam off-axis. μ_0 is the coefficient along the axis of the machine.

The depth-dose, output factor and off-axis ratio are measured routinely for symmetrical fields. The above analysis shows that the only other data needed to compute the dose distribution for asymmetrical fields is the measurement of the in-air POCR and the angular dependence of the beam hardening.

Thomas and Thomas (1990) measured the necessary data and showed that these two equations predicted the dose accurately when compared with measurements for a Brown–Boveri CH8 8 MV linac. The beam hardening was expressed by

$$\mu (\phi) = 0.037 + 0.020\phi \tag{7.8}$$

with μ in cm^{-1} and ϕ in radians.

Independently movable jaws crossing the mid-line have also be used to generate a wedged field (Levene *et al* 1978) similar to that produced by the 'dynamic wedge'.

7.8. TWO-DIMENSIONAL TISSUE COMPENSATORS

A two-dimensional tissue compensator is a material placed in the path of a beam, usually in the 'blocking tray' whose thickness varies across its area in a way designed to overcome the effects of obliquity of the beam, the non-flat patient contour and internal tissue inhomogeneities. A wedge is a particularly simple form of such a compensator and correspondingly achieves a simple distribution of dose. In contrast, a tissue compensator can be designed to give a uniform dose on a specified *plane* within the patient. This has related objectives to conformal therapy. For example, combination of several beams, each with its own compensator, would be a desirable objective (see later this section).

For the moment consider a single beam and the objective of obtaining a uniform dose to a specified plane, which, in general, need not be normal to the beam axis. Mageras *et al* (1991) have described a method of computing the shape of the compensator.

The compensator is specified by its transmission $T (x, y)$. The thickness of the compensator is then

$$t (x, y) = - \ln T (x, y) / \mu \tag{7.9}$$

where μ is the linear attenuation coefficient of the compensator material. For the purposes of computation, the transmission matrix T is imagined to be centred at the isocentre and perpendicular to the z central axis of the beam. The required plane of constant dose is shown shaded in figure 7.6. A point in this plane is labelled x_p, y_p and a ray from the source to this point intersects the transmission matrix at x, y. The process of computing $T (x, y)$ is iterative:

• Compute the dose to x_p, y_p from

$$D_{x_p, y_p} = D_{ref} T A R \left(d, W_p \times H_p\right) O C R \left(d, x_p, y_p, W_p, H_p\right) (s_{ref}/s)^2 C \tag{7.10}$$

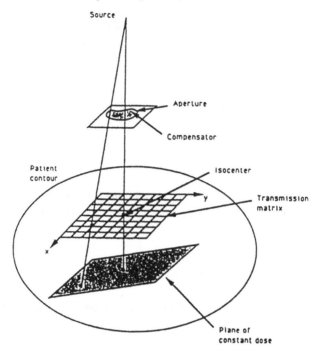

Figure 7.6. *The geometry for computing the thickness of a 2D compensator needed to generate a uniform dose distribution over the dark-shaded plane. The compensator is shown in the blocking tray. The aperture is defined by the beam's-eye-view of the target and the transmission matrix is imagined to be at the isocentre for the purposes of computation (see text). (From Mageras et al (1991).)*

where

D_{ref} is the reference dose at distance s_{ref} from the source,

d is the water-equivalent depth of x_p, y_p from the source (computed from 3D CT data),

s is the distance of the point from the source,

$TAR\left(d, W_p \times H_p\right)$ is the tissue-air ratio at depth d,

W_p, H_p is the field size at x_p, y_p,

OCR is the off-centre ratio, and

C is a correction factor which accounts for the change in the fluence (and hence dose) due to the compensator (and blocks etc).

- C is a ratio of two doses at the point x, y, d, in a flat-surfaced uniform phantom, one with the fluence ϕ_{comp} modified by the compensator transmission and the other with the fluence ϕ_{open} being the uniform fluence within the beam's-eye-view of the plane; i.e.

$$C = \frac{\int\int \phi_{\text{comp}}\left(x', y'\right) k\left(x - x', y - y', d\right) dx' dy'}{\int\int \phi_{\text{open}}\left(x', y'\right) k\left(x - x', y - y', d\right) dx' dy'}. \tag{7.11}$$

The function k is the pencil-beam dose distribution.

- For the first iteration the fluence ϕ_{comp} is assumed to be unmodified ($=\phi_{\text{open}}$) and $C = 1$. Then the first estimate of the transmission is

$$T^{(1)}(x, y) = D^{(1)}_{\text{const}}/D^{(1)}(x, y) \tag{7.12}$$

where the superscript $^{(1)}$ has been added to show this is the first estimate. $D^{(1)}_{\text{const}}$ is chosen so the maximum transmission is unity.

- Now the process is repeated, except that this time the fluence ϕ_{comp} in the numerator of C is obtained from the matrix of transmissions T. C now departs from unity and a new dose $D^{(2)}$ is computed from which the transmissions are adjusted to (x, y dependence not explicitly shown)

$$T^{(2)} = T^{(1)} D^{(2)}_{\text{const}}/D^{(2)}. \tag{7.13}$$

The calculations are of course made for each point in the plane where it is required to have uniform dose. After several iterations, (and Mageras *et al* (1991) suggest only some two or three are required in practice), the transmission matrix becomes:

$$T^{(k)} = T^{(k-1)} D^{(k)}_{\text{const}}/D^{(k)}. \tag{7.14}$$

Because of the use of the pencil-beam kernel, this iterative process has accounted for more than just the effect of changed primary fluence on the dose and included the three-dimensional transport of radiation within the patient. Mageras *et al* (1991) used this technique to compute the thicknesses of compensator and to drive an automatic milling machine to fabricate the compensator. Film dosimetry then verified that the method could accurately flatten the dose in a plane to within a few percent. Renner *et al* (1989) tackle this problem similarly, but do not use the convolution method of including energy transport.

Ideally one wishes to combine beams, each equipped with its own compensator, such that the resulting combination of doses generates a distribution which is optimal. For conformal radiotherapy this means creating as uniform a target dose as possible whilst minimizing the dose to other healthy tissue. Djordjevich *et al* (1990) formulated the solution to this problem. Each beam is divided into a number of small elementary beams; the problem is to compute the weight of these elementary beams. Let the weight of the jth elementary beam be w_j and t_{ij} indicate the contribution to the ith target point from the jth elementary beam. The total dose to the ith target point from N elementary beams is then

$$^{T}d_i = \sum_{j=1}^{N} t_{ij} w_j, i = 1, 2 \ldots m_i \tag{7.15}$$

for m_i target points. Similarly, let u_{ij} indicate the contribution to the ith point in an organ at risk from the jth elementary beam. The total dose to the ith point in an organ at risk from N elementary beams is then

$$^u d_i = \sum_{j=1}^{N} u_{ij} w_j, i = 1, 2 \ldots m_u \tag{7.16}$$

for m_u organ at risk points. Suppose the prescribed dose to the ith target point is r_i, then the problem becomes how to minimize

$$\chi^2 = \left(^T D - R\right)^t \left(^T D - R\right) + \lambda Y^t \left(^u D\right) \tag{7.17}$$

where ^{t} signifies transpose, bold capitals indicate vectors of the corresponding lower case components and Y is a vector which weights the importance of each healthy (organ at risk) point. λ is a multiplier which may be varied to control the relative importance of obtaining a homogeneous dose distribution in the target, compared with the importance of minimizing the dose to organs at risk. A further constraint to control the range of weights was applied; for example, weights clearly must be positive. Djordjevich *et al* (1990) show that minimizing χ^2 is a well known problem in quadratic programming, and they obtain solutions for model planning problems.

Although formalism of this kind has been discussed in terms of designing a set of two-dimensional tissue compensators, the same arguments apply to designing the weights of elemental sub-beams within an irregularly shaped field determined by a multileaf collimator. The solution of the inverse problem 'given the required dose constraints, find the intensity distribution across the open field of the multileaf collimator' has been given by (among others) Boyer *et al* (1991) and Webb (1991, 1992a,b) (see chapter 2). Boyer *et al* (1991) used an analytic inversion technique based on deconvolving the point-spread function to obtain photon fluence; Webb (1991, 1992a,b) optimized a quadratic cost function with prescribed doses to target volume and organs at risk by the technique of simulated annealing.

7.9. SUMMARY

Historically, some of the earliest approaches to conformal radiotherapy were based on treatment machines with isotope sources. One such machine, the RFH tracking cobalt unit, has been in development and use for nearly four decades. It is still in regular use. Other centres have developed tracking techniques based around a linear accelerator. Attention now seems to be more focused on shaping fields with a multileaf collimator. The technology was covered in chapter 5 and planning techniques were covered in chapter 2. Meanwhile, linear accelerators have acquired a variety of new features, such as the universal wedge, the dynamic wedge, independent asymmetric collimators and two-dimensional tissue compensators. These are sometimes called 'accessories', in that the machine can be operated without them, but they are really an integral part of the capability of the modern machine to perform conformal radiotherapy.

REFERENCES

Bjärngard B, Kijewski P and Pashby C 1977 Description of a computer-controlled machine *Int. J. Rad. Oncol. Biol. Phys.* (Supplement 2) **2** 142

Blake T, Hickling P A, Lane R A, Nahum A E, Rosen I I and Rosenbloom M E 1987 Assessment of the Philips SL-25 linac for conformation therapy *The use of computers in radiation therapy* ed I A D Bruinvis *et al* (Proc. 9th ICCRT) pp 537–540

Boyer A L, Desobry G E and Wells N H 1991 Potential and limitations of invariant kernel conformal therapy *Med. Phys.* **18** (4) 703–712

Brace J A 1985 Computer systems for the control of teletherapy units *Progress in medical radiation physics* 2 ed C G Orton (New York: Plenum) pp 95–111

Brucer M 1956 An automatic controlled pattern cesium 137 teletherapy machine *Am. J. Rontg.* **75** 49–55

Cheng C W and Chin L M 1987 A computer-aided treatment planning technique for universal wedges *Int. J. Rad. Oncol. Biol. Phys.* **13** 1927–1935

Chin L M, Kijewski P K, Svensson G K, Chaffey J T, Levene M B and Bjärngard B E 1981 A computer-controlled radiation therapy machine for pelvic and para-aortic nodal areas *Int. J. Rad. Oncol. Biol. Phys.* **7** 61–70

Chui C and Mohan R 1986 Off-center ratios for three-dimensional dose calculations *Med. Phys.* **13** 409–412

Chung-Bin A and Wachtor T 1979 Applications of computers in dynamic control of linear accelerators *Proc. 3rd Int. Computer Software and Applications Conf.* (New York: IEEE) pp 270–275

Davy T J 1985 Physical aspects of conformation therapy using computer-controlled tracking units *Progress in medical radiation physics 2* ed C G Orton (New York: Plenum) pp 45–94

Davy T J 1987 *Conformation therapy methods and systems* IAEA-SM-298/87

Djordjevich A, Bonham D J, Hussein E M A, Andrew J W and Hale M E 1990 Optimal design of radiation compensators *Med. Phys.* **17** (3) 397–404

ICRU 1976 Determination of absorbed dose in a patient irradiated by beams of x or gamma rays in radiotherapy procedures *ICRU (Washington, DC) report number 24*

Jennings W A 1985 The Tracking Cobalt Project: From moving-beam therapy to three-dimensional programmed irradiation *Progress in medical radiation physics 2* ed C G Orton (New York: Plenum) pp 1–44

Khan F M, Gerbi B J and Deibel F C 1986 Dosimetry of asymmetric x-ray collimators *Med. Phys.* **13** 936–941

Leavitt D D, Martin M, Moeller J H and Lee W L 1990 Dynamic wedge field techniques through computer-controlled collimator motion and dose delivery

Med. Phys. **17** (1) 87–91

Levene M B, Kijewski P K, Chin L M, Bjärngard B E and Hellman S 1978 Computer controlled radiation therapy *Radiology* **129** 769–775

Loshek D D and Keller K A 1988 Beam profile generator for asymmetric fields *Med. Phys.* **15** 604–610

Mageras G S, Mohan R, Burman C, Barest G D and Kutcher G J 1991 Compensators for three-dimensional treatment planning *Med. Phys.* **18** (2) 133–140

Mansfield C M, Suntharalingam N and Chow M 1974 Experimental verification of a method for varying the effective angle of wedge filters *Am. J. Radiol.* **120** 699–702

McParland B J 1990 The effect of a dynamic wedge in the medial tangential field upon the contralateral breast dose *Int. J. Rad. Oncol. Biol. Phys.* **19** 1515–1520

—— 1991 The effects of a universal wedge and beam obliquity upon the central axis dose buildup for 6-MV x-rays *Med. Phys.* **18** (4) 740–743

Nakagawa K, Aoki Y, Sakata K, Karasawa K, Muta N, Kojima K, Onogi Y, Hosoi Y, Akanuma A and Ito M 1987 Dynamic therapy utilizing CT-Linac on line system *The use of computers in radiation therapy* ed I A D Bruinvis *et al* (Proc. 9th ICCRT) pp 541–544

Ochran T G, Boyer A L, Nyerick C E and Otte V A 1992 Dosimetric characteristics of wedges mounted beyond the blocking tray *Med. Phys.* **19** (1) 187–194

Petti P L and Siddon R L 1985 Effective wedge angles with a universal wedge *Phys. Med. Biol.* **30** 985–991

Phillips Medical Systems Division 1983 *Product data* 764 (Eindoven: Philips)

Renner W D, O'Connor T P and Bermudez N M 1989 An algorithm for design of beam compensators *Int. J. Rad. Oncol. Biol. Phys.* **17** 227–234

Richter J, Klemm P, Neumann M, Nowak G and Sauer O 1987 Computer control of a linac for dynamic treatment *The use of computers in radiation therapy* ed I A D Bruinvis *et al* (Proc. 9th ICCRT) pp 545–548

Saunders W and Chin L M 1986 Innovative techniques: dynamic therapy and utilisation of non-coplanar beams *A categorical course in Radiation therapy treatment planning* ed B R Paliwal and M L Griem (Oak Brook, IL: Radiological Society of North America) pp 123–128

Schlegel W, Boesecke R, Bauer B, Alandt K and Lorenz W J 1987 Dynamic therapy planning *The use of computers in radiation therapy* ed I A D Bruinvis *et al* (Proc. 9th ICCRT) pp 361–365

Shentall G S and Brace J A 1987 Fully automated planning for conformation radiotherapy of the spinal axis *The use of computers in radiation therapy* ed I A D Bruinvis *et al* (Proc. 9th ICCRT) pp 367–370

Tatcher M 1970 A method for varying the effective angle of wedge filters *Radiology* **97** 132

Thomas S J and Thomas R L 1990 A beam generation algorithm for linear accelerators with independent collimators *Phys. Med. Biol.* **35** 325–332

Webb S 1991 Optimisation by simulated annealing of three-dimensional conformal treatment planning for radiation fields defined by a multileaf collimator *Phys. Med. Biol.* **36** 1201–1226

—— 1992a Optimisation of three dimensional treatment planning for volumes with concave outlines, using a multileaf collimator *Proc. ART91 (Munich)* (abstract book p 66) *Advanced Radiation Therapy: Tumour Response Monitoring and Treatment Planning* ed A Breit (Berlin: Springer) pp 495–502

—— 1992b Optimisation by simulated annealing of three-dimensional conformal treatment planning for radiation fields defined by a multileaf collimator. 2: Inclusion of two-dimensional modulation of the x-ray intensity *Phys. Med. Biol.* **37** 1689–1704

Zwicker R D, Shahabi S, Wu A and Sternick E S 1985 Effective wedge angles for 6 MV wedges *Med. Phys.* **12** 347–349

CHAPTER 8

IMAGING FOR CONFORMAL RADIOTHERAPY PLANNING

8.1. INTRODUCTION

It is axiomatic that fully three-dimensional images are required for planning conformal radiotherapy. This restricts the useful imaging modalities to just four: x-ray CT, MRI, SPECT and PET. Whilst the many other imaging modalities play major roles in diagnosing disease, they have little if any part in planning conformal radiotherapy (Webb 1988). Throughout this text we have been referring to the *use* of medical images. In order not to detract from the main thrust of presenting the physics of conformal radiotherapy, the presentation of how these images are *obtained* has been left to this point. We now fill in these details, keeping in mind the application of imaging to therapy. In section 8.2, principles common to transmission and emission tomography are reviewed. Later sections discuss some of the important features of these imaging modalities applied to planning conformal radiotherapy.

Table 8.1 is an attempt to summarize some of the features of the imaging modalities which play a role in radiotherapy. Most of the questions addressed cannot be precisely quantified, as values vary from machine to machine. The main purpose of the table is to show the relative merits of the techniques at a glance and to provide a feel for how they are used for non-diagnostic purposes.

The simulator plays a major role in 'conventional' treatment planning. An extensive review has been provided by Taylor (1988). In chapter 1 we have already discussed how, for conformal radiotherapy, many groups are working towards using the 3D images in a 3D computer-based planning system to take over many of the traditional roles of the simulator, including the definition of shaped, possibly non-coplanar, fields and block design. If this trend continues, the simulator will become the instrument to check the decisions from such methodology. Simulator (planar) images will remain important for comparison with megavoltage portal images (see chapter 6). CT images can also be performed on a simulator and, whilst this has not yet progressed to generating 3D data-sets, the subject is reviewed in section 8.8.

Table 8.1. *Methods of 3D medical imaging†.*

Method	X-ray CT	MRI	SPECT‡	PET
Property being imaged (A=anatomy; F=function)	A	A+F	F	F
Are multisections gathered simultaneously? (Y=yes; N=no)	N	Y	Y	Y
Best spatial resolution (mm)	1	1	9	6
In-plane resolution same (S) or different (D) from slice width?	D	S	D	S
Typical data acquisition time (for multisections) (min)	2	20	15	20
Typical data reconstruction time (for multisections) (min)	2	2	10	10
Main application site (H=head; B=body)	H+B	H+B	H+B	H+B
Is modality useful for treatment planning? (C=common; R=rare; S=sometimes used)	C	S	R	R
Is modality useful for monitoring effects of treatment? (C=common; R=rare; S=sometimes used)	C	S	S	S
Cost of imaging equipment (M=moderately expensive; E=expensive; V=very expensive)	E	V	M	V
How common is equipment in the UK? (C=common; R=rare)	C	C	C	R
Is the physical basis of the modality still under active development? (Y=yes; O=already optimized; C=close to optimized)	O	Y	C	Y

† This summary does not imply that other methods of medical imaging have no role in radiotherapy. However, these, even when they are tomographic, are 'single-section' imagers and so cannot truly guide 3D conformal radiotherapy.
‡ By gamma camera.

8.2. PRINCIPLES OF IMAGING BY COMPUTED TOMOGRAPHY

X-ray CT scanning relies on the principle of reconstruction from projections. Three-dimensional images may be thought of as a set of stacked two-dimensional tomograms. The CT scanner records data which can be expressed as line integrals of the spatially varying x-ray linear attenuation coefficient for a large set of directions through the slice. The earliest so-called 'first-generation' CT scanners recorded this data by translating a narrow parallel beam of x-rays impinging on a detector the other side of the slice (figure 8.1). Each projection comprised measurements with the line between the source and the detector at a fixed orientation to the slice. A large set of projections were formed by stepping this orientation through a range of 180°. 'Second-generation' CT scanners were a variation on this theme, with a narrow fan of radiation falling on a few detectors. 'Third-generation' CT scanners had the x-ray beam collimated to a wide fan, spanning the patient, impinging on a detector array facing the source and co-rotating with it. Modern 'fourth-generation' CT scanners have a ring of detectors in a complete circle round the slice. The x-ray beam is collimated to a wide fan, spanning the patient, impinging on the detector. A complete projection data-set is recorded by rotating the source through 180° plus the fan angle. The recorded data in all these geometries are proportional to the intensity of the radiation beam emerging from any particular line through the patient. If the equivalent measurement is made for a beam passing through air only, the transmission measurements can be converted to a measure of the line integral of x-ray attenuation (Webb 1987, 1990a, Swindell and Webb 1988).

Creating the tomographic map of linear attenuation coefficient involves solving the inverse problem: given the 1D line integrals of a 2D spatially varying property, compute the distribution of that property throughout the section. This subject is sufficiently important to warrant extensive study and there is an enormous literature on reconstruction theory (Herman 1980, Natterer 1986). Arguably the best known method is that of convolution and backprojection (CBP) (Ramachandran and Laksminarayanan 1971). The 1D projections are convolved with a filter and the convolved data are backprojected along the lines joining the detection point to the source for each projection element. The task of the filter is twofold; it compensates for the blurring which would be introduced by simple backprojection, and it has variable features controlling noise. In the earliest days of x-ray CT scanning, iterative methods were used (Hounsfield 1968, Gilbert 1972). These started with an initial guess at the reconstructed section and then refined it until the projections matched the recorded data in some clearly defined sense. These methods are still used in emission tomography, where an analytic solution may not be possible. Iterative methods have been substantially developed beyond the form in which they first appeared in the late 1960's (Todd Pokropek 1989).

Worldwide there are many well known reconstruction computer packages which emanate from university departments and are much used by constructors of special-purpose imaging equipment (Herman *et al* 1975, 1989, Huesman *et al* 1977). Commercial manufacturers of x-ray CT and SPECT scanners tend to develop their own in-house reconstruction software, much of it hard-wired and some of it restricted by commercial secrecies. This is generally considered to be less of a problem now than it was in the 1970's, when these disciplines were relatively young.

To a first approximation, the principles of SPECT are very similar. A gamma

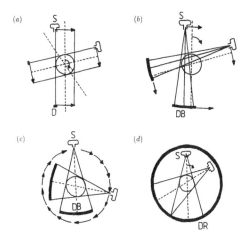

Figure 8.1. *Showing the 'generations' of CT scanner technologies. (a) A 1st generation CT scanner. The source S and single detector D face each other and the line between them translates across the reconstruction circle. This movement is repeated at each angle in a range of 180°. (b) A 2nd generation CT scanner. The source S emits a narrow fan of x-rays in the plane of the section which impinges on a small detector bank DB. The source and the detector bank translate together across the field for a 180° range of angles. The angular sampling can now be coarser than in arrangement (a), reducing scanning time. (c) A 3rd generation CT scanner. The source emits a wide fan of x-rays in the plane of the slice which impinges on a large bank of detectors DB opposite. The source and coupled detector bank rotate together around the reconstruction circle. The angular coverage need only be 180° + the fan angle, although there is no reason why more data from a full circle could not be used to improve image quality. The major disadvantage of this geometry is the generation of circular artefacts. (d) A 4th generation CT scanner. As in the 3rd generation CT scanner, the source S emits a wide fan of x-rays and rotates through at least 180° +the fan angle. This time the detector ring DR is static and, as the source rotates, different detecting elements are called into play. This eliminates circular artefacts.*

camera rotates around the patient through 360°, recording at fixed orientations a planar image into a square digital matrix (figure 8.2). Each line of data, in the plane of rotation of the camera, for each planar image is a 1D projection of the activity in the corresponding slice. The situation is actually rather more complicated than this because the line integral really records *attenuated* activity and special steps have to be taken during reconstruction to cope with this. Once again, a range of geometries is possible. The most common method has a parallel-hole collimator attached to the camera and data are recorded analogous to 'first-generation' CT (without the need for translation, of course). Fan-beam collimators mimic 'third- or fourth-generation' CT. There is no emission equivalent of 'second-generation' CT. A single rotation of the gamma camera records all the data needed for a complete

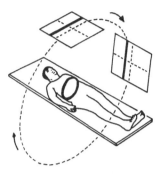

Figure 8.2. *Showing the principle of SPECT. The patient lies horizontally. If the images are to be correlated with x-ray CT and MRI images, the couch must be flat. A gamma camera rotates around the patient through 360°, usually in a series of discrete angular increments. At each orientation the data are recorded into a square matrix as shown. (The camera is usually circular and is not shown for clarity.) One particular line of data on the camera in the plane of rotation (shown as a solid line) contributes to the reconstruction of one particular slice in the patient (as shown). Unlike x-ray CT, all the projection data for a series of slices are recorded in one single rotation of the camera.*

set of tomographic slices. Data are generally reconstructed into transaxial cross sections and re-ordered into sagittal and coronal slices. Emission SPECT is much slower than transmission CT. Typically 15 m may be needed for a complete study, whereas each x-ray CT set of data are taken in a few seconds (Webb *et al* 1985a).

Positron emission tomography can operate on similar principles of reconstruction. When a positron emitter decays, two 511 keV gammas are emitted more or less back-to-back. They are detected in coincidence, thus defining the line along which the decay must have occurred (figure 8.3). Some PET systems have a ring of scintillation detectors surrounding the patient and the recorded data can be arranged into projections at a large set of orientations, from which the emission tomogram may be recorded in the same way as a SPECT image. Other PET systems make use of rotating area-detectors. For these it is more appropriate to backproject each recorded coincidence event into the reconstruction space to create a blurred 3D image. The point-spread function of the imaging system is then deconvolved to produce a sharp image. In this deconvolution stage, appropriate windowing can control the noise. The usual trade-off arises between controlling the effects of noise and the spatial resolution obtainable (Phelps 1986).

Although magnetic resonance imaging, following exactly the principles of x-ray CT, can be performed, until recently this method has been thought too slow to be practical. Hence MRI tomograms are generally recorded differently. NMR signals are radiofrequency emissions resulting from the relaxation of spins which have been excited by radiofrequency pulses. The exact location from which each signal arises is determined by the frequency of emission, which in turn is made to vary spatially over the region to be imaged by applying magnetic gradients in several orthogonal directions superposed to the main, large magnetic field (of the

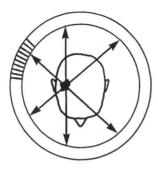

Figure 8.3. *Schematic showing the principle of* PET *scanning. The patient is surrounded by a ring of detectors (only a few are shown) collimated to a thin plane parallel to the plane of the paper. Coincidence circuits (not shown) link the detectors. A region in the brain of the patient (shown dark) is emitting positrons and three coincidence paths are shown. (In practice coincidence paths emanate in all directions in the plane.) A single tomographic plane is reconstructed from this data. To obtain more planes, more coincidence rings are required. Alternatively, 'area detectors' may be used.*

order 2 T). MRI tomograms can be made to reflect different physical properties by varying the 'pulse sequences' which capture the data. This is a complex subject and the above is a big simplification (Chen and Hoult 1989).

There are many physics and engineering design considerations in optimizing these medical-imaging modalities and whole volumes have been devoted to each. It would not be appropriate here to oversimplify these important considerations and readers wishing to know more might consult one of the following: Moores *et al* (1981), Hamilton (1982), Wells (1982), Gifford (1984), Guzzardi (1987), Aird (1988), Webb (1988), Chen and Hoult (1989). The next sections focus on some of the aspects of these imaging technologies which are particularly relevant to conformal radiotherapy.

8.3. X-RAY COMPUTED TOMOGRAPHY

X-ray computed tomography has a long and complex history (Webb 1990a). However, the important event for clinical practice occurred in April 1972 when Sir Godfrey Hounsfield unveiled the world's first commercial CT machine (from the EMI Company) at the 30th British Institute of Radiology Annual Congress (Hounsfield 1973). It was not long before CT images were routinely forming the basis of radiotherapy treatment planning. The present state of the art was reached with amazing rapidity by the mid to late 1970's (Süsskind 1981, Blume 1992), and whilst small improvements continue to be made to CT scanners, no further quantum leaps are expected.

The main requirements for therapy CT scanning include the need for the patient to lie on a flat couch in the treatment position so that planning is based on the

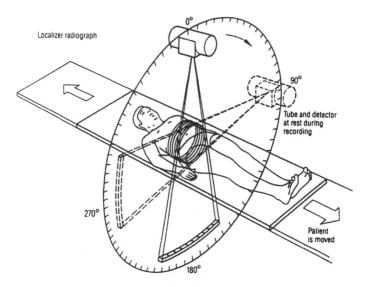

Figure 8.4. *Schematic showing how a topogram is formed. The patient lies on a flat couch and is translated through the beam, which is collimated to a narrow fan. In the picture the beam is in either of two positions, labelled as 0° and 90°. A digital radiograph is formed which shows dotted lines indicating the location of the CT slices. (From Krestel (1990).)*

geometry as it will be at treatment time. The patient quietly respires, and time taken to assure patient comfort is time not wasted in this respect. Radio-opaque markers are placed on selected skin tatoos to act as landmarks. A 'scout view' (or 'topogram') is recorded to determine the necessary range of travel of the couch. This is formed by taking the patient through the CT scanner with the fan beam and detectors stationary, thus generating a kind of planar radiograph (figure 8.4). This has the characteristic feature that the beam diverges normal to the direction of travel and is parallel in the orthogonal direction. In this sense, such a radiograph could not be recorded by a 2D detector and an unmoving x-ray source (figure 8.5). Sections are taken from a few cm below the target volume to a few cm above and registered on to the 'scout view'. These sections are generally spaced 5 mm apart in the body, but may be as close as 2 mm in the head or in the periphery of fields in the body where beam penumbra will occur. Contrast media are used for selected examinations. Reconstructed CT numbers convert to tissue electron densities. Image data are generally transferred to the planning computer by magnetic tape.

There have been countless reviews of the impact of x-ray CT on therapy planning. The issues are discussed well by Goitein (1983) and Hogstrom (1983). Reviews by Webb (1990b) and Dobbs and Webb (1988) included a discussion of special-purpose CT scanners (see also section 8.8).

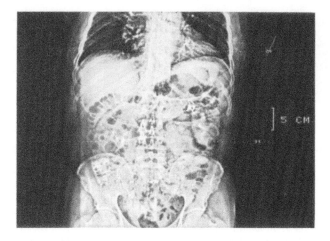

Figure 8.5. *Topogram to facilitate the localization of CT slices within the torso. The dotted lines give the slice numbers. (From Krestel (1990).)*

8.4. MAGNETIC RESONANCE IMAGING

At first encounter, magnetic resonance tomography and x-ray computed tomography appear to have similar roles. They both produce high-resolution fully three-dimensional tomographic data-sets. Both have transformed the practice of diagnostic radiology. However, important differences between them ensure that they play complementary rather than competing roles in diagnostic radiology. CT had not been in use for diagnosis for long before the images were being exploited to assist radiotherapy planning. Hence, when diagnostic MRI became widely available, the same questions were asked concerning its suitability for assisting radiation therapy planning. Early experience suggested the disadvantages in MRI were insurmountable (Henkelman *et al* 1984, Sontag *et al* 1984). This view has been modified with the passage of time. Experience has shown that, for this application also, the role of MRI is complementary to that of CT (Coffey *et al* 1984, Ten Haken *et al* 1991).

There are important differences which ensure that this complementarity is likely to remain. Whilst both modalities have good spatial resolution, MRI scan times are much longer than for CT. Hence there is a danger of artefacts from patient movement and also reduced patient throughput. MRI cannot give electron-density images needed for x-ray therapy planning, whereas the electron density of tissues directly follows from the image of CT numbers representing x-ray linear attenuation coefficient. In its favour, MRI directly generates non-transaxial data-sets, including sagittal, coronal and oblique. X-ray CT is geometrically very accurate, whereas image distortion can be a problem for MRI. High-sensitivity MRI sometimes makes use of head and body coils which restrict access to and positioning of the patient. In its early days, x-ray CT had a limited patient aperture, but this is less of a problem with the latest CT machines. For both modalities the

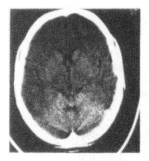 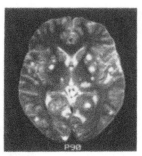 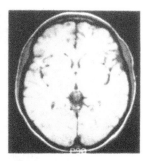

Figure 8.6.   *Brain metastases from a primary breast tumour which are not visible on the CT slice (left). MRI show the metastases with high tumour-specific contrast on both a T_2-weighted sequence (centre), and to a lesser extent on a T_1-weighted sequence (right). In the central image the lesions show as bright spots; in the right-hand image they show as dark spots. (From Henkelman (1992).) (Reprinted with permission from Pergamon Press Ltd, Oxford, UK.)*

patient should be imaged lying on a flat couch with quiet respiration. Finally, one of the strengths in diagnostic work, the ability of MRI to image different physical properties within any slice by varying the pulse sequences, might be thought a limitation for its use in therapy planning. The MRI images may become too dependent on the choice of acquisition parameters, whereas there is a certain inherent consistency in how CT images are formed (Fraass *et al* 1987, McShan and Fraass 1987).

The ability of MRI to display functional differences between benign and malignant tissue, which often extend beyond the anatomical differences imaged by x-ray CT, is the driving force behind efforts to harness it into the planning process. For example, figure 8.6 shows brain metastases from a primary breast tumour which are invisible on a CT scan but show up on MRI. Tumours can have characteristically longer relaxation times than the corresponding normal anatomy, providing the physical basis for tumour contrast. Sometimes the MR sensitivity to flow can be used to generate image contrast (Henkelman 1992). Shuman *et al* (1985) reported a study of 30 patients who were imaged with both CT and MRI. Radiation therapy planning was initially based on the CT data alone and the effect of having the additional MRI information was systematically assessed. In over 50% of the cases the radiation therapy plan was changed as a result of having the MRI data. In many other cases, whilst no changes were made, confidence was increased in the CT-based plan. Whilst the authors of this study admitted that there might be some element of bias in the study, in that patients were selected where MRI data might be expected to provide additional information, an alternative way of looking at this situation is that the new (MRI) imaging modality was being used appropriately. MRI can also be a useful tool for monitoring the efficacy of therapy by imaging changing tissue function. Indeed Henkelman (1992) argues that we should not expect further great improvements in geographical definition of structures, but look to improvements in monitoring physiological function.

The literature on the use of MRI in radiotherapy planning largely concentrates on how to overcome the difficulties (Henkelman *et al* 1984). One of the main criticisms of the use of MRI in radiotherapy planning is that images may be distorted. Since conformal therapy requires exquisite precision, this feature appears contrary to aims. The distortion may arise in one or both of two ways. Geometric distortion may arise from inhomogeneities in the magnetic field, particularly at the edge of body images. This is a smaller, if non-existent, problem in head scanning. Further distortions may arise due to the patient perturbing the field. The question arises whether these distortions are constant with time and from patient to patient. If not, individual corrections are required. Phantoms of known geometry, e.g. a matrix of fluid-filled tubes, can be imaged to characterize the distortion from the former cause. Unwarping algorithms may then restore the geometric consistency of the data (Fraass *et al* 1987). With long scan times, patient movement may lead to artefacts and/or distortion.

MRI and CT images require to be registered so that the electron density data from CT can be mapped on to the corresponding MRI images where the target volume may be more easily determined. Henkelman *et al* (1984) showed a complete lack of correlation between electron density and MRI image data; registering MRI and CT is the way to overcome the problem. It may not be possible to immobilize patients identically for both imaging sessions (e.g. because of the coils mentioned earlier). CT immobilization methods involving ferromagnetic materials are also excluded for MRI. Hence the many registration techniques discussed in chapter 1 are called into play. In particular, the use of external markers can form the basis of correlation. Fraass *et al* (1987) comment that a source of inaccuracy arises because of the slippery nature of the oil-filled markers used in MRI. Also, these need to be arranged so that they show as points in all sections, including those which are non-transaxial. Figure 8.7 illustrates this.

After registration, images are presented to the 3D planning system. A difficulty which may arise here is that the pulse sequences for orthogonal slices may have been different, leading to structures appearing with different grey levels in each slice. When these are merged into an axonometric display, the effect may be confusing. To overcome this, some workers prefer to show MRI and CT data separately, but alongside each other, and transfer structures from one data-set to the other via linked cursors and similar tools.

Three-dimensional planning systems, traditionally CT-based, may not be able to accept 3D MRI data, especially from non-transaxial planes. Some systems, such as UMPLAN, are specifically designed to use MRI and CT data interchangeably. The beam algorithms are written to yield the dose distributions in such planes and to display them accordingly.

Bearing in mind these difficulties, MRI is least problematic in the head, where distortions may be minimal, and MRI is indicated because of its ability to show the tumour and oedema better than x-ray CT. MRI presents a particularly strong case for routine quality assurance.

In summary, MRI is not yet routinely used for planning radiotherapy, but this is

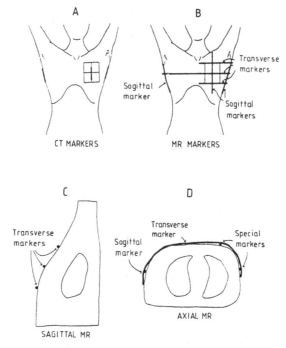

Figure 8.7. *(A) shows three line markers used in a multisection CT study. These will show as dots in transaxial tomograms. (B) shows how orthogonal line markers may be used in MRI, showing as points in sagittal slices (such as (C)) and either points or a line in transaxial slices (such as (D)). The intersections between the lines can define points in the two data-sets, which may be used for image registration. (From Fraass et al (1987).) (Reprinted with permission from Pergamon Press Ltd, Oxford, UK.)*

expected to change with the passage of time.

8.5.   ULTRASOUND IMAGING

Despite its enormous importance for diagnosis, ultrasound imaging plays virtually no part in planning conformal radiotherapy. Ultrasound is reflected at barriers of high acoustic-impedance mismatch. It is difficult for ultrasound to penetrate bone or gas, so ultrasonic imaging is unable to generate 3D tomographic data-sets for treatment planning. Ultrasound has been used to determine chest wall thickness for breast treatment planning (Lock and Coules 1986, Taylor 1988). Comparison of the external contour and lung position located by ultrasound with CT and a life-size pantograph showed that the contour points from the ultrasound scan were never more than 3 mm from the corresponding points determined by the comparative method. The ability to obtain a contour in a non-transaxial plane was an advantage for ultrasound.

With the passing of the old compound ultrasound imaging systems and the emergence to total predominence of the hand-held real-time scanners, any last semblance of a coordinate registration system has been lost. Unless these features were added to modern ultrasound scanners it would not be possible to use them for planning in the same way as CT or MRI. This is not impossible, but it has not been done.

This situation may change soon, since there is currently considerable activity in 3D ultrasound data collection. Scanners which automatically generate a sequence of parallel-slice images over a pre-defined volume in about 5 s have been designed. Early commercial versions already exist. Research programmes are studying options for real-time 3D scanning using parallel processing systems. These will still be restricted to applications where the presence of air and bone are not limitations (Bamber 1992).

8.6.   SINGLE-PHOTON EMISSION COMPUTED TOMOGRAPHY—SPECT

Both SPECT and PET provide three-dimensional functional images (Ahluwalia *et al* 1989). Since functional defects may not necessarily correlate with anatomical defects, images can be useful for planning therapy. In chapter 1 we looked at methods by which SPECT and PET could be registered and entered into the planning process. The use of emission tomography to assess the effectiveness of radiotherapy is equally important and calls for the same precision in image registration. It is required to superimpose three-dimensional dose distributions on to three-dimensional functional images.

SPECT is characterized by a fairly poor spatial resolution if a parallel-hole collimator is used (Webb *et al* 1985a). This depends on the reconstruction filter, but might be typically as poor as 1.5 cm. For correlating body images, a flat couch should be used during scanning. Since scan times are typically 15 m, the patient is not suspending respiration, so there is the possibility of generating artefacts in thoracic images. SPECT images are generally reconstructed into cubic arrays with cubic voxels, since the planar images, taken as the camera rotates, are square with square pixels. The main advantage of SPECT is that a wide variety of radiopharmaceuticals may be imaged (Ott *et al* 1988). Some 'single-section' emission tomographic devices can provide images with improved spatial resolution, but, unless the patient is translated through the machine to obtain multiple slices, these images would be less useful for assisting conformal therapy.

8.7.   POSITRON EMISSION TOMOGRAPHY—PET

Positron emission tomography opens up the possibility of imaging radiolabelled elements naturally ocurring in the body, such as hydrogen, oxygen and carbon. For those radionuclides with a short half-life, an on-site cyclotron is essential and many 'PET centres' have the cyclotron and imaging technology close to each other. However, PET without a cyclotron is possible using, for example, ^{68}Ga-labelled radiopharmaceuticals.

The spatial resolution of PET images is generally rather better than SPECT images. Precise values depend on the radionuclide and on the method of detecting the photons, but spatial resolution of the order 6 mm is possible. Machines fall into two categories: 'ring systems' in which there are a number of rings of scintillation detectors and 'area-detector' systems. The former produce exquisite functional images, but with a small number of slices covering a limited axial extent. The latter generate as many slices as a SPECT system, covering a larger axial extent; in principle a bonus for assisting conformal radiotherapy (Ott *et al* 1985). Research in imaging technology is still very active, with the supporters of area-detector systems working hard to emulate the good sensitivity achievable with ring systems.

8.8.   CT IMAGING ON A RADIOTHERAPY SIMULATOR

A number of groups worldwide have investigated the extent to which the imaging requirements for radiotherapy planning may be met by the development of non-standard CT scanners. Some of these groups have constructed apparatus based on a radiotherapy simulator gantry. Others have built special-purpose CT devices with radioactive sources, but these will not be reviewed as in general they were early developments not intended for clinical use. The major use of simulator-based machines has been in planning radiotherapy in the upper thorax, where tissue-inhomogeneity corrections are important.

After the first commercial CT scanner was announced in 1972 (New Scientist 1972, Ambrose and Hounsfield 1972), it was not long before body scanners were developed (Ledley *et al* 1974, Ledley 1976). Both head and body scanners were used to generate 'therapy CT scans', images of internal tumours and anatomy, useful for radiotherapy treatment planning (Parker and Hobday 1981, Dobbs and Webb 1988).

Full 3D diagnostic-quality CT data-sets are essential for planning most conformal radiotherapy. However, a small body of people developed simpler CT machines to assist with *specific* treatment planning problems and these are briefly reviewed here. Planning to conform the high-dose volume to the breast, avoiding irradiating too much lung, is one such application. A full review has been written by Webb (1987, 1990b). If multi-section CT could be achieved with a simulator-based scanner, this would be very helpful to 3D planning conformal therapy and greatly speed up patient throughput.

8.8.1.   *Systems for non-diagnostic computed tomography (NDCT)*

The term NDCT is here used to mean all those *single-slice* computed tomography machines whose aim is to produce CT cross sections of lower quality— i.e. poorer spatial and electron density resolution—than obtained by state-of-the-art commercial CT scanners. The latter were of course developed for diagnostic purposes, their use in treatment planning being inspired lateral thinking subsequently. An abundance of terms appears in the literature for NDCT machines, including 'simple CT scanners', 'alternative CT scanners', 'non-commercial CT scanners' etc. The term NDCT is prefered because it immediately differentiates the purpose of such machines and is a valid generic term. It removes the sense that NDCT scanners are in some way inferior. They can and do provide enough data for the purpose for which they were designed and engineered.

Most NDCT machines were designed and built in university and hospital departments, rather than by commercial companies. Hence, in reviewing their construction and capabilities we are faced with a miscellany of technology. No two machines are alike, although some use common detector technology.

8.8.2. *NDCT on a simulator with an image intensifier as detector*

A number of workers built NDCT systems based around the use of an image intensifier as x-ray detector. The idea appears to have been first suggested in a German patent by Dummling (1970). Baily (1977,1979), Baily *et al* (1976a,b) and Kak *et al* (1977) were among the first to investigate this possibility, coupling an image intensifier to a vidicon TV tube and recording real-time fan-beam x-ray projections. The method took account of the non-linear logarithmic response of the image-intensifier detector. The data were obtained in fan-beam geometry by collimating the x-ray source to a narrow fan spanning just eight TV lines. At the time of this work (circa 1975) reconstruction from fan-beam projections had received little attention and Kak *et al* (1977) decided to treat the beams as if they were parallel.

Harrison and Farmer (1976, 1978) in Newcastle used a similar technique to image body cross sections. They took the work into the clinic, based around a radiotherapy simulator and were well aware of the potential for improving radiotherapy planning (Harrison 1981). A 48.5^2 cm^2 fluorescent screen was viewed by a low light-level Intensifier Isocon television camera. This detector was mounted on a TEM Ximatron Mark 1 gantry. Image data were reconstructed off-line by an IBM-370 computer using a hybrid convolution and backprojection (CBP) algorithm. This treated the projection data as if collected in parallel geometry for the convolution part but then backprojected along the correct fan-beam lines. The early papers from this group noted that the achievable spatial accuracy of some ± 5 mm was consistent with getting dosimetry right to within 2–5% for ^{60}Co, 4 MV and 8 MV x-radiotherapy.

The NDCT system at Newcastle is one of the few examples of a system which has been heavily used clinically (Kotre *et al* 1984, 1986, Kotre and Harrison 1988) for both planning external radiotherapy of thoracic regions and assisting localizing the sources in brachytherapy. It does suffer from the

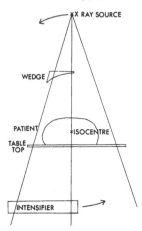

Figure 8.8. *The 'half-fan' method of projection data capture with an offset image intensifier, as used by several groups. The source and detector must make a complete 360° rotation in order to capture the data. The intensifier is offset to increase the field of view. The wedge close to the x-ray source is to compensate (approximately) for the changing shape of the patient so that the data do not have too large dynamic range. (From Redpath and Wright (1985).)*

major disadvantage of poor field of view (30 cm diameter reconstruction circle) necessitating combining NDCT data with conventional lead-wire outlining for large patients. This is because the image intensifier is not offset to one side as in some other developments.

A team from Edinburgh overcame the field-of-view limitation of body-section imaging using a radiotherapy simulator by introducing a lateral offset for the position of the image intensifier of some 15 cm at a distance of 35 cm from the isocentre (figure 8.8). This 'half-fan mode' of data capture necessitates a full rotation through 360° followed by combining half projections into half the number of full projections (Redpath and Wright 1985, Redpath 1988a,b,c, Wright *et al* 1984). It was found most convenient to re-bin the fan-beam data into equivalent parallel projections prior to reconstruction.

The use of a television system coupled to a fluorescent screen and image intensifier on a rotating gantry proved a popular way of creating a NDCT scanner.As early as 1978, one company, Oldelft (N V Optische Industrie 'de Oude Delft', Delft Holland) produced a commercial machine (Duinker *et al* 1978). This was subjected to a rigorous experimental investigation by Kijewski *et al* (1984) in Boston. The scanner was utilized in two quite separate modes. Firstly, it was used, as the manufacturer intended, to generate images by analogue reconstruction means. Secondly, the projection data were digitized and reconstructed on a digital computer. The images obtained in the two ways were compared quantitatively.

The data capture system comprised a flat fluorescent screen of gadolinium oxysulphide coupled by mirrors and lenses to an image intensifier and an Isocon television. For analogue mode, the detector was offset to collect 1500 half-projections in 360° (reconstruction circle 40 cm diameter). For digital mode the offset was eliminated and the reconstruction circle shrank to 20 cm diameter. The slice width for the analogue mode was 7 mm (16 TV lines) and 3.5 mm (8 TV lines) for the digital mode. Digital projection data were obtained (143 frames in 360°) using a Colorado Video Corporation model 270A video digitizer interfaced to a VAX 11/780 computer, the digitization taking 2.5 h. Digital reconstruction used a fan-beam CBP algorithm with a ramp-Hanning windowing filter.

The results of experiments carried out both ways showed that the spatial and density resolutions of the analogue images were 3.1 mm and 3.5% respectively. The corresponding figures for the digital reconstruction were 1.1 mm and 15.4%. Naturally these latter figures could be changed by implementing a different reconstruction filter. Kijewski *et al* (1984) found that when a blurring filter was chosen which yielded the same (3.1 mm) spatial resolution as for the analogue method, the noise reduced to 3.2 %. Further, they noted that analogue reconstruction made use of some 10 times the number of projections and that had this number been used for digital reconstruction the noise value would have fallen even further to 1.8% at this spatial resolution.

At London's Hammersmith Hospital a system was constructed using half-fan-beam projections recorded by a Siemens Siretom RBV 30H triplex image intensifier with a caesium iodide screen of curved cross section coupled to a Siemens Videomed N vidicon TV camera (Arnot *et al* 1984). These were mounted on a TEM Mark 5 Ximatron gantry. A distinguishing feature of this development was the attention paid to designing a hardware electronic filtering circuit which approximated to the Shepp–Logan filter. Spatially filtered data were backprojected in a fast-arithmetic array processor (Cynosure Imaging Systems Inc). This filter used only nine data points (four each side of the element being filtered) and was implemented in a binary-word shift register with bit shifting and add/subtract circuitry. The filtering took $\leqslant$ 40 ms per projection and so the reconstruction became essentially real time.

This group corrected for the problems of image intensifier non-uniformity and the effect of the earth's magnetic field. They concluded that no figure for spatial resolution could be given because the image of a fine wire varied with position in the field of view. Density resolution was of the order 3–5 %. The performance of the system was satisfactory for delineating skin contours, bony tissue, abnormal and normal lung fields and was thus suitable for localization and inhomogeneity correction for radiotherapy. It would appear however that this group had loftier hopes because they found that soft-tissue discrimination was not acceptable and it was, for example, not possible to differentiate ventricles nor tumour in the brain. Figure 8.9 shows a head and chest transverse section to illustrate the quality of data from this machine.

Cynosure Imaging Systems Ltd went on to develop their array processor so that analogue television signals from a rotating image intensifier could be digitized in the AP and then reconstruction took place by digital convolution and digital

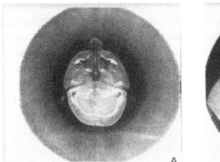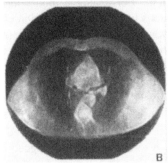

Figure 8.9. *Images of transverse sections of (a) head and (b) chest of normal subjects taken at nominal 125 kVp and 4.2 mA on the Hammersmith Hospital system, which utilized an image intensifier detector. (From Arnot et al (1984).)*

fan-beam backprojection. The whole process took only 2.5 s (Cynosure 1985).

One problem with all such systems based on an image intensifier is the scatter contribution to the image. The systems did not employ collimation at the detector (unlike commercial DCT systems) and this was a limitation.

Very recently a commercial system (the Ximatron CT option) has been marketed by Varian Associates (Varian 1992) (figure 8.10). The detector is an offset image intensifier with photodiode outrigger. A collimator at the detector assists scatter rejection. 500 projections are recorded with 512 elements each. Data are reconstructed by a 2D fast Fourier transform method into an image matrix of 512^2 pixels. A spatial resolution of $\leqslant 2$ mm and density resolution of $\leqslant 1\%$ is achieved. This is very close to the performance of a diagnostic CT scanner. It takes 1 minute to take the data and a further 1.5 minutes to reconstruct the image (figure 8.11). The reconstruction circle is 50 cm in diameter and the aperture is 80 cm.

8.8.3. NDCT with multiwire or scintillation detector

A group from the University of California have reported two machines. The first (Smith *et al* 1976a,b, 1977, 1978, 1980) employed a multi-element xenon detector replacing the image intensifier on a radiotherapy simulator. This detector had a collection efficiency of 99% at the energy in question, comprised 121 radially oriented wires subtending a 40° arc at the x-ray source. Reference detectors were used to compensate for the fluctuating x-ray intensity and software normalization between channels ensured that the output from each channel was stable. The projection data (typically 180 views) were re-ordered into parallel geometry and reconstructed by CBP algorithms employing either the Ram and Lak filter or the Shepp and Logan filter. Spatial resolution of some 7.2 mm and density resolution of 4.1% were achieved. Live subjects were scanned, but the small patient aperture of only 30 cm limited the device principally to head scanning,

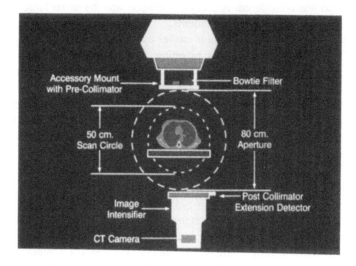

Figure 8.10.   *The geometry of the Varian CT simulator. The image intensifier is offset and there is an additional post-collimation extension detector (outrigger). At the x-ray head is a precollimator, with a reference detector and a 'bowtie' filter to reduce the dynamic range of the output values. The image intensifier is coupled to a camera which is different from the one used for routine screening. The source and detector must make a complete rotation about the patient to capture the data. (From Varian product literature.)*

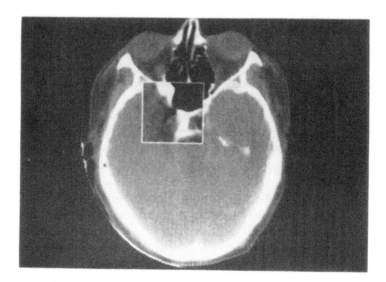

Figure 8.11.   *CT scan of a head taken with the Varian CT simulator with a magnifying glass, showing detail of the sinus. (From Varian product literature.)*

although the possibility of improving breast radiotherapy by imaging in the 'arms up' position was considered. A second limitation was the rather large dead space (some 7 mm) between detector elements.

Smith *et al* (1982) built on their earlier experience to construct a simulator-based CT scanner for radiotherapy planning. They used a bank of 256 solid-state cadmium tungstate crystals coupled to silicon photodiodes and mounted these on a separate C-arm gantry which sat at the opposite end of the simulator couch from the simulator gantry. Using data capture electronics with 17-bit dynamic range, their system performance ($\frac{1}{2}$% density and 2.3 mm spatial resolution) was closer to that of a diagnostic CT scanner than to most other NDCT scanners reviewed here.

8.8.4. The Royal Marsden Hospital CT Simulator

The author and colleagues worked on the development of NDCT scanners between 1976 and 1987. The work progressed in three main stages, the first two being test-bed arrangements leading to the development of the RMH 'CT Simulator'.

Work began in 1976 on a system for recording transmission projection data on film in a special cassette holder which locked on to (and thus replaced as detector) the image intensifier of a radiotherapy simulator (Webb *et al* 1977a,b, 1978, 1981, Webb and Leach 1981).

This group also constructed a NDCT scanner based on a modified J and P Engineering gantry and having a ^{137}Cs source (Webb *et al* 1981, Webb and Leach 1981, Leach and Webb 1979, 1980, Leach *et al* 1982, 1984a, and Flatman *et al* 1982).

The instrument designed for patient scanning (the 'CT Simulator') arose from these studies. Construction began in 1980 and was completed in 1983. (Bentley *et al* 1987, Dobbs and Webb 1988, Leach and Webb 1982, Leach *et al* 1984b, 1985a,b, Maureemootoo *et al* 1988, Toms *et al* 1986a,b, Webb 1982, Webb *et al* 1984a,b,c, 1985b, 1986, 1987a,b, 1988, and Webb and Long 1985).

The machine comprises a detector of novel design, a bar of plastic scintillator viewed at each end by photomultiplier tubes. The detector is collimated first by a narrow slit (3 mm wide), to define the section to be reconstructed from the x-ray fan-beam projections. Additionally, the plastic scintillator of circular section is collimated by a pair of rotating lead cylinders, one inside the other and rotating in the opposite direction to each other. Cut into these cylinders are helical slots, also 3 mm wide and spiralling in opposite ways. The effect of rotating these two helical slots is that a diamond-shaped aperture to the scintillator 'scans' from one end to the other, once per rotation of the cylinders. As soon as the aperture reaches one end, it 'reappears' at the other end. The detector collimators rotate once every 3.31 s continuously whilst the gantry supporting the detector and the source rotate continuously through 360° in about 5 minutes. By sampling the summed output from the photomultiplier tubes every 100 ms, 'projections' (of length 331 elements—but see below) are continuously recorded. The novel collimation thus

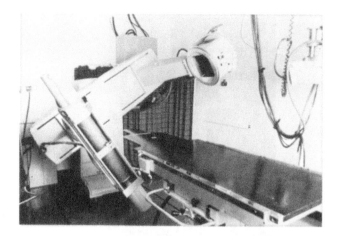

Figure 8.12. *View of the Royal Marsden Hospital 'CT Simulator'. This was built from a Fairey Engineering simulator with excellent geometric bearings. The source collimation was re-engineered and the special one-dimensional detector was added opposite the source. A completely new isocentric couch facility was engineered.*

gives the detector its positional sensitivity. Additionally, the detector is a photon counter rather than a recorder of electrical current.

This is not an efficient method of collecting data because the x-ray beam is not steered with the moving aperture. The method was adopted because the technology of photon counting was well known and reliable and the detector was inexpensive to construct. The delivered dose, despite being largely wasted dose, is not prohibitive for patients receiving radiotherapy. The projections recorded in this way are in a unique geometry. Instead of being able to think of a small finite number of projections at some angular orientation of the gantry with many projection elements, one needs to think of projections as being only single elements but at an enormous number of orientations. The problem of reconstructing this data was solved by Herman (1982) who showed that the data could be reorganized into an equivalent small number (of the order 100) of 'conventional' fan-beam projections.

The scanner is controlled by a PDP1105 computer through a NASCOM Z80 micro, data being captured direct to disc. The data are then transferred via floppy disc to a VAX 750 for image reconstruction. Typically 4.1 mm spatial and 13% density resolution is achieved with a dose to the sternum of 55 mGy. The slice width is 4.6 mm. The scanner is shown in figure 8.12 and the detector in close up in figure 8.13. The patient aperture was some 95 cm and the reconstruction circle 46 cm, entirely adequate for even the largest patient.

The RMH CT Simulator was used to study the importance of including cross-sectional imaging data in the treatment planning process for post-operative breast radiotherapy. It was soon recognized that images taken at the mid-line, upper

Figure 8.13. *View of the detector for the RMH CT Simulator from the source side. The rotating lead cylinders visible in figure 8.12 are below the extra slit collimator. The two mu-metal-shielded photomultiplier tubes are in the brass tubes at each end of the cylindrical plastic scintillator.*

and lower borders of the treatment field were very different with respect both to external contour and internal lung spaces (see figure 8.14). Conventional treatment planning based on simple external contour localization by lead-wire ran the risk of compressing the breast, as well as not providing for accurately tailoring the treatment to the individual chest-wall thickness and lung size. The study showed that significant under and overdosage to breast tissue could occur in these circumstances, that lung tissue was being unnecessarily irradiated by conventional non-conformation techniques and that both could have serious effects on local control in the breast and pulmonary complications.

8.8.5. Cone-beam x-ray CT on a simulator

If CT performed on a simulator is ever to become a true contender for assisting planning conformal radiotherapy for sites other than the breast, multiple-slice data must be obtained. Presently it takes of the order of a minute to image one slice, limited by the speed of rotation of the gantry. One might envisage an area detector which would allow more slices to be simultaneously imaged, but as yet this has not been engineered, although reconstruction techniques are well understood for this geometry (Feldkamp *et al* 1984, Webb *et al* 1987c).

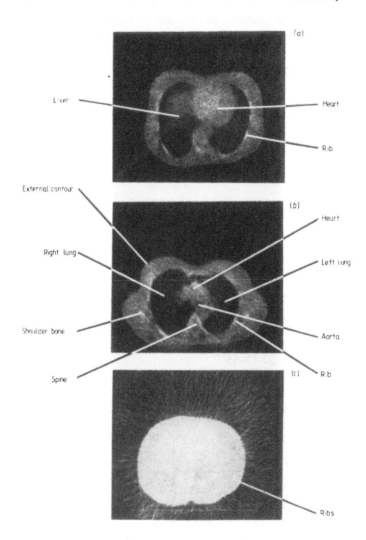

Figure 8.14. *CT images of the breast (with the patient in the treatment position) taken with the RMH CT Simulator (a) mid-field, (b) upper border, (c) lower border. This illustrates the large changes in both internal and external contours across the radiation field and indicates the need for fully three-dimensional imaging to be engineered on a simulator. As yet no one has achieved this.*

8.9. SUMMARY

Medical imaging 'drives' conformal radiotherapy. For true 3D conformation, only those imaging modalities which are truly three-dimensional can play a major role. These are x-ray CT, MRI, SPECT and PET. There are special imaging requirements when these are used to assist radiotherapy, rather than for diagnosis. Imaging serves a dual role: planning and monitoring treatment. For both it is necessary for 3D image data-sets to be registered so regions of interest in one modality (for example x-ray CT with a superposed dose distribution) can be transferred to another modality (for example PET monitoring changed function). Because of the importance of the simulator for planning, the development of x-ray CT on a simulator may play an increasing role, but single images are not enough and a digital area detector is needed to generate 3D images with the patient in the treatment position.

REFERENCES

Ahluwalia B D (ed) 1989 *Tomographic methods in nuclear medicine: physical principles, instruments, and clinical applications* (Boca Raton: Chemical Rubber Company)

Aird E G A 1988 *Basic physics for medical imaging* (Oxford: Heinemann)

Ambrose J and Hounsfield G N 1972 Computerized transverse axial tomography *Brit. J. Radiol.* **46** 148–9

Arnot R N, Willetts R J, Batten J R and Orr J S 1984 Investigations using an x-ray image intensifier and a TV camera for imaging transverse sections in humans *Brit. J. Radiol.* **57** 47–55

Baily N A 1977 Computerized tomography using video techniques *Optical Engineering* **16** 23–27

—— 1979 The fluoroscopic image as input data for computed tomography *IEEE Trans. Nucl. Sci.* **NS26** 2707–2709

Baily N A, Keller R A, Jakowatz C V and Kak A C 1976a The capability of fluoroscopic systems for the production of computerized axial tomograms *Invest. Rad.* **11** 434–439

Baily N A, Lasser E C and Keller R A 1976b Tumour localisation and beam monitoring-electrofluorotomography *Med. Phys.* **3** 176–180

Bamber J 1992 private communication

Bentley R E, Webb S, Leach M O, Maureemootoo K, Yarnold J R, Toms M A and Gardiner J 1987 A CT scanner, based on a radiotherapy simulator, as an aid to planning the conservative treatment of the breast *Proc. 24th Kongress der Gesellschaft Fur Medizinishe Radiologie der DDR Mit Internationaler Beteiligung (Leipzig)* Vol 1 paper V160

Blume S S 1992 *Insight and industry: On the dynamics of technological change in medicine* (Camb Mass: MIT Press)

Chen C N and Hoult D I 1989 *Biomedical magnetic resonance tomography* (Bristol: Adam Hilger)

Coffey C W, Hines H C, Wang P C and Smith S L 1984 The early applications and potential usefulness of NMR in radiation therapy treatment planning *Proc.*

8th Int. Conf. on the Use of Computers in Radiation Therapy (Toronto, 1984) (IEEE Computer Society, Toronto) pp 173–180

Cynosure 1985 *Cynosure 2000-CT package for trans-axial computed tomography on radiotherapy simulator gantries* (Cynosure Imaging Systems Inc, Northleach, Gloucester, UK)

Dobbs H J and Webb S 1988 Clinical applications of x-ray computed tomography in radiotherapy planning *The Physics of Medical Imaging* ed S Webb (Bristol: Adam Hilger) pp 128–141

Duinker S, Geluk R J and Mulder H 1978 Transaxial analogue tomography *Oldelft Sci. Eng. Quarterly* **1** 41–66

Dummling K 1970 Einrichtung zur Anfertigung von röntgenologischen Körper-Querschnittsaufnahmen *German patent no* 2050825

Feldkamp L A, Davies L C and Kress J W 1984 Practical cone-beam tomography *J. Opt. Soc. Am.* A **1** (6) 612–619

Flatman W, Leach M O and Webb S 1982 The use of a median window filter with a simple C T scanner *Proc. 6th ICMP (Hamburg, 1982)* paper 20:12

Fraass B A, McShan D L, Diaz R F, Ten Haken R K, Aisen A, Gebarski S, Glazer G and Lichter A S 1987 Integration of magnetic resonance imaging into radiation therapy treatment planning: 1. Technical considerations *Int. J. Rad. Oncol. Biol. Phys.* **13** 1897–1908

Gifford D 1984 *Handbook of physics for radiologists and radiographers* (Chichester: Wiley)

Gilbert P 1972 Iterative methods for three-dimensional reconstruction of an object from projections *J. Theor. Biol.* **36** 105–117

Goitein M 1983 Impact and use of CT in planning treatment *Advances in Radiation Therapy Treatment Planning* ed A E Wright and A L Boyer (New York: American Institute of Physics) pp 310–320

Guzzardi R 1987 *Physics and engineering of medical imaging* (Dordrecht: Martinus Nijhoff) (in collaboration with NATO Scientific Affairs Division)

Hamilton B 1982 (ed) *Medical diagnostic imaging systems: technology and applications* (New York: Frost and Sullivan Press)

Harrison R M 1981 Potential applications of computerized tomography to treatment planning using a radiotherapy simulator *Computerized Tomographic Scanners in Radiotherapy in Europe (Brit. J. Radiol. Supplement 15)* chapter 48

Harrison R M and Farmer F T 1976 Possible application of a radiotherapy simulator for imaging of body cross-sections *Brit. J. Radiol.* **49** 813

—— 1978 The determination of anatomical cross-sections using a radiotherapy simulator *Brit. J. Radiol.* **51** 448–453

Henkelman R M 1992 New imaging technologies: prospects for target definition *Int. J. Rad. Oncol. Biol. Phys.* **22** 251–257

Henkelman R M, Poon P Y and Bronskill M J 1984 Is magnetic resonance imaging useful for radiation therapy planning? *Proc. 8th Int. Conf. on the Use of Computers in Radiation Therapy (Toronto, 1984)* (IEEE Computer Society,

Toronto) pp 181–185

Herman G T 1980 *Image reconstruction from projections: The fundamentals of computed tomography* (New York: Academic)

—— 1982 Reconstruction algorithms for non-standard CT scanner designs *J. Med. Sys.* **6** 555–568

Herman G T, Hinds J A, Peretti R W and Rowland S W 1975 SNARK-75- a programming system for the reconstruction of objects from shadowgraphs *Technical report* number 96 (State University of New York at Buffalo, Dept. of Computer Science)

Herman G T, Lewitt R M, Odhner D and Rowland S W 1989 SNARK-89- a programming system image reconstruction from projections *Report number* MIPG160 (Medical Image Processing Group, Philadelphia, Pennsylvania)

Hogstrom K R 1983 Implementation of CT treatment planning *Advances in Radiation Therapy Treatment Planning* ed A E Wright and A L Boyer (New York: American Institute of Physics) pp 268–281

Hounsfield G N 1968 A method of and apparatus for examination of a body by radiation such as X or gamma radiation *UK Patent no* 1283915

—— 1973 Computerised transverse axial scanning (tomography): Part 1 Description of system *Brit. J. Radiol.* **46** 1016–1022

Huesman R H, Gullberg G T, Greenberg W L and Budinger T F 1977 RECLBL Library users manual: Donner algorithms for reconstruction tomography *Lawrence Berkeley Laboratory Publication* 214

Kak A C, Jakowatz C V, Baily N A and Keller R A 1977 Computerized tomography using video recorded fluoroscopic images *IEEE Trans. on Biomed. Eng.* **BME 24** (2) 157–169

Kijewski M F, Judy P F and Svensson G K 1984 Image quality of an analog radiation therapy simulator-based tomographic scanner *Med. Phys.* **11** (4) 502–507

Kotre C J and Harrison R M 1988 Clinical use of simulator-based computed tomography in Newcastle upon Tyne *Brit. J. Radiol.* **61** 561

Kotre C J, Harrison R M and Ross W M 1984 A simulator-based CT system for radiotherapy treatment planning *Brit. J. Radiol.* **57** 631–635

Kotre C J, Lambert G D and Dawes P J D K 1986 The use of a simulator-based computed tomography in iridium-192 dosimetry *Brit. J. Radiol.* **59** 1035–1036

Leach M O, Flatman W, Webb S, Flower M A and Ott R J 1984a The application of variable median window filtering to computed tomography *Information processing in medical imaging* ed F Deconinck (Dordrecht: Martinus Nijhoff) pp 151–168

Leach M O and Webb S 1979 Application of a radiotherapy simulator for imaging body cross-sections *Proc. HPA Ann. Conf. (Bristol, 1979)*

—— 1980 A simple CT scanning system suitable for radiotherapy treatment planning: recent results *Brit. J. Radiol.* **53** 1024–1025

—— 1982 A dedicated CT scanner using a position sensitive detector for radiotherapy treatment planning *Proc. 6th ICMP (Hamburg, 1982)* paper 20:19

Leach M O, Webb S and Bentley R E 1982 A rotate translate CT scanner providing cross-sectional data suitable for planning the dosimetry of radiotherapy treatment *Med. Phys.* **9** (2) 269–275

—— 1985a An x-ray detector system and modified simulator providing CT images for radiotherapy dosimetry planning *Phys. Med. Biol.* **30** 303–311

Leach M O, Webb S, Bentley R E and Yarnold J R 1984b The application of a radiotherapy simulator-based CT scanner for radiotherapy treatment planning *Phys. Med. Biol.* **30** 94

Leach M O, Webb S, Bentley R E, Yarnold J R and Toms M A 1985b Radiotherapy dosimetry planning using data from a radiotherapy simulator based CT scanner *71st RSNA Conf. (Radiology* **157** (P) 311)

Ledley R S 1976 Introduction to computed tomography *Comp. Biol. Med.* **6** 239–246

Ledley R S, Wilson J B, Golab T and Rotolo L S 1974 The ACTA scanner: the whole body computerized tomograph *Comp. Biol. Med.* **4** 145–155

Lock V and Coules D 1986 Improved localisation for radical treatment of breast carcinoma using CT and ultrasound *Radiography* **52** (606) 281–286

Maureemootoo K, Webb S, Leach M O and Bentley R E 1988 The performance characteristics of a simulator-based CT scanner *IEEE Trans. Med. Imag.* **7** 91–98

McShan D L and Fraass B A 1987 Integration of multi-modality imaging for use in radiation therapy treatment planning *Computer Assisted Radiology* ed H U Lemke, M L Rhodes, C C Jaffee and R Felix (Berlin: Springer) pp 300–304

Moores B M, Parker R P and Pullan B R (eds) 1981 *Physical aspects of medical imaging* (Chichester: Wiley)

Natterer S 1986 *The mathematics of computerised tomography* (New York: Wiley)

New Scientist 1972 *X-ray diagnosis peers inside the brain* April 27th 1972

Ott R J, Batty V, Bateman T E, Clack R, Flower M A, Leach M O, Webb S and McCready V R 1985 Low cost positron computed tomography *Recent developments in medical imaging* ed R P Clark and M R Goff (London: Taylor and Francis) pp 126–131

Ott R J, Flower M A, Babich J W and Marsden P K 1988 The physics of radioisotope imaging *The physics of medical imaging* (Bristol: Adam Hilger) pp 142–318

Parker R P and Hobday P A 1981 CT scanning in radiotherapy treatment planning: its strengths and weaknesses *Computerised axial tomography in oncology* ed J E Husband and P A Hobday (Edinburgh: Churchill Livingstone) pp 90–100

Phelps M E 1986 Positron emission tomography: principles and quantitation *Positron emission tomography and autoradiography* ed M E Phelps, J C Mazziotta and H R Schelbert (New York: Raven)

Ramachandran G N and Lakshminarayanan A V 1971 Three-dimensional reconstruction from radiographs and electron micrographs: applications of convolutions instead of Fourier transforms *Proc. Natl Acad. Sci., USA* **68** 2236–2240

Redpath A T 1988a Clinical use of the Edinburgh simulator computed tomography system *Brit. J. Radiol.* **61** 561

—— 1988b An analysis of the changes to radiation dose distributions resulting from the use of simulator computed tomography *Brit. J. Radiol.* **61** 1063–1065

—— 1988c *Proc. HPA Ann. Conf. (Brit. J. Radiol.* **61** 561)

Redpath A T and Wright D H 1985 The use of an image processing system in radiotherapy simulation *Brit. J. Radiol.* **58** 1081–1089

Schuman W P, Griffin B R, Haynor D R, Johnson J S, Jones D C, Cromwell L D and Moss A A 1985 MR imaging in radiation therapy planning *Radiology* **156** 143–147

Smith V and Baker R J 1976b A proposed scheme to implement a transverse section transmission scanner for radiotherapy treatment planning *Computer applications in radiation oncology* ed E S Sternick pp 420–423

Smith V, Boyd D P, Kan P T 1977 A computerized transverse section transmission scanner for radiation therapy treatment planning *Phys. Med. Biol.* **22** 128

Smith V, Boyd D P, Kan P T and Baker R J 1976a A computerised transverse section transmission scanner for radiation therapy treatment planning *Proc. 4th ICMP Ottawa (Phys. Can.* **32** paper 26:6) (Ottawa: Canadian Association of Physicists)

Smith V, Parker D, Peschmann K, Phillips T, Wang E and Jimbo M 1982 An improved simulator-based CT scanner for radiation therapy treatment planning *Proc. 6th ICMP (Hamburg, 1982)* paper 20:18

Smith V, Parker D L, Stanley J M, Phillips T J, Boyd D P and Kan P T 1980 Development of a computed tomographic scanner for radiation therapy treatment planning *Radiology* **136** 489–493

Smith V, Stanley J H, Parker D L 1978 Progress towards a CT scanner for radiation therapy treatment planning *Med. Phys.* **5** 339–340

Sontag M R, Galvin J M, Axel L and Bloch P 1984 The use of NMR images for radiation therapy treatment planning *Proc. 8th Int. Conf on the Use of Computers in Radiation Therapy (Toronto, 1984)* (Toronto: IEEE Computer Society) pp 168–172

Süsskind C 1981 The invention of computed tomography *History of Technology 1981* ed A R Hall and N Smith (London: Mansell Publishing Co) pp 39–80

Swindell W and Webb S 1988 X-ray transmission computed tomography *The physics of medical imaging* ed S Webb (Bristol: Adam Hilger) pp 98–127

Taylor J 1988 *Imaging in radiotherapy* (London: Croom Helm)

Ten Haken R K, Kessler M L, Stern R L, Ellis J H and Niklason L T 1991 Quality assurance of CT and MRI for radiation therapy treatment planning *Quality assurance in radiotherapy physics* ed G Starkschall and J Horton (Madison, WI: Medical Physics Publishing) pp 73–103

Todd Pokropek A 1989 Theory of tomographic reconstruction *Tomographic methods in nuclear medicine: physical principles, instruments and clinical applications* ed B D Ahluwalia (Boca Raton: Chemical Rubber Company) pp 3–33

Toms M A, Gardiner J, Webb S, Leach M O, Maureemootoo K and Yarnold J R 1986a Clinical dosimetry for breast radiotherapy based on imaging with the prototype Royal Marsden Hospital CT Simulator *Proc. 5th Annual ESTRO Meeting (Baden-Baden, 1986)* pp 207

—— 1986b Clinical dosimetry for breast radiotherapy based on imaging with the prototype Royal Marsden Hospital CT Simulator *1st Meeting of the British Oncological Assoc. (Imperial College, 1986)*

Varian 1992 *Varian Ximatron Simulator/ CT Option* Varian Associates technical literature

Webb S 1982 A modified convolution reconstruction technique for divergent beams *Phys. Med. Biol.* **27** 419–423

—— 1987 A review of physical aspects of x-ray transmission computed tomography *Proc. IEE* **134** Pt A (2) 126–135

—— (ed) 1988 *The physics of medical imaging* (Bristol: Adam Hilger)

—— 1990a *From the watching of shadows: the origins of radiological tomography* (Bristol: Adam Hilger)

—— 1990b Non-standard CT scanners: their role in radiotherapy *Int. J. Rad. Oncol. Biol. Phys.* **19** 1589–1607

Webb S, Flower M A, Ott R J, Leach M O, Fielding S, Inamdar R, Lowry C and Broderick M D 1985a A review of studies in the physics of imaging by single photon emission computed tomography *Recent developments in medical imaging* ed R P Clark and M R Goff (London: Taylor and Francis) pp 132–146

Webb S and Leach M O 1981 Alternative tomographic systems for radiotherapy planning *Computerized axial tomography in oncology* ed J Husband and P A Hobday (Edinburgh: Churchill Livingstone) pp 154–169

Webb S, Leach M O, Bentley R E, Maureemootoo K, Nahum A, Toms M A, Gardiner J, Parton D, Unwin S and Yarnold J R 1987a The role of the Royal Marsden Hospital computed tomography simulator in planning radiotherapy treatment for breast cancer *Brit. J. Radiol.* **61** 562

Webb S, Leach M O, Bentley R E, Maureemootoo K, Toms M A and Yarnold J R 1986 Clinical experience with the Royal Marsden Hospital CT Simulator applied to planning radiotherapy treatment for breast cancer *Brit. J. Radiol.* **59** 831

Webb S, Leach M O, Bentley R E, Maureemootoo K, Yarnold J R, Toms M A Gardiner J and Parton D 1987b *Phys. Med. Biol.* **32** 835–845 and (corrigendum) 1116–1117

Webb S, Leach M O, Bentley R E and Yarnold J R 1984a Computer aided imaging of the breast using a purpose-built CT scanner *Proc. 6th Nordic Conf of the Biological Engineering Society (Aberdeen, 1984)* paper CIG3:3

—— 1985b Physical performance of the Royal Marsden Hospital CT Simulator applied to planning the conservative radiotherapy of breast cancer *Med. Biol. Eng* **23** Suppl 2, 955–956

Webb S, Leach M O and Herman G T 1984b Reconstruction from a non-standard CT scanner *IEEE Trans. Med. Im.* **MI3** 193–196

Webb S, Leach M O and Newbery P 1988 CT scanner and detector therefor *UK Patent no* 2170980, *US Patent no* 4769829, *European Patent no* 0194743

Webb S, Leach M O and Yarnold J R 1984c CT of the breast achieved in the treatment position using a CT facility developed for a radiotherapy simulator *Proc. 3rd Annual ESTRO Meeting (Jerusalem, 1984)* p 215

Webb S, Lillicrap S C, Steere H and Speller R D 1977a Application of a radiotherapy simulator for imaging of body cross-sections *Brit. J. Radiol.* **50** 152–153

—— 1977b Application of a radiotherapy simulator for imaging body cross-sections *Brit. J. Radiol.* **51** 158

Webb S, Lillicrap S C, Steere H, Speller R D and Jones J 1978 Reconstruction of body cross-sections for use in radiotherapy treatment planning *Computers in Radiation Therapy* ed U Rosenow pp 749–753

Webb S and Long 1985 The effect of reversing projection data in reconstruction tomography *Phys. Med. Biol.* **30** 81–83

Webb S, Speller R D and Leach M O 1981 The generation of CT data for radiotherapy planning using devices other than commercial scanners *Computerized Tomographic Scanners in Radiotherapy in Europe (Brit. J. Radiol. Supplement 15)* pp 178–181

Webb S, Sutcliffe J, Burkinshaw L and Horsman A 1987c Tomographic reconstruction from experimentally obtained cone-beam projections *IEEE Trans. Biomed. Imag.* **MI6** (1) 67–73

Wells P N T (ed) 1982 *Scientific basis of medical imaging* (Edinburgh: Churchill Livingstone)

Wright D H, Redpath A T, Jarvis J G H and Harris J R 1984 Image processing techniques applied to fluoroscopic x-ray pictures obtained from a radiotherapy simulator *Proc. 8th Int. Conf on the Use of Computers in Radiation Therapy (Toronto, 1984)* (IEEE Computer Society, Toronto) pp 343–349

EPILOGUE

The goal of conformal radiotherapy is clearly not new, but the technology allowing progress towards achieving this goal is now under serious active development and the subject has a rejuvenated topicality. Physicists and engineers are now able to manufacture equipment for the practical and accurate delivery of radiation, and computer science has matured to the point that systems for planning radiotherapy, controlling radiation delivery and verifying its positional accuracy are available. Medical imaging 'drives' conformal radiotherapy planning and recent improvements in 3D imaging have led naturally into improvements in conformal radiotherapy. As might be expected, progress has inevitably spawned new questions. For example, the prediction of biological effect is still in its infancy by comparison. Ironically there is also concern that too precise a definition of target volume might not include microscopic peripheral spread of malignant disease. This serves to emphasize that the physicist is not the person to solve all the problems. There is a need for close collaboration with radiobiologists, diagnostic radiologists, radiotherapists and other cancer medicine clinicians.

Achieving conformal radiotherapy relies on optimizing a chain of events; any weakness in one element of the chain weakens the whole.

At the heart of planning conformal radiotherapy are computer systems, able to accept, register, display and analyse 3D medical images from different modalities. In the 1980's these were largely to be found as 'one-off' developments in university hospitals, but the 1990's have seen a number of systems become available commercially. They are not cheap, but one may expect the cost to fall even as their features expand.

Techniques to optimize dose are actively under development. Some are realizable in practical computational times; others giving 'more conformal' 3D dose distributions take longer and some invoke technology which is not readily available, such as intensity-modulated fields.

The availability of the multileaf collimator opens up new and exciting possibilities for treatment delivery and should be able to make the repetitive casting of shielding blocks redundant. Under computer control it creates new opportunities for dynamic field shaping which are quite unachievable with block casting. It may even solve the problem of delivering fields with spatially varying beam intensities. Dynamic therapy with moving gantry, couch and collimator jaws will undoubtedly continue, but it will be interesting to see to what extent the multileaf collimator takes over as the preferred method of tailoring the geometry of the fields.

Stereotactic radiation of brain disorders, malignant and benign, will continue apace with linear accelerators. Enough work has now been done to demonstrate that with just a few arcs, very satisfactory dose distributions can be obtained. It will be interesting to see whether the gamma knife continues to compete. Perhaps the problems will divide into two classes appropriate to each radiation treatment machine?

Electronic portal imaging is the key to quantifying patient registration between imaging, treatment planning and the time of treatment delivery. One may expect that soon it will become routine to check each field and, if necessary, reposition the patient. Several studies are addressing this problem and more will undoubtedly do so.

The use of protons continues to be rare, despite their obviously superior physical properties. The reason is largely one of cost. The first purpose-built, as opposed to shared, proton treatment facility has just come on stream and the community looks to evaluate its success for the justification of future proton irradiation facilities.

The application of the physical sciences to medicine has undoubtedly been a major contribution to radiotherapy and will continue to be so, as far as developing technology is concerned. Much has been achieved, but much remains to be done and this book has attempted to cover the major areas of activity. It cannot, however, cover all the problems and, in particular, it has not attempted to be too definitive where precise knowledge is not to hand. In particular, the translation of dose to biological effect is currently based on less data than one would like, and a developing area will undoubtedly be the collaboration of physicists and clinical partners to put right this deficiency. Finally, all sponsors expect a return on their investments and there are still unanswered questions concerning the efficacy of conformal radiotherapy and some doubts about the best moment in the course of research to mount the necessary exercises in assessment.

NUMERICAL QUESTIONS

The following numerical questions may be helpful when teaching from this book. To answer the questions will require data which can be found in tables of physical constants such as Hubbell (1969,1982), Hubbell *et al* (1980), Duck (1990) and Kaye and Laby (1986). All the relevant theory is covered in this book. Some calculations require writing a short piece of computer code.

Chapter 1

1. The tumour control probability (TCP) for a tissue is specified by an α value from a Gaussian distribution of mean 0.35 and standard deviation 0.08. The clonogenic cell density is 10^7 per cc. If a target volume of 160 cc is irradiated to a uniform dose of 60 Gy, what is the predicted TCP with the cell-kill model? What is the TCP for the same volume at a dose of 65 Gy? What is the TCP for a larger volume of 320 cc at the first dose of 60 Gy? (You may need to write a computer program to do this question.)

2. The normal tissue complication probability (NTCP) for a uniform irradiation of the full volume of an organ at risk is 0.2 (i.e. 20%). What would be the NTCP if only half the organ were raised to the same dose? What would be the NTCP if instead 0.1 of the organ were raised to the same dose?

3. The full volume of an organ at risk is raised to 60 Gy, giving a normal tissue complication probability of 0.1 (10%). What would be the equivalent dose delivered to half the organ to give the same normal tissue complication probability? (Assume $k = 10$ for this organ.)

4. A volume containing an organ at risk in a radiotherapy treatment has a (0.5) 50% normal tissue complication probability when irradiated uniformly to a dose of 50 Gy. What would be the NTCP if the dose were (i) 40 Gy (ii) 30 Gy? (Assume $k = 10$ for this organ.)

5. A cumulative dose–volume histogram is specified by the formula

$$V(d) = 1 - \left\{ \frac{1}{1 + [(121 - d)/41]^{10}} \right\}$$

where d is the dose in Gy (in the range 1 to 120 Gy) and $V(d)$ is the volume raised to at least this dose. Plot a graph of $V(d)$ versus d. What is the effective fractional volume raised to the maximum dose of 120 Gy according to the Kutcher–Burman reduction method? (Take $n = 0.1$ and you may need to write a computer program to do this question.)

6. Two imaging modalities specified on a rectangular cartesian coordinate system are related as follows:

- a rotation by 90° about the x axis,
- a rotation by 90° about the y axis,
- no rotation about the z axis,
- an isotropic scaling by a factor 0.5,
- a shift by 1 cm in the x direction, by 2 cm in the y direction and by 3 cm in the z direction.

Write down the coordinate transformation between the two imaging modalities as a matrix multiplication only.

Chapter 2

7. A radially symmetric dose distribution (see equation (2.9)) is given by rotating a non-uniform beam profile specified by a cosine dependence along a radius through the origin of rotation coordinates up to a radius of 5 cm and zero thereafter (specified by equation (2.22)). Assume that μ=0.02 cm^{-1} for the radiation. Write a small piece of computer code to plot a graph of the dose distribution as a function of radius up to a distance of 10 cm. Discuss how the dose distribution falls off for radial distances beyond 5 cm. Why is the dose not zero in this region?

8. Use the code from question 7 to vary the radius beyond which the beam profile is zero inwards from 5 cm to (sequentially) 3 cm, 1 cm, 0.1 cm. What do you notice about the dose distribution up to a radial distance of 10 cm as this is done? What is the name given to the last of these distributions? How is it used in 'convolution dosimetry'?

9. A rotationally symmetric dose distribution is generated in which the dose is zero inside a radius of 4 cm and uniform from 4 to 8 cm. Plot the intensity profile at the source in the region x lying between 4 and 8 cm required for radiation with $\mu = 0.02$ cm^{-1}. Assume the phantom is a circle of radius 15 cm with its centre at the axis of rotation.

10. A rotating beam of the type shown in figure 2.19 can be considered to deliver a dose of 0.1 arbitrary units under the central portion between a distance of ±1 cm and 1 unit under the region between ±(1–6) cm. Plot the radial distribution of the circularly-symmetric dose distribution for radii 0 to 10 cm.

11. What is the advantage in terms of speed for computing a 3D convolution of two 2048^3 matrices in Fourier space compared with real space? What is the corresponding advantage for a 2D convolution of two 512^2 matrices?

Chapter 4

12. A narrow beam of protons diverges to irradiate a circular area of radius 4 cm at a dose-measurement plane. A simple beam-stopper is placed on axis such that its projection on the dose-measurement plane fills a circle of radius $A_1 = 1$ cm. Plot the distribution of proton intensity in the measurement plane if the proton beam has been scattered twice, firstly by a foil with $R_1 = 1.7A_1$ and secondly by a foil with $R_2 = 1.4A_1$ just behind the beam-stopper. (You will probably need to write a short computer code for this question.)

13. A narrow beam of protons diverges to irradiate a circular area of radius 4 cm at a dose-measurement plane. An annular beam-stopper is placed on axis such that its projection on the dose-measurement plane fills an annulus of outer radius $A_1 = 1$ cm and inner radius $A_2 = 0.38$ cm. Plot the distribution of proton intensity in the measurement plane if the proton beam has been scattered twice, firstly by a foil with $R_1 = 1.7A_1$ and secondly by a foil with $R_2 = 1.0A_1$ just behind the beam stopper. (You will probably need to write a short computer code for this question.)

14. Given that a 100 Mev proton has a range in tissue of about 8 cm, what would be the approximate energies of an alpha particle and a deuteron with the same range?

15. If a proton beam facility can produce 10^{-10} A cm^{-2}, what dose rate would be delivered to the last 1 cm of the proton range?

chapter 5

16. A multileaf collimator comprises 40 pairs of leaves. Each leaf is 5 mm wide. The leaves are 40 cm from the x-ray target, can open to a maximum of 10 cm on the leaf-side of mid-line and can over-run the mid-line in either direction by 6 cm. What is the largest field that can be set at the isocentric distance of 1 m?

17. A multileaf collimator is made of tungsten leaves 5 cm thick. What fraction of 2 MeV monoenergetic photons 'leaks' through any closed portion of the collimator (ignoring the possible hairline gap between any two adjacent leaves)? If in addition a back-up collimator 2 cm thick resides behind the leaves, to what fraction is the leakage reduced at this energy?

18. A multileaf collimator is made of tungsten leaves 5 cm thick and they are 'stepped' at a depth of 2.5 cm along the edges of the leaves. What leakage radiation could be expected at the step for monoenergetic 2 MeV photons if there were no back-up collimator?

chapter 6

19. Monoenergetic x-rays at 2 MeV are incident normally on the surface of a 25 mm thick slab of the scintillator bismuth germanate. Calculate the fraction

of the x-ray flux which is stopped by the scintillator. What fraction of the flux would be stopped if the scintillator were zinc tungstate instead? Calculate the same fraction for these two materials if the x-ray energy were increased to 6 MeV. How thick should a scintillator of the material caesium iodide be to stop 75% of x-ray photons of energy 1 MeV? What are the implications for megavoltage imaging?

20. The detector for megavoltage imaging comprises a 1 mm thick plate of copper bonded to a 300 mg cm^{-2} thick scintillator of gadolinium oxysulphide. Calculate the fraction of 1 MeV monoenergetic x-rays which stop in the copper plate. What would be the consequences, of making the copper plate thicker, for detective quantum efficiency and spatial resolution?

21. A fluoroscopic megavoltage imaging system images the light from a scintillator of area 400 mm^2 on to a CCD camera with a face size of 4 mm^2 via an optical lens with F-number 0.9 and optical transmission coefficient 0.8. If the photon detection efficiency of the CCD camera is 70% and 2×10^4 optical photons are emitted by the scintillator for each detected x-ray photon and the detective quantum efficiency of the metal-plate converter is 3%, assuming unity refractive index for the scintillator, by how much is the signal-to-noise ratio of the input x-ray flux degraded by the input stage of the CCD camera?

22. A digital portal-imaging system based on a fluoroscopic detector creates images with an improved signal-to-noise ratio by frame averaging. By how much does the signal-to-noise ratio improve if (i) 16, (ii) 256 frames are averaged? What are the consequences for static and dynamic imaging?

23. In a fluoroscopic megavoltage imaging system, should the F-number of the optical component of the system be as small as possible or as large as possible, and why?

24. In a megavoltage imaging system comprising a scanning array of photodiodes, the photodiodes are encapsulated in lead 1.5 mm thick. Estimate the fraction of monoenergetic x-rays of energy 4 MeV which are detected.

25. What is the dose contrast on a film introduced by an extra thickness of 2 mm of water-equivalent tissue for 1 MeV monoenergetic radiation if the scatter-to-primary ratio is 30%? What is the film density contrast if the γ of the film is 2.7 in the region where these intensities occur?

26. Contrast improvement in portal imaging is to be achieved by 'gamma multiplication' using three films. If the first film has a gamma of 2.5, the third has a gamma of 3.2, what should be the gamma of the second (middle) film if the overall gamma multiplication is to be 22.5?

27. An image intensifier is used for capturing an image of a patient being positioned for radiotherapy on a simulator for a square field whose side is 20 cm long. If the displacement δ of an image point from its true position is given by the

formula

$$\delta = \alpha R^2 + \beta R^4$$

where R is the true distance of that point from the origin, what is the maximum displacement which a point suffers if $\alpha = 6 \times 10^{-4}$ pixels^{-1} and $\beta = 4 \times 10^{-8}$ pixels^{-3} and the pixel size is 2 mm?

REFERENCES

Duck F A 1990 *Physical properties of tissue: A comprehensive reference book* (London: Academic)
Hubbell J H 1969 Photon cross sections, attenuation coefficients and energy absorption coefficients from 10 keV to 100 GeV *Report number* NSRDS-NBS 29, US Government Printing Office, Washington DC
Hubbell J H 1982 Photon mass attenuation and energy absorption coefficients from 1 keV to 20 MeV *Int. J. Appl. Rad. Isot.* **33** 1269–1290
Hubbell J H, Gimm H A and Øverbø 1980 Pair, triplet and total atomic cross sections (and mass attenuation coefficients) for 1 MeV–100 GeV photons in elements $Z = 1$ to 100 *J. Phys. Chem. Ref. Data* **9** (4) 1023–1147
Kaye G W C and Laby T H 1986 *Tables of physical and chemical constants* (London: Longmans) 15th edn

APPENDIX B

GLOSSARY OF TERMS

This glossary also includes some widely used terms in medical imaging, which is so important to conformal radiotherapy.

ANGER CAMERA Gamma-ray detector, named after its inventor, in which a large-area scintillation crystal fitted with a collimator is viewed by a matrix of photomultiplier tubes with electronics to give position sensitivity.

ARTEFACT An unwanted structure in an image generally due to some unsatisfactory (but often insurmountable) feature in imaging equipment.

AUTOTRACKING Method of obtaining the contour of a specific tissue from images automatically by giving the computer features of the tissue which enable it to be differentiated from its surroundings (e.g. its linear attenuation coefficient).

BANDLIMIT The maximum frequency present in an image. Real information at frequencies above the limit fail to be recorded.

BEAM PROFILE The shape of the intensity distribution in a radiation field having an intensity varying in one or two dimensions.

BEAM SPACE Mathematical space in which beams are defined, possibly with spatially varying intensities (see also DOSE SPACE).

BEAM-STOPPER Material placed in a charged-particle beam and which completely absorbs the beam. A combination of a beam-stopper with an annular shape and a double-scattering system can create a large-area field for radiotherapy.

BEAM'S-EYE-VIEW The view of the patient anatomy as seen through the radiation collimator by an imaginery observer at the source location. The view is generally constructed from tomographic image data. It is particularly helpful when planning to avoid irradiating an organ-at-risk.

BEAM-WEIGHT Rather loose term applied to radiation fields delivered by a linear accelerator, meaning the number of monitor units per field or part of a field if this has a spatial variation.

BIOLOGICAL RESPONSE The response of normal and pathological tissues to radiation. Calculating this requires a model for the damage produced by a known radiation dose.

BLOCKED FIELDS Radiation fields defined by a collimator to which additional lead blocks have been added to create a non-rectangular-shaped field.

BLOCKING TRAY Tray in which blocks are secured at the linear accelerator during photon treatment.

BODY SCANNER Rather loose term usually referring to a machine for computed tomography of the body (rather than the head).

BOLUS Material placed next to the patient to change the contour as it appears to the radiation beam.

BRAGG PEAK The region of high-energy deposition at the end of the track of a particle (e.g. a proton) slowing down in a tissue.

BUILD-UP REGION A volume in a patient where the electrons from photon interactions are not in equilibrium. This may be near the patient surface, or an interface between two materials with widely varying electron densities.

CCD CAMERA Television camera using a solid-state charged-coupled device. These are found in some megavoltage imaging systems and also in mechanisms for verifying the leaf positions of a multileaf collimator.

CLASSICAL TOMOGRAPHY Generic term for all non-computed or so-called 'blurring tomography'

CLINICAL TRIAL Comparison of the effectiveness (with some clearly specified criteria) of a new method of radiotherapy against some standard and commonly accepted technique. Patients should be randomized into two arms of the trial, one set of patients being treated with the 'old' method and the other set receiving the new technique. Ideally neither the patients nor their clinicians should know which patients are in which arm until the trial is analysed (double blind trial).

CLONOGENIC CELLS Tumour cells which are reproducing.

COINCIDENCE DETECTION Form of imaging wherein two gamma rays are detected simultaneously (or nearly so). The gammas may be those from a positron-emitting radionuclide or an isotope emitting two gammas in cascade.

COLOUR WASH Method of shading grey-scale anatomical tomograms with colours to indicate the value of dose at each region.

COMPENSATOR Material placed in the path of radiation before it enters a patient, designed to compensate for the effect of tissue inhomogeneity in the path of the radiation. Compensators are constructed on different principles for photons and charged particles.

COMPUTED TOMOGRAPHY (CT) Section imaging in which the required image must be reconstructed from projection measurements, usually using a digital computer.

CONFORMAL THERAPY In the old Japanese definition, term for rotation radiotherapy; in the modern definition, term for radiotherapy with the high-dose volume geometrically tailored to the target.

CONVOLUTION AND BACKPROJECTION Mathematical technique for reconstructing CT data from (filtered, or convolved) projections.

CONVOLUTION DOSIMETRY Method of computing dose distributions based on a knowledge of the elemental dose distribution from radiation interactions at a point.

COPLANAR FIELDS Radiation fields with central axes all lying in a common plane.

CORONAL SLICE Tomographic section bordered by superior, inferior, right and left of the patient (i.e. face-on view); generally not obtained by direct reconstruction, but by reorganizing transaxial sections.

COST FUNCTION Mathematical function parametrizing the effect of arranging beams in some particular way. For example, a simple cost function could be the RMS difference between the prescribed dose and the delivered dose. More complicated functions could include biological models. The aim of optimization would be to minimize the cost function, possibly subject to constraints.

COUCH TWIST When the axis of rotation of a couch coincides with the axis of rotation of a radiotherapy gantry, a radiation field which does not have its axis in the transaxial plane of the patient may be achieved by rotating the couch so it is not perpendicular to the plane of rotation of the gantry. This is colloquially referred to as couch twist. It allows non-coplanar beams to be arranged.

CROSSTALK Information leaking from one place to another (generally adjacent) place; for example, when an x-ray impinges on one element of a matrix of radiation detectors, another element may record a smaller but non-zero signal.

CYCLOTRON Circular particle accelerator which may be used for the production of certain radionuclides and for charged particles for radiotherapy.

DECONVOLUTION Inverse process to convolution. Deconvolution of a dose kernel from a dose distribution can yield the irradiation density for example.

DIGITAL RADIOLOGY Form of radiology (tomographic or non-tomographic) in which the image is captured (often by discrete detectors) and stored as a digital matrix of numbers.

DIGITAL RECONSTRUCTED RADIOGRAPH Planar radiograph constructed by ray-tracing from the position of an x-ray source through the 3D matrix of CT data. This image can be compared with a portal image at the time of treatment or with the image from a simulator of the patient at treatment simulation.

DOSE-DIFFERENCE PLOT Map of the difference between the desired dose distribution (the dose prescription) and the dose distribution resulting from some particular arrangement of beams.

DOSE KERNEL The distribution of dose around a point at which a photon has interacted, computed from an ensemble of all the possible interactions.

DOSE SPACE Mathematical space in which dose is defined. The radiation planning process relates dose space to beam space.

DOSE–VOLUME HISTOGRAM (DVH) Histogram showing the fraction of the target volume raised to a particular dose value (differential DVH) or the fraction of the volume receiving the specified dose value or greater (integral DVH).

DOUBLE FOCUSING Arrangement by which the leaves of a multileaf collimator are engineered. The tangents to all sides and ends of leaves converge to the source of radiation. This limits penumbra.

DOUBLE-SCATTERING SYSTEM Method by which a narrow proton beam is broadened to a wide area useful for radiotherapy. The technique uses two thin metal foils for scattering.

DYNAMIC THERAPY Form of radiotherapy with the beam moving relative to the patient during the irradiation (opposite of STOP-AND-SHOOT). Sometimes this is engineered by moving the source. Sometimes, instead, the couch is translated relative to the source.

DYNAMIC WEDGE The creation of a wedged field by moving one jaw of an accelerator collimator during irradiation.

EGS4 The electron gamma shower computer package much used for radiation dose computations in benchmark situations.

ELECTRONIC PORTAL IMAGING Recording the exiting radiation at the time of treatment with an electronic method without the use of film (sometimes also called megavoltage portal imaging (MVI)).

EMISSION TOMOGRAPHY Body-section imaging of the distribution of uptake of a radiopharmaceutical or some other property of its metabolism.

EMISSION COMPUTED TOMOGRAPHY (ECT) Emission tomography requiring the use of a digital computer to reconstruct digital tomograms. When based on the detection of single photons, the technique is known as SINGLE-PHOTON EMISSION COMPUTED TOMOGRAPHY (SPECT). When based on the coincidence detection of high-energy photons from positron emission, the technique is known as POSITRON EMISSION TOMOGRAPHY (PET).

ERROR PLAN A dose plan in which the effect of certain known or estimable sources of error (such as tissue movement) is taken into account. (Not a plan with errors!)

FIDUCIAL A marker fixed to the patient which will show as a point in a medical image. Multiple fiducials play an important part in image registration.

FIELD OF VIEW Region in space which is viewed at all times during tomography.

FILM Generally here meaning film sensitive to x-rays, not visible light.

FILM GAMMA The slope of the linear part of the curve expressing film density as a function of dose.

FILTERED BACKPROJECTION The technique (or the image created thereby) by which CT data are reconstructed from filtered projections.

FIRST-GENERATION CT Form of rotate–translate CT using a single detector (see ROTATE–TRANSLATE CT).

FLUENCE Photons per unit area.

FLUOROSCOPIC SCREEN Detector which emits light photons when x-ray photons are incident upon it.

FOURIER TRANSFORM Mathematical transformation, named after its inventor, which converts data from a function of some variable (e.g. time) into the corresponding function of the inverse variable (e.g. frequency).

FOUR VECTOR Vector with four components. Rotations, scaling and translations can all be handled by matrix multiplications if points in space are represented as four vectors.

FOURTH-GENERATION CT Form of rotate-only CT in which the detectors comprise a stationary ring and the x-ray source only rotates.

FRACTIONATION Process of delivering radiation in a series of small doses spaced out in time.

FRAME AVERAGING Finding the mean image of a number of (generally sequential) video frames to reduce the effect of noise.

FRAME GRABBER Computer-based device for creating a digital image from a video image.

GAMMA CAMERA Usually means Anger camera (see above).

GAMMA KNIFE Specialized irradiation device designed by Leksell for stereotactic radiosurgery. It contains a large number of cobalt sources focused to a point.

GAMMA MULTIPLICATION The method of increasing the contrast on a film by making sequential copies on to films with a film gamma greater than one.

GRAVITY-ASSISTED BLOCKING Method by which a block is suspended in the field from a rotating treatment machine. The block shields specified volumes of tissue, whatever the orientation of the gantry.

HAT–HEAD METHOD Technique for registering two image data-sets of the head. Surfaces are constructed from each and fitted like a hat to a head to minimize the space between.

HEAD SCANNER Rather loose term, usually referring to a machine for computed tomography of the head (rather than the body).

HEAD TWIST Colloquial term for rotating the collimator of a treatment machine so the sides of a square field would not lie parallel to, and orthogonal to, the plane of rotation of the gantry.

HELMET Collimator for the gamma knife.

IMAGE OF REGRET A colour-washed grey-scale image of anatomy, wherein the colour codes the inability of a technique to produce a dose distribution matching the prescription (regret).

IMAGE INTENSIFIER Optoelectronic device creating an optical image from an x-ray flux impinging upon it.

INDEPENDENT COLLIMATOR X-ray collimator on a linear accelerator with the ability for each jaw to move independent of its paired jaw.

INTENSITY MODULATION FUNCTION Mathematical function describing the variation of radiation intensity in beam space.

INVERSE COMPUTED TOMOGRAPHY Somewhat loose term meaning the process of computing the optimum distribution of beam intensities which will combine to yield some specified dose distribution. When formulated in two-dimensions, the problem has certain similarities with the mirror of the method of reconstructing CT slices from 1D projection data, hence the name.

INVERSE RADON TRANSFORM The mathematics embodied in reconstruction theory for computed tomography.

IRREGULAR FIELD Rather loose term, generally meaning radiation field which is not rectangular and which can therefore not be collimated by a pair of simple jaws without the addition of blocks.

ISOCENTRE The axis of rotation of a gantry, for example that of a CT scanner or a machine delivering radiotherapeutic x-radiation.

ISODOSE LINE/SURFACE Line (in a 2D plan) or surface (in a 3D plan) connecting points where the dose is the same.

ISOEFFECT FORMULA Equation giving the relationships between quantities such as time, dose per fraction and number of fractions, which may be varied to give the same biological effect.

KERMA Kinetic energy released per unit mass. Kerma includes energy released by charged particles only. It excludes radiation transport from scattered photons. Kerma has the same dimensions as dose (compare with TERMA).

KICKER Device for extracting a charged particle beam from an accelerator.

LANDMARK-BASED CORRELATION Method of registering two (generally 3D) image data-sets by identifying common (generally internal anatomical) landmarks in the two images.

LARGE-SCREEN DIGITIZER Device for projecting an image of a patient cross section and some means of pointing at the image. By tracing a feature of interest, the required contour is input to the computer.

LEAKAGE RADIATION Unwanted radiation for example passing through the gaps between closed collimators or adjacent leaves of a multileaf collimator.

LOCAL CONTROL Control of the primary tumour without metastatic spread.

MATCH-LINE PROBLEMS Potential sites of overdose or underdose where two adjacent radiation fields meet.

MATRIX IONIZATION CHAMBER A type of megavoltage portal imager.

MEGAVOLTAGE COMPUTED TOMOGRAPHY (MVCT) CT performed on a treatment machine at the time of treatment, used to verify the patient positioning.

MEGAVOLTAGE PORTAL IMAGING (MVI) Making a picture of the radiation emerging from the patient at the time of treatment.

METAL SCREEN Thin sheet of metal placed in front of a film or scintillator to convert an x-ray flux into a detectable electron flux.

MODALITIES Techniques whereby medical images may be obtained by physical probes.

MONTE CARLO METHODS Calculation techniques using a computer, whereby the macroscopic effect of an irradiation is constructed as the ensemble average of the individual effects of following the scattering history of discrete photons or particles. Millions of histories need to be computed for accurate results. The accuracy of the results is limited only by the input data, the sophistication of the modelling of the interactions and the amount of available computer time.

MORBIDITY Treatment-related complications which adversely affect the quality of life of a patient.

MULTICRYSTAL PET SCANNER Form of positron tomography device using many scintillation crystals usually arranged in a circle.

MULTILEAF COLLIMATOR X-ray collimator comprising many 'leaves' and thus able to create an irregularly shaped window for the radiation to pass through. Typically there might be some 40 pairs of these leaves with each leaf projecting to a width of about 1 cm at the isocentre.

MULTIPLE-BEAM RADIOTHERAPY Radiotherapy of tumours by combining one or more x-ray beams from different directions relative to the patient.

MULTIPLE SCATTERING The processes by which photons or charged particles lose energy in traversing tissue. Photons may only scatter a few times before being annihilated, whereas electrons and protons undergo thousands of small scattering events.

MULTIWIRE PROPORTIONAL CHAMBER (MWPC) Form of gamma-ray detector based on proportional chamber theory.

NORMAL TISSUE COMPLICATION PROBABILITY (NTCP) Graph of the probability of inducing a complication in a normal healthy tissue as a function of the applied radiation dose. It is necessary to specify the 'end-point', the specific complication appropriate to the curve (e.g. rectal bleeding).

OBLIQUE TOMOGRAPHY Form of tomography wherein arrangements are made for the slice in sharp focus to lie in some plane other than transaxial, coronal or sagittal.

OPTIMIZATION Technique to establish the best arrangement of radiation fields to maximize uncomplicated tumour control. The word is widely used, with varying shades of interpretation.

ORGAN-AT-RISK Sensitive structure requiring shielding during radiotherapy (e.g. the spinal cord).

OVER-RUNNING Ability for the leaves of a multileaf collimator to be set past the axis of rotation of the treatment machine, i.e. the beam port is entirely to one side of the axis of rotation.

PARALLEL-BEAM CT Form of computed tomography in which all the ray-paths to projection elements (at each orientation) are parallel (i.e. 1st generation CT).

PARALLEL-HOLE COLLIMATOR Collimator for an Anger camera, also enabling SPECT in parallel (as opposed to fan- or cone-) beam geometry.

PARALLEL-OPPOSED FIELDS Radiation fields sharing a common axis and pointing towards each other.

PARTIAL VOLUME EFFECT The effect whereby the value of a physical parameter ascribed to a voxel is some average value due to the voxel spanning tissues of different types.

PENCIL-BEAM DOSE DISTRIBUTION The distribution of dose surrounding a narrow parallel beam of radiation with an infinitessimal cross section. This forms the elemental kernel for some methods of dose calculation.

PENUMBRA In radiology, term meaning the part shadows caused by (among other things) finite source size and collimation. In radiotherapy, term meaning the region at the edge of a field where the dose rapidy decreases, but does not become zero due to scatter, as well as the above.

PHANTOM Object generally comprising tissue substitute materials used to simulate a patient or part thereof.

PHOTOSTIMULABLE PHOSPHOR PLATE Radiation detector able to store an image for later digital readout.

PIN-CUSHION DISTORTION Geometrical distortion, symmetrical about the detector centre, misplacing the detected event on an image intensifier.

PIXEL Picture element in a digital image matrix (for example, a CT scan).

POINT IRRADIATION DISTRIBUTION Distribution of dose at a point resulting from the superposition of many narrow radiation fields converging on that point, for example by rotation therapy.

POINT-SPREAD FUNCTION The response of an imaging system to a point source of radiation.

PORT The opening, usually defined by a collimator, through which radiation emerges to impinge on a patient. On the distal side of the patient the emerging radiation can be recorded by 'portal imaging'.

POSITION-SENSITIVE DETECTOR A radiation detector in which the position of the x-ray interaction can be very precisely determined. Some position-sensitive detectors are multi-element (e.g. crystal scintillators); other continuous detectors use position-sensing electronics (e.g. as in the Anger or gamma camera).

POSITRON EMISSION TOMOGRAPHY (PET) Form of sectional imaging of a radiopharmaceutical, in which the radionuclide is a positron emitter (in contrast to a single-gamma emitter).

PROJECTION The sum along a particular (usually straight) line path through a patient of some physical property of biological tissue. In transmission computed tomography the projections are sums of x-ray linear attenuation coefficient. In emission computed tomography the sums are complicated functions of both the uptake of a radiopharmaceutical in body tissues and the photon attenuation properties of the biological tissue.

PROJECTION FILTERING A vital stage in the mathematical process of reconstructing cross-sectional images from projections (CT). After projection filtering, data are back-projected into the reconstruction space.

PROTON CT Form of computed tomography using protons as probe.

QUANTUM EFFICIENCY Fraction of photons which impinge on a detector which contribute to the image (for example, in megavoltage imaging).

RADIODENSITY Loose term for x-ray linear attenuation coefficient (which is related to the electron density of tissue).

RADIONUCLIDE Isotope of an element which is radioactive.

RADIOPHARMACEUTICAL Pharmaceutical, element or compound labelled with a radionuclide.

RADIOTHERAPY TREATMENT FRACTIONS The delivery of part of a radiotherapy treatment via a series of visits for the patient.

RANGE The maximum distance a charged particle will travel in a tissue.

RANGE MODULATION The process of reducing the range in a patient of a charged-particle beam by placing absorbing material in its path upstream of the patient.

RANGE-MODULATING WHEEL OR PROPELLOR Rotating wheel comprising a fan of modulators, each of different thicknesses, which, when placed in a monoenergetic proton beam, generates a beam with a mixture of ranges. The carefully calculated geometry of the wheel creates an approximately flat spread-out Bragg peak.

RECONSTRUCTION ALGORITHM The mathematical method by which an image may be created from measurements of some physical property of biological tissues (such as x-ray linear attenuation coefficient or the uptake of a radiopharmaceutical) along a large number of different paths through the patient.

REDUCTION SCHEME method of converting the dose–volume histogram of a volume which receives an inhomogeneous radiation dose to either (*a*) an equivalent volume receiving a homogeneous single dose or (*b*) an equivalent single dose to the whole volume.

REGISTRATION Process of aligning two (generally 3D) image data-sets so that a region in one exactly corresponds to the same region in the patient in the other.

REGRET Inability to meet a dose specification (see VOLUME OF REGRET and IMAGE OF REGRET).

RISK CURVE Loose term for TCP or NTCP curve.

ROTATE-ONLY CT Form of computed tomography in which an x-ray source delivering a fan-beam of radiation on to a bank of detectors (the fan completely spanning the patient) rotates (with no translation) about the patient; if the source and detectors both rotate then this is also known as 3rd generation CT scanning. If only the source rotates and the detectors comprise a stationary ring, this is known as 4th generation CT.

ROTATE–TRANSLATE CT Form of computed tomography in which a single x-ray source and single detector (or small bank of detectors) are translated in a straight line relative to a patient, with subsequent rotation of the line of translation through a large number of orientations relative to the patient. The PROJECTIONS so recorded form the basis of reconstruction. The form of CT with a single detector is also known as 1st generation CT; the form with a bank is known as 2nd generation CT.

SAGITTAL SLICE Tomographic section bordered by superior, inferior, anterior and posterior of the patient (i.e. side-on view); generally not obtained by direct reconstruction but by re-organizing transaxial sections.

SCATTER Radiation which results from the interaction of primary radiation, for example by the Compton effect in tissue.

SCINTILLATION DETECTOR Form of x-ray detector in which x-ray energy is converted into optical photons.

SCOUT VIEW A digital two-dimensional planar radiograph formed using a CT scanner, operated such that the source and detector do not rotate about the patient, who is translated relative to the x-ray beam (used for localization or patient set-up).

SCREEN-FILM X-ray detector comprising both an intensifying screen and a film (opposite, NON-SCREEN FILM).

SECOND-GENERATION CT Form of rotate–translate CT using a bank of detectors (see ROTATE–TRANSLATE CT).

SECONDARY COLLIMATION Usually refers to a simple set of jaws in a linear accelerator in addition to a multileaf collimator.

SECTION IMAGING Tomography.

SHADED-SURFACE DOSE DISPLAY Computer display of dose. Isodose surfaces are shown as if they were solid bodies illuminated by light.

SIMULATED ANNEALING An optimization technique which uses principles modelled on crystal annealing. Such processes converge to a global minimum result.

SIMULATOR Machine which emulates the geometry of a treatment machine, but which uses diagnostic energy x-rays to take images of the patient in the treatment position.

SINGLE-PHOTON EMISSION COMPUTED TOMOGRAPHY (SPECT) See EMISSION COMPUTED TOMOGRAPHY

SINOGRAM The image formed by stacking projections, one beneath the other, taken by detectors at sequential (and generally equal) angular increments around a patient. An off-axis point would describe a sine wave in the sinogram, hence its name.

SLICE THICKNESS The thickness of a tomographic section.

SMART Stereotactic multiple-arc radiotherapy.

SPILL The burst of charged particles from an accelerator.

SPOT SCANNING Method of creating a large-area proton field by electro-magnetically changing the position of a small-area beam.

SPREAD-OUT BRAGG PEAK (SOBP) The wide region of high-dose deposition in the graph of energy deposition versus depth for a charged-particle radiation. The SOBP is formed by a range-modulating wheel.

STEREOTACTIC RADIOTHERAPY Radiotherapy with the patient's head firmly constrained to a precise geometry in relation to the radiation. This may be by a stereotactic frame bolted to the calvarium, or by some less invasive frame.

STEREOTACTIC RADIOSURGERY As stereotactic radiotherapy, except the aim is to sterilize the target volume in one or two fractions.

SYNCHRONOUS SHIELDING Method by which a block is suspended in the field from a rotating treatment machine. The block shields specified volumes of tissue, whatever the orientation of the gantry.

SYNCHRONOUS FIELD SHAPING Method of conforming the radiation aperture to the projected size of the target in rotation therapy by gravity-suspended collimators.

TARGET VOLUME The full extent of the volume to be treated, including the tumour, marginal spread of disease and a 'safety' margin. The 'biological target volume' (BTV) and 'mobile target volume' (MTV) are defined in chapter 1. A new report (ICRU50) to be published in late 1992 will redefine targets in terms of 'gross target volume' (GTV), 'clinical target volume ' (CTV) and 'planning target volume' (PTV).

TERMA Total energy released per unit mass. This include all radiation transport, both particles and photons. Terma has the same dimensions as dose (compare with KERMA).

THERAPEUTIC RATIO Ratio of the probability of tumour control to the probability of normal tissue complication.

THERAPY CT SCAN CT scan (for localization purposes) taken with the patient in the same position as they will be set up for subsequent radiotherapy.

THIRD-GENERATION CT See ROTATE-ONLY CT

THREE-DIMENSIONAL TREATMENT PLANNING SYSTEMS (3DTP) Computer packages which are able to use 3D image data for planning radiation therapy, calculating dose distributions and evaluating and displaying these in 3D.

TISSUE INHOMOGENEITY CORRECTIONS Part of the radiotherapy planning process which takes account of the varying x-ray attenuation of different body tissues. CT data are often used to assist this process.

TRACKING THERAPY Method of dynamic therapy whereby the radiation field is constantly adjusted to conform to the projected area of the target. This may be arranged by moving the patient relative to a stationary source, or vice versa. The machine at London's Royal Free Hospital is called the TRACKING COBALT UNIT.

TRANSIT DOSIMETRY Calculation of the distribution of dose actually received by a patient, starting with the portal image.

TRANSMISSION COMPUTED TOMOGRAPHY (TCT) X-ray CT as, for example, distinguished from emission CT (ECT).

TOMOGRAPHY Since 1962 the term has been decreed by ICRU to be applied to *all* methods of body-section imaging, howsoever performed.

TOMOGRAM Term used for the body section generated by tomography.

TOPOGRAM Same as scout view.

TUMOUR CONTROL PROBABILITY (TCP) The graph of the probability of killing the tumour as a function of the applied radiation dose.

UNIVERSAL WEDGE Fixed-angle wedge in a linear accelerator. By combining two fields—one with the wedge and one without the wedge—of different durations, any arbitrary wedge angle may be achieved.

VIRTUAL SIMULATION Method of creating images of the radiation portals which will be used at treatment. Instead of these being physically created on a simulator, they are constructed computationally from tomographic data. The radiation planning system able to do this is called a VIRTUAL SIMULATOR.

VOLUME OF REGRET A tissue volume receiving more or less dose than required by prescription.

VOXEL Volume element within the patient, in which some physical property is measured and displayed.

WEDGE Generally a piece of metal shaped to a wedge with a variable radiation transmittance along its length. This creates a non-zero angle between each isodose line and the normal to the beam axis. The same effect can be obtained by moving accelerator jaws, called a DYNAMIC WEDGE.

WEIGHT Rather loose term, applied to radiation fields delivered by a linear accelerator, meaning the number of monitor units per field.

WELL-DIFFERENTIATED DISEASE Primary tumour which can be visualized in images and separated from surrounding normal healthy tissue.

XERORADIOGRAPHY X-ray imaging technique in which a latent image is formed by the radiation selectively discharging an electrostatically charged plate. The image is formed by powder deposition.

APPENDIX C

IMPORTANT DEVELOPMENTS IN 'SUPERVOLTAGE' RADIOTHERAPY FOR CONTEXTUAL FRAMING OF CONFORMAL RADIOTHERAPY

Date	Development
1924	The idea of a linear accelerator was first suggested by Gustaf Ising in Sweden.
1925	Sorenson (California Institute of Technology) created a transformer capable of producing a peak potential of 1 MV.
1927	Coolidge produced a set of cascaded tubes capable of accelerating electrons to 900 keV.
1928	Rolf Wideroe first postulated the concept of the betatron accelerator.
1928	Lauritsen used the Sorenson transformer array with a tube to generate 750 keV x-rays at Caltech.
1929	van de Graff built his prototype generator at Princeton.
1930	The first patients were treated with 600 keV radiation from the Lauritzen x-ray tube at Caltech.
1931	Lawrence accelerated protons to 1 MeV in the cyclotron.
1931	Failla and Quimby at Memorial Hospital (New York) treated patients with 700 keV radiation from a two-section cascaded Coolidge tube.
1933	Patients were treated with 1 MeV x-rays at Caltech in a laboratory funded by W K Kellogg 'the breakfast cereal king'.
1934	van de Graff generator, capable of achieving a potential of 7 MV, developed at Round Hill, Massachusetts.
1935	1 MeV x-rays were produced from a van de Graaff generator at MIT.
1937	Metropolitan Vickers 1 MV supervoltage unit with Cockroft–Walton high-voltage supplies installed at St Bartholomew's Hospital London. In 1936 the Metropolitan Vickers unit was operated at 700 kV, rising to 1 MV by 1939. The cost of the unit and building was £12 000.
1937	Steenbeck patented the idea for a betatron, but this was not built.

Date	Development
1937	Development of the klystron by the Varian brothers (Russell and Sigurd).
1938	Development of the rhumbatron by W W Hansen at Stanford University.
1939	Randall and Boot developed a magnetron with 1 MW power.
1940	Development of the betatron by D W Kerst at the University of Illinois.
1941	Radioactive cobalt-60 first produced in a cyclotron.
1943	Kerst 20 MeV betatron available for medical use.
1946	Mayneord brought three disks of cobalt-59 to North America for activation to cobalt-60. Two were inserted into the AEC Chalk River reactor and the other was inserted into the Oak Ridge National Laboratory reactor.
1946	Mitchell (Cambridge) first published a proposal for the use of cobalt-60 as a therapy source.
1948	L G Grimmett (M D Anderson Hospital) produced first practical design for a cobalt-60 treatment unit.
1948	Although Kerst treated one patient in his laboratory with a betatron, the first clinical radiation programme was started at Saskatoon by Harold E Johns with an exact copy of the original 2.3 MeV betatron. The same year, a second Allis–Chalmer's (A–C) betatron was being used clinically by John Laughlin at the University of Illinois in Chicago.
1949	William Hansen, linac inventor, died, aged only 40.
1949	First commercially available betatron.
1951	Cobalt-60 radiotherapy started at Saskatoon Clinic (Aug), Ontario Institute of Radiotherapy (Oct) and at the M D Anderson Hospital, Houston.
1951	Leksell introduced radiosurgery, initially performed with x-ray equipment.
1952	A home-made cobalt-60 unit was operating at the Los Angeles Tumour Institute.
1952	The Eldorado Cobalt-60 unit was commercially manufactured.
1953	First medical linear accelerator (8 MV) installed at the Hammersmith Hospital, London.
1953	Linear accelerators installed at Newcastle General Hospital and at the Christie Hospital, Manchester.
1954	First (Varian) medical linear accelerator installed at Stanford after early design work by inventors Hansen and Ginzton.
1954	Vickers 15 MeV linac installed at St Bartholomew's Hospital Medical College at Charterhouse Square.
1955	First medical proton irradiation at the Lawrence Berkeley Laboratory, University of California.
1956	First radiotherapy treatment in the US with a linear accelerator at Stanford.

Date	Development
1958	The first treatment-planning systems, which were batch programs with punch cards and line-printer output, by Wood in Cardiff and Tsien in Philadelphia.
1959	First conformation therapy with a tracking cobalt unit (London UK).
1959	Gscheidlen's patent for a prototype multileaf collimator.
1959	Russell Varian died.
1961	Sigurd Varian died in a plane crash off the Pacific Coast of America.
1961	K C Tsien produced a world survey of teletherapy units. There were 1120 worldwide; 47 in Britain (for other totals see Grigg p 313).
1962	First Varian clinical accelerator capable of a full 360 degree rotation installed at the UCLA medical centre.
1962	First electronic megavoltage portal image by Benner.
1965	Takahashi's famous book on rotation therapy published.
1966	The first interactive treatment-planning system, the Programmed Console in St. Louis. Development started in 1966–7, but the system was not available clinically until about 1969–70.
1968	First gamma knife at the Sophiahemmet Hospital, Stockholm, Sweden.
1971	The RAD-8 (Royal Marsden Hospital planning system) was brought into clinical use in 1971–2 and remained in use for about 15 years.
1980	First modern equipment for electronic portal imaging.
1984	First (modern) multileaf collimator.

This table does not list the many 'firsts' in different countries, or by different manufacturers. Grigg (1965) provides the best comprehensive source of such information.

REFERENCES

Bentley R E 1992 private communication

Brecher R and Brecher E 1969 *The rays; a history of radiology in the United States and Canada* (Baltimore, MD: The Williams and Wilkins Co)

Grigg E R N 1965 *The trail of the invisible light; from X-strahlen to radio(bio)logy* (Springfield, IL: Charles C Thomas)

Innes G S 1988 The one million volt x-ray therapy equipment at St Bartholomew's Hospital, 1936–1960 *Megavoltage Radiotherapy 1937–1987 (Brit. J. Radiol. Supplement 22)* 11–16

Jones A 1988 The development of megavoltage x-ray therapy at St Bartholomew's Hospital *Megavoltage Radiotherapy 1937–1987 (Brit. J. Radiol. Supplement 22)* 3–10

Laughlin J S 1989 Development of the technology of radiation therapy *Monograph Issue: the technical history of radiology. Radiographics* **9** (6) 1245–1266

Figure C.1. *Russell and Sigurd (standing) Varian with an early klystron.
(Courtesy Varian Associates.)*

Sisterson J 1992 (ed) Worldwide charged particle patient totals *Particles Newsletter* no 9, January 1992

Takahashi S 1965 Conformation radiotherapy: rotation techniques as applied to radiography and radiotherapy of cancer *Acta Radiol.* Suppl 242

Varian (undated) *Varian Associates: An early history* (Palo Alto: Varian Assoc)

FURTHER READING FOR PRE-SUPERVOLTAGE RADIOTHERAPY

Burroughs E H 1986 *Pioneers and early years: A history of British radiology* (Alderney: Colophon Ltd)

Mould R F 1980 *A history of x-rays and radium* (Sutton: IPC Press)

Pallardy G, Pallardy M-J and Wackenheim A 1989 *Histoire Illustrée de la radiology* (Paris: Les Editions Roger Dacosta)

Webb S 1990 *From the watching of shadows: the origins of radiological tomography* (Bristol: Adam Hilger)

INDEX

Numbers in **bold** refer to **figures** and those in *italics* refer to *tables*.

Printed and bound by CPI Group (UK) Ltd, Croydon, CR0 4YY
01/05/2025
01858532-0001